An Introduction to Statistics for Appraisers

Readers of this text may also be interested in the following publications from the Appraisal Institute:

- *The Appraisal of Real Estate*, 13th edition
- *Scope of Work*
- *Mathematics for Real Estate Appraisers*
- *The Appraiser's Workbook*, second edition

An Introduction to Statistics for Appraisers

by Marvin L. Wolverton, PhD, MAI

Appraisal Institute • 550 West Van Buren • Chicago, IL 60607 • www.appraisalinstitute.org

The Appraisal Institute advances global standards, methodologies, and practices through the professional development of property economics worldwide.

Reviewers: Kenneth M. Lusht, MAI, SRA
Jeffrey A. Johnson, MAI
Thomas O. Jackson, MAI

Chief Executive Officer: Frederick H. Grubbe
Director, Marketing & Membership Resources: Hope Atuel
Senior Manager, Publications: Stephanie Shea-Joyce
Senior Technical Writer/Book Editor: Michael D. McKinley
Technical Book Editor: Emily Ruzich
Manager, Book Design/Production: Michael Landis

For Educational Purposes Only

The materials presented in this textbook represent the opinions and views of the author. Although these materials may have been reviewed by members of the Appraisal Institute, the views and opinions expressed herein are not endorsed or approved by the Appraisal Institute as policy unless adopted by the Board of Directors pursuant to the Bylaws of the Appraisal Institute. While substantial care has been taken to provide accurate and current data and information, the Appraisal Institute does not warrant the accuracy or timeliness of the data and information contained herein. Further, any principles and conclusions presented in this publication are subject to court decisions and to local, state and federal laws and regulations and any revisions of such laws and regulations.

This textbook is sold for educational and informational purposes only with the understanding that the Appraisal Institute is not engaged in rendering legal, accounting or other professional advice or services. Nothing in these materials is to be construed as the offering of such advice or services. If expert advice or services are required, readers are responsible for obtaining such advice or services from appropriate professionals.

Nondiscrimination Policy

The Appraisal Institute advocates equal opportunity and nondiscrimination in the appraisal profession and conducts its activities in accordance with applicable federal, state, and local laws.

Printed in the United States of America

Library of Congress Cataloging-in-Publication Data

Wolverton, Marvin L.
An introduction to statistics for appraisers / by Marvin L. Wolverton.
p. cm.
ISBN 978-1-935328-05-6
1. Real property–Valuation–Statistics. 2. Real property–Valuation–Mathematical models. I. Title.
HD1387.W56 2009
519.5–dc22

2009028114

Table of Contents

1.	Introduction	1
2.	Mathematics Review	17
3.	Working with Tables and Charts	41
4.	Describing Numerical Data	69
5.	Probability	109
6.	Research Design, Hypothesis Testing, and Sampling	139
7.	Inferences About Population Means and Proportions	165
8.	Nonparametric Tests	215
9.	Simple Linear Regression Analysis	241
10.	Multiple Linear Regression Analysis	291
	Solutions to Practice Problems	349
	Useful Symbols and Formulas	367

Appendices

A.	Standard Normal Probabilities	375
B.	Student's *t* Values for One-Tailed Tests	377
C.	Upper Critical Values of the *F* Distribution	380
D.	Wilcoxon Signed Ranks Test	393
E.	Mann-Whitney U Statistic	394
F.	Wilcoxon Sum of Ranks Test	395
G.	89015 Townhouse Data	397
	Bibliography	401

About the Author

Marvin L. Wolverton, PhD, MAI, is a practicing real property valuation theorist and consultant currently employed as a Senior Director in the national Dispute Analysis and Litigation Support practice at Cushman & Wakefield, where he engages in litigation consulting and expert witness services. Dr. Wolverton is also an emeritus professor, and the former Alvin Wolff Distinguished Professor of Real Estate, at Washington State University. He is a state-certified general appraiser and has been a member of the Appraisal Institute since 1985.

Dr. Wolverton is a current member of the Appraisal Journal Review Panel. He has also served as editor of the *Journal of Real Estate Practice and Education* and on the editorial boards of the *Journal of Real Estate Research* and *The Appraisal Journal.* He has authored more than 40 articles in refereed and professional journals including the *Journal of Real Estate Research, Real Estate Economics, Journal of Real Estate Finance and Economics, The Appraisal Journal, Assessment Journal, Journal of Real Estate Portfolio Management, Journal of Property Valuation and Investment,* and *Journal of Property Research.* He has edited and written books and chapters of books on valuation theory and specialized appraisal topics, and he teaches appraisal courses on behalf of the Appraisal Institute.

His formal education includes a Bachelor of Science in Mining Engineering from New Mexico Tech, a Master of Science in Economics from Arizona State University, and a Doctor of Philosophy from Georgia State University specializing in Real Estate and Decision Science. He and his wife, Mimi, currently reside in Leander, Texas.

Acknowledgments

The Appraisal Institute was supportive and responsive when I first contacted them about writing this book many moons ago. Michael McKinley, senior technical writer/book editor at the Appraisal Institute, was immensely helpful both as a copyeditor and as a highly capable mathematics problem solution checker. Words are insufficient to express my gratitude for Michael's assistance with this project.

Several people read all or parts of early versions of the book and provided useful comments and words of encouragement including my son, Steve Wolverton, PhD, who is a professor at the University of North Texas; J. Andrew Hansz, PhD, CFA, at the University of Texas at Arlington; William G. Hardin III, PhD, at Florida International University; and Jim Sanders, SRA. In addition, the anonymous reviewers selected by the Publications Review Panel of the Appraisal Institute provided many useful and instructive comments.

I thank you, one and all, and recognize that despite all of our efforts an undertaking of this magnitude is bound to have retained some arithmetic or typographical errors, for which I am solely and ultimately responsible.

Foreword

The use and misuse of statistics are growing concerns for all real property professionals. In the past, some in the appraisal profession have been reluctant to engage fully with statistical models, seeing them more as direct competition for traditional fee appraisal services than as tools to add greater mathematical rigor and statistical significance to their value conclusions. Now, however, new educational criteria require that appraisers learn more about statistics and modeling and demonstrate competence in their use. To that end, the Appraisal Institute is proud to present *An Introduction to Statistics for Appraisers*, a book that will both challenge readers and reward them with a new understanding of a complex subject.

The basis of statistics is mathematics, and the road to statistical competence begins with a review of essential mathematical skills. Readers will encounter increasingly complex material as they proceed through the text, starting with advice on the responsible use of graphic illustrations of data and then delving into the concept of probability and its application in statistical analysis. More powerful tools such as falsification in testing research hypotheses and sophisticated statistical tests of the variance and reliability of data samples are also explored. The reader's final destination is a rigorous discussion of statistical model building using simple and multiple linear regression.

Like the popular Appraisal Institute workbook *Capitalization Theory and Techniques Study Guide*, which has helped generations of appraisers learn how to use discounted cash flow analysis with confidence, this new publication should serve the appraisal profession for years to come as a valuable educational resource. *An Introduction to Statistics for Appraisers* was written by a real estate appraiser and designed to provide a logical starting point for the study of statistics. Readers who take up the intellectual challenge and diligently work through the exercises in the book will be rewarded with a deeper understanding of their profession and a new set of tools to ensure their continued competence and success.

Jim Amorin, MAI, SRA
2009 President, Appraisal Institute

1 Introduction

Real property appraisal has evolved over the years as appraisal professionals have confronted and adapted to new tools, techniques, and theories. Although applied statistics have been used to describe and analyze data for many decades, widespread practical application of statistical tools in business, economics, and appraisal is a relatively recent occurrence. The technological revolution of the 1980s and 1990s is the primary reason for the widespread use of statistical tools today–particularly the advent of the personal computer and user-friendly software.

Currently, statistical applications are an essential element of the practicing appraiser's tool kit. The extent to which statistics are employed in appraisal practice depends on a number of circumstances, such as the scope of work of the assignment, the statistical competence of the analyst, and the availability of data. Nevertheless, it is likely that most appraisal analyses could benefit from inclusion of applied statistics. Simple descriptive tools like charts and tables or measures of central tendency and dispersion can provide more effective appraisal communication. In addition, more complex inferential tools and analyses, while more difficult to master, provide new and effective ways to solve many traditional valuation problems.

Applying statistical tools to valuation problems is much more than employing and understanding automated valuation models (AVMs). In a narrow sense, AVMs are systems that answer questions by applying hidden, and often undisclosed, algorithms to interpret proprietary data. Users of AVMs are not generally expected to know much about how to actually perform the underlying mathematical analyses, and they typically do not gather, numerically code, and maintain the AVM data. AVM users can be characterized more as operators of the AVM than as applied statisticians. They are generally more interested in the output of the AVM than in how the output was derived.

This book has not been written with a goal of making you a better user of AVMs, although studying the book will more than likely have this effect. Rather, the book is intended to help you develop statistical competence at whatever level you aspire to. If you seek the ability to use descriptive statistical tools as ways to improve communication in your appraisal practice, you will find much useful information here. If you want to take your education further, the book can help you become more competent at statistical inference. If you are even more ambitious, the book will function as an important first foundational step toward becoming a highly skilled valuation statistician.

The book is designed to capture the essence of a first college course in statistics while focusing on essential tools for appraisers and omitting material not vital for the day-to-day practice of appraisal. Therefore, the book's contents should not be viewed as providing everything needed to become a statistician. Rather, it is an initial step that may be more than sufficient for some and insufficient for others. If you are part of the latter group, then additional study probably lies ahead, perhaps including university course work in applied statistics or econometrics.

Organization

The book's chapters are organized in a logical sequence, beginning with a mathematics review and ending with multiple linear regression. Each chapter includes the mathematics behind most of the statistical procedures you will be studying, example problems and solutions that illustrate the statistical procedures, as well as practice problems to test your progress through the book. (The solutions to practice problems are provided at the back of the book.) The examples and practice problems, to the extent possible, are written in a real property context to help you make the connection between the techniques being studied and their application in your professional practice. In addition to the example and practice problems, several chapters also contain an ongoing analysis of actual townhouse data from zip code 89015. Where appropriate, the real world data is analyzed at the end of the chapter to demonstrate the application of some of the chapter's tools and to give you an opportunity to "play with" real data without forcing you to assemble an actual data set.

Chapter 2 reviews all of the mathematics required to understand and master the material in the book. Topics include working with positive and negative numbers; the order of mathematical operations; fractions, decimals, and percents; exponents and roots; logarithms; summation notation; equations; factorials and combinations; and types of data. If you are comfortable with most of these topics, you may want to skim over this chapter. If your math skills are rusty, then Chapter 2 will provide plenty of exercises designed to get you up to speed.

Chapters 3 and 4 cover the essential elements of descriptive statistics. Chapter 3 deals with tables and charts, using them to present numerical and categorical data. It also discusses the crucial issue of ethics in charting. Chapter 4 deals with the description of numerical data, discussing measures of central tendency, dispersion, shape, and correlation between numerical measures.

Chapters 5 and 6 discuss essential theory necessary for competence in statistical inference–that is, reaching a conclusion about an unknown characteristic of a population based on sample data. Inferred conclusions are never certain, so an understanding of probability is essential to developing a credible statistical inference. Chapter 5 introduces the topic of probability, the laws of probability, and two important probability distributions. It also includes a detailed discussion of probability in the context of the Normal Probability Distribution. Chapter 6 deals with how to design and implement inferential statistical research. The topics in this chapter include hypothesis construction, reliability and validity in research, sampling, setting significance levels, and sample size.

The remainder of the book explains how to apply the inferential theories found in Chapters 5 and 6 to problem solving. Chapter 7 covers the process of making inferences about population means and proportions based on an underlying normality assumption–including a discussion of the all-important Central Limit Theorem. The construction of confidence intervals is introduced along with methods of hypothesis testing. Comparison of multiple sample means and proportions is also included in Chapter 7. Chapter 8 provides nonparametric tests analogous to the statistical tests presented in Chapter 7. Nonparametric tests are applicable to situations where a normal distribution of data cannot be assumed. In other words, nonparametric tests generally apply to the types of small data sets that appraisers often encounter.

Chapters 9 and 10 cover linear regression analysis. Chapter 9 discusses simple linear regression, where the model includes a single response variable and a single explanatory variable. Chapter 9 lays the foundation for the chapter that follows and is intended to be read in tandem with Chapter 10, the discussion of multiple linear regression. Chapter 10 deals with models that have a single response (outcome) variable and numerous explanatory (predictor) variables. These two chapters explain how to run and interpret linear regression models as well as the assumptions underlying the models and how to examine the models for assumption violations. Other topics include curve-fitting, variable transformations, and model building. At the end of Chapter 10, the 89015 zip code data set is analyzed to illustrate the realities of building a valuation model from actual sales data.

Where appropriate, the book includes tips on how to conduct the various analyses in Excel, Minitab, and SPSS (the "Statistical Package for the Social Sciences"). Note that not all of the procedures shown in the book are available in each software package. Many other statistical analysis programs exist, many of which are capable of doing much more than these three. The choice to include "how to" suggestions for Excel, SPSS, and Minitab was based on the widespread use of these programs and does not constitute an endorsement of any of them. The choice of one software package over another ultimately depends on the types of analyses you plan to do and the cost associated with software investments.

More About Software

As stated above, many of the statistical routines presented in the book are standard functions of software marketed under the brand names Excel, Minitab, and SPSS. As you probably know, Excel is principally a spreadsheet program having some statistical analysis capabilities. Minitab and SPSS are dedicated statistics programs that read data organized in spreadsheet form. Consequently, Minitab and SPSS include more statistical procedures and model building tools than Excel. Nevertheless, much of what you will encounter in the book can be done in Excel.

The exhibits that follow are designed to help you develop a basic understanding of where the commands can be found in the various programs. In later chapters, examples of software routines or procedures contain much more detail than the basics included here. If you are experienced at working with one or more of these software packages, you can probably skip the following section.

Excel

The statistical tools in Excel 2003 are found by clicking the **Tools** tab in the ribbon at the top of the spreadsheet and then selecting **Data Analysis** in the drop-down menu. **Figure 1.1** shows a typical spreadsheet with the **Tools** menu exposed and **Data Analysis** highlighted. The format will differ slightly depending on which version of Excel you are running, but the selection routine is similar for all versions. For example, in Excel 2007 the **Data Analysis** command is found under the **Data** tab.

If you select **Tools** and do not find **Data Analysis** in the drop-down menu, then one of two procedures is necessary. First, click on the double-down arrows at the bottom of the drop-down menu to ensure that all of the menu's options are being revealed. **Data Analysis** may have been hidden due to lack of use. If **Data Analysis** is still

Figure 1.1 Typical Excel Spreadsheet

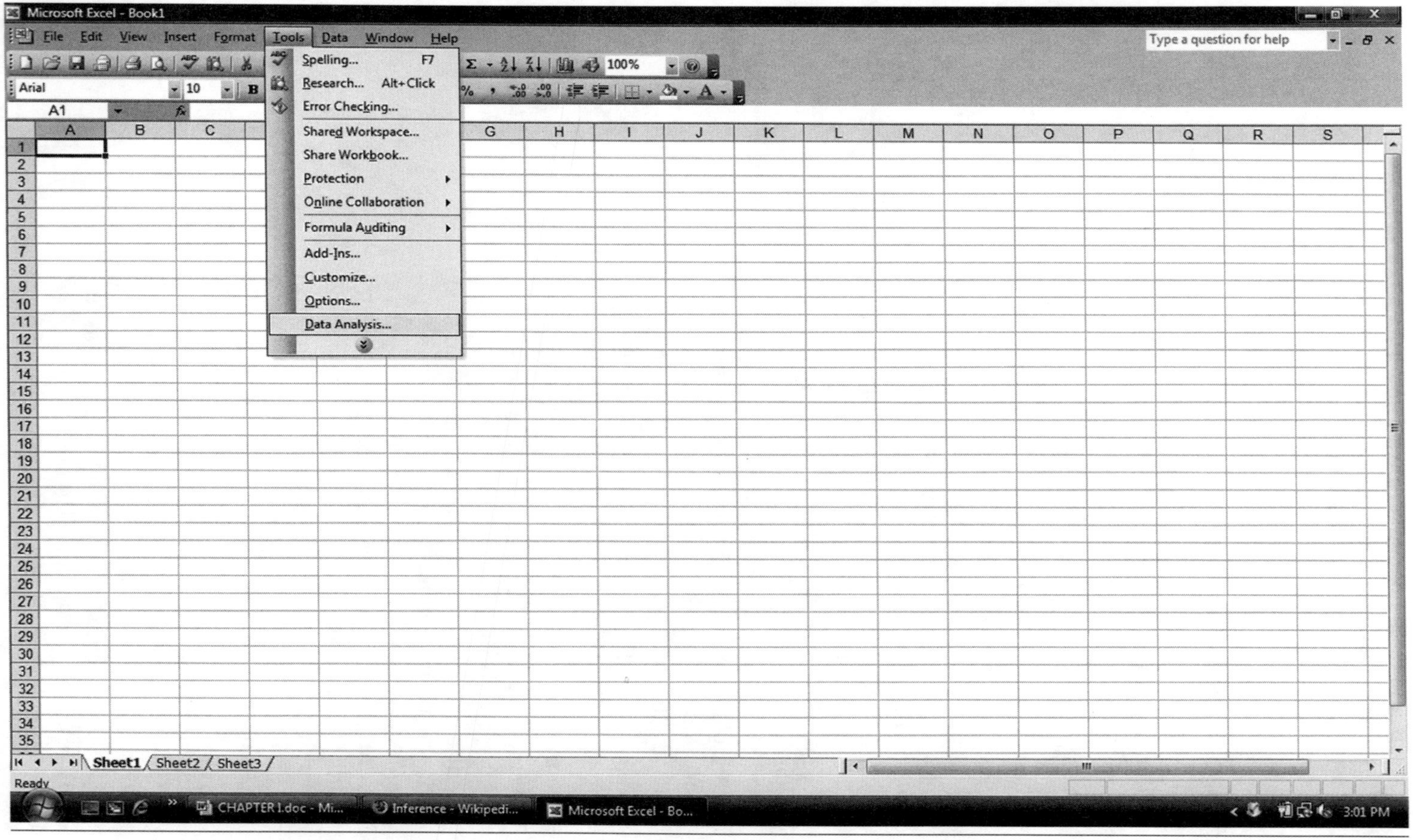

Microsoft Excel - Book1
File Edit View Insert Format Tools Data Window Help
Type a question for help
Arial 10 B
100%
A1
Spelling... F7
Research... Alt+Click
Error Checking...
Shared Workspace...
Share Workbook...
Protection
Online Collaboration
Formula Auditing
Add-Ins...
Customize...
Options...
Data Analysis...
A B C G H I J K L M N O P Q R S
1 2 3 4 5 6 7 8 9 10 11 12 13 14 15 16 17 18 19 20 21 22 23 24 25 26 27 28 29 30 31 32 33 34 35
Sheet1 Sheet2 Sheet3
Ready
CHAPTER 1.doc - Mi... Inference - Wikipedi... Microsoft Excel - Bo... 3:01 PM

not listed, you will need to install the **Data Analysis** add-in. Select **Add-Ins** from the **Tools** drop-down menu. (In Excel 2007, you will find **Add-Ins** under **Excel Options**.) The **Add-Ins** window shown in **Figure 1.2** should open, listing available Add-Ins. Select **Analysis ToolPak** and **Analysis ToolPak–VBA** from the list shown and click **OK**. Excel's statistical data analysis tools should either load automatically or you will be prompted to insert the program disk containing the Excel software, depending on the version of Excel you are running and how it was originally installed.

After the Analysis ToolPak has been installed, the window shown in **Figure 1.3** will open when you select **Data Analysis** from the **Tools** drop-down menu. This exhibit shows a partial list of the statistical routines programmed into Excel. Movement of the slide bar to the right of the list will reveal the remaining routines.

Excel also includes several useful statistical function "macros." (A macro is a single computer instruction that takes the place of a sequence of operations.) If you select **Insert** and select **Function** from the drop-down menu, the function window shown in **Figure 1.4** will appear. This window lists macros that can be applied to spreadsheet data, including medians, averages, minimums, maximums, normal probabilities, t-distribution probabilities, *F*-distribution probabilities, standard deviations, and the like. (Don't worry about what these distributions are at this time–we get into this much later in the book.)

The standard deviation macro is highlighted in Figure 1.4. Notice that the portion of this window imme-

Figure 1.2 Excel Add-Ins Window

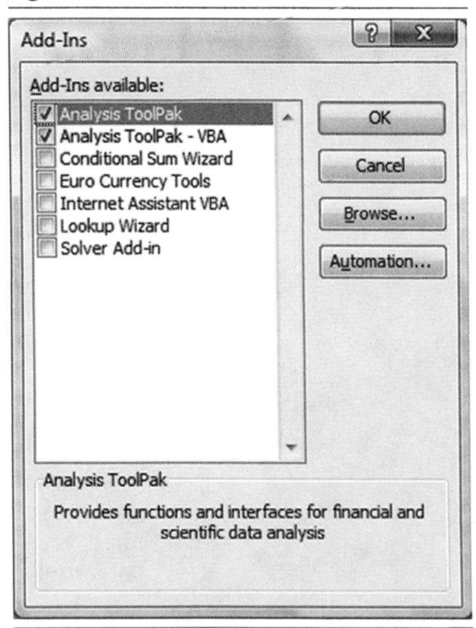

Figure 1.3 Excel Data Analysis Window

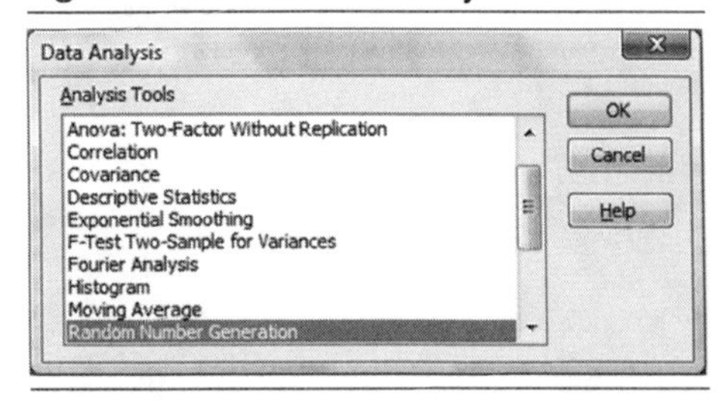

Figure 1.4 Excel Insert Function Window

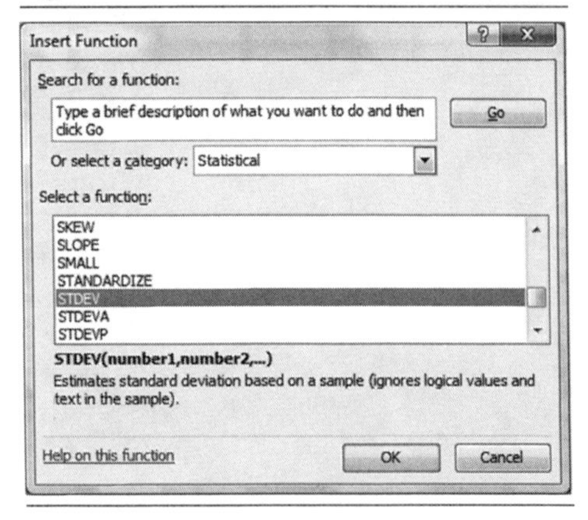

diately below the macro list describes what the macro does and identifies the type of data used in the macro. For example, the standard deviation macro shown here will run when "=STDEV(*list of numbers*)" is entered into a cell, where *list of numbers* denotes the cells containing the sample. If the sample numbers were to be, for instance, listed in cells A1 through A15, then entering "=STDEV(A1: A15)" in cell A16 would cause the standard deviation calculation for the 15 numbers in the list to appear in cell A16.

Now would be a good time to open Excel and try some of the procedures introduced above. Look for the **Tools** menu and find the **Data Analysis** link. Look through the list and become familiar with some of the statistical routines available to you in Excel. Enter some numbers into a spreadsheet and use the **Insert Function** window to select macros to perform calculations you are familiar with such as calculating a minimum, maximum, and average (arithmetic mean).

SPSS

Two windows are available when SPSS is first opened–a **Data View** window and a **Variable View** window. The **Data View** window shows a spreadsheet with a menu bar across the top. This menu bar is used to employ the program's statistics and graphing capabilities. The **Variable View** window allows you to name variables and select variable attributes such as type of data, number of decimals to show in the data view, value labels, and the like. **Figure 1.5** shows the **Variable View** window in SPSS. Note that the menu bar in the **Variable View** window is the same as the menu bar in the **Data View** window. This allows you to run analyses from either window. (You can move between the **Data View** and **Variable View** windows using the tabs in the lower lefthand corner of the spreadsheet.) Once an analysis has been run in SPSS it will appear in a new **Output** window.

Figure 1.6 shows the **Data View** window in SPSS with the **Analysis** drop-down menu open. SPSS provides numerous analytical routines and subroutines. Notice that there are seven available subroutines associated with the **Descriptive Sta-**

Figure 1.5 SPSS Variable View Window

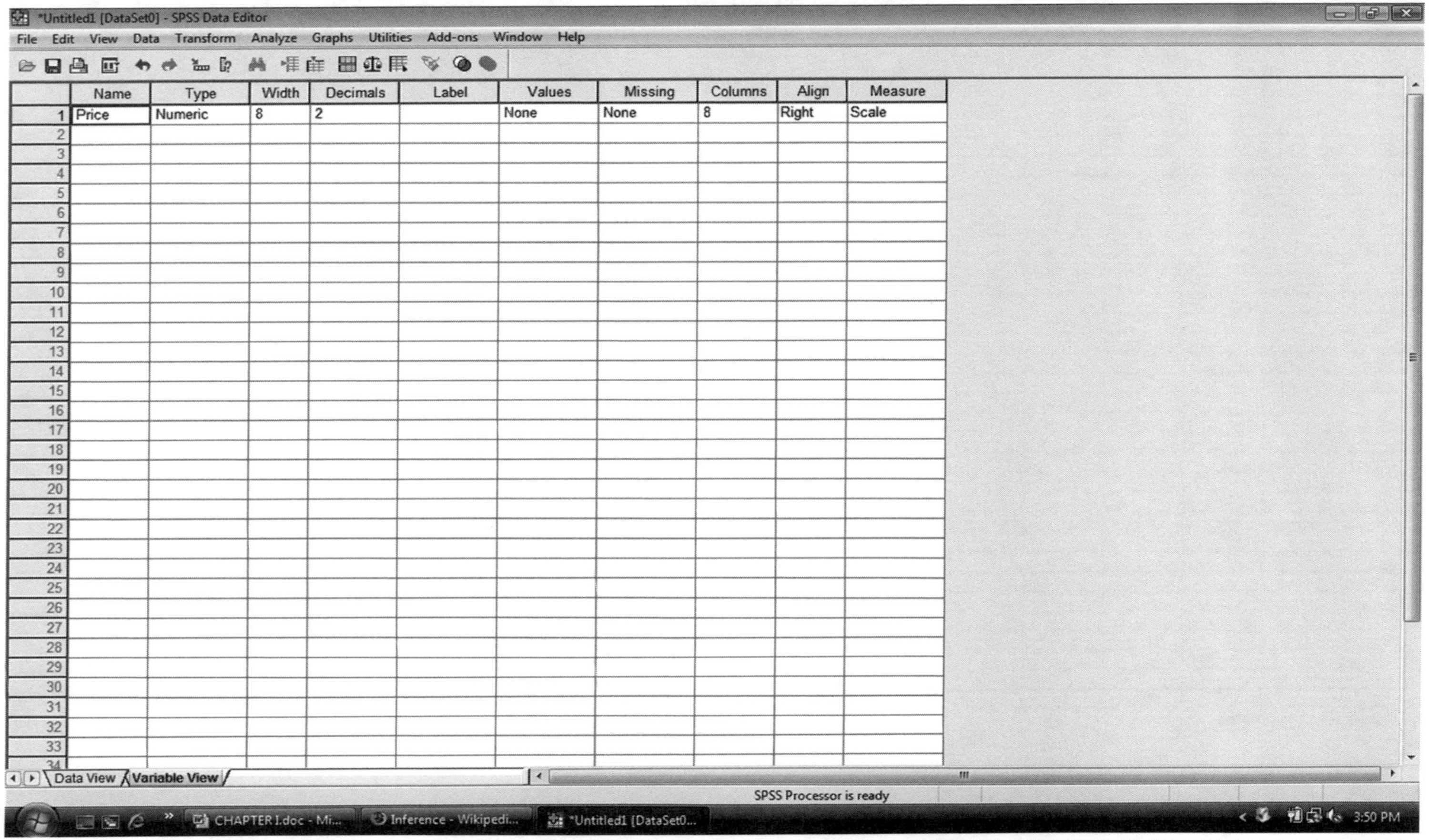

*Untitled1 [DataSet0] - SPSS Data Editor

File Edit View Data Transform Analyze Graphs Utilities Add-ons Window Help

	Name	Type	Width	Decimals	Label	Values	Missing	Columns	Align	Measure
1	Price	Numeric	8	2		None	None	8	Right	Scale
2										
3										
4										
5										
6										
7										
8										
9										
10										
11										
12										
13										
14										
15										
16										
17										
18										
19										
20										
21										
22										
23										
24										
25										
26										
27										
28										
29										
30										
31										
32										
33										
34										

Data View / Variable View

SPSS Processor is ready

CHAPTER I.doc - Mi... Inference - Wikipedi... *Untitled1 [DataSet0... 3:50 PM

Figure 1.6 SPSS Data View Window

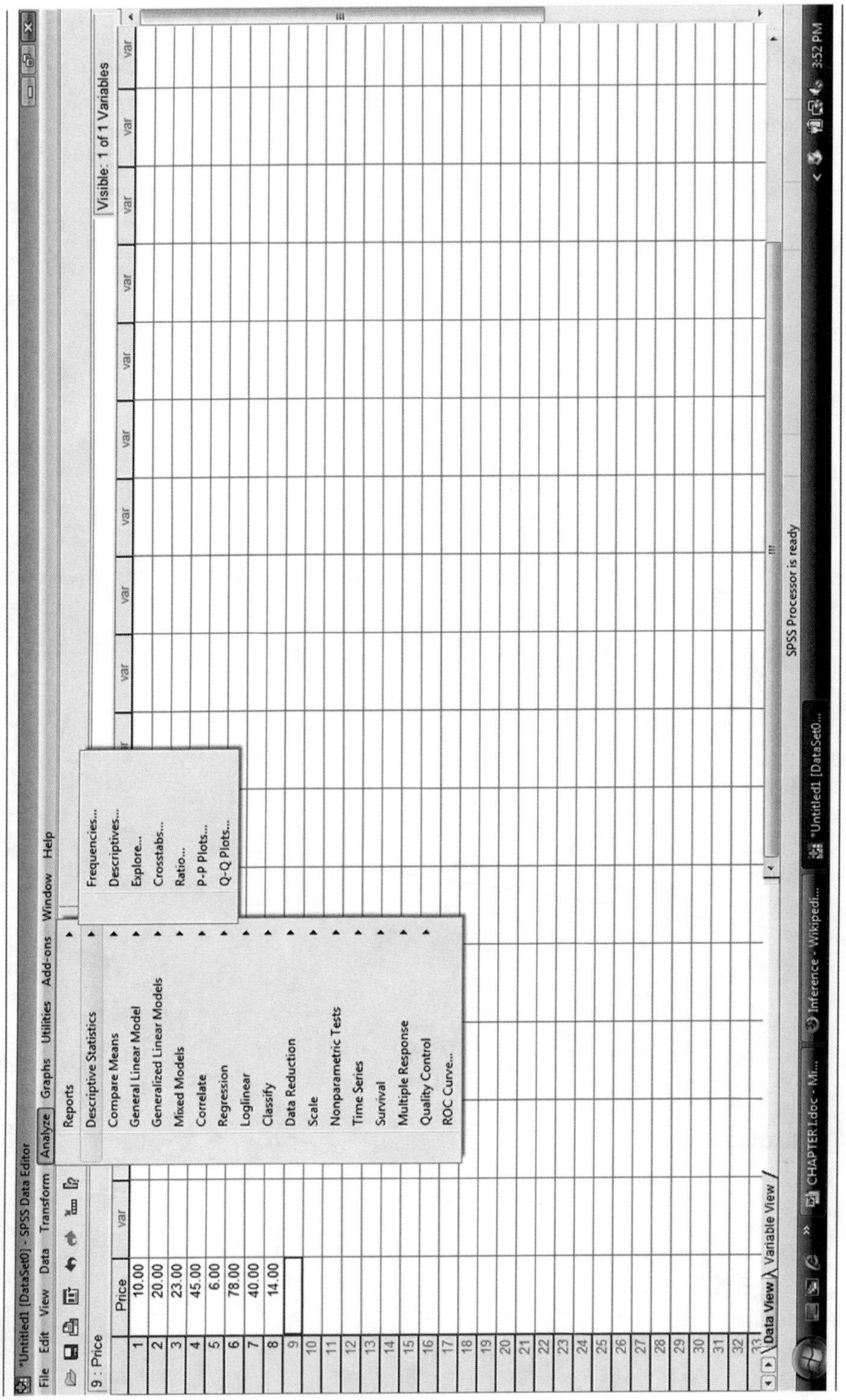

*Untitled1 [DataSet0] - SPSS Data Editor

File Edit View Data Transform Analyze Graphs Utilities Add-ons Window Help

Reports
Descriptive Statistics
Compare Means
General Linear Model
Generalized Linear Models
Mixed Models
Correlate
Regression
Loglinear
Classify
Data Reduction
Scale
Nonparametric Tests
Time Series
Survival
Multiple Response
Quality Control
ROC Curve...

Frequencies...
Descriptives...
Explore...
Crosstabs...
Ratio...
P-P Plots...
Q-Q Plots...

9 : Price

Visible: 1 of 1 Variables

	Price	var
1	10.00	
2	20.00	
3	23.00	
4	45.00	
5	6.00	
6	78.00	
7	40.00	
8	14.00	
9		
10		
11		
12		
13		
14		
15		
16		
17		
18		
19		
20		
21		
22		
23		
24		
25		
26		
27		
28		
29		
30		
31		
32		
33		

Data View Variable View

SPSS Processor is ready

CHAPTER I.doc - Mi... Inference - Wikipedi... *Untitled1 [DataSet0... 3:52 PM

tistics item located in the drop-down list. Figure 1.6 also shows a variable called "Price" entered into the SPSS spreadsheet. Data can be entered into the SPSS spreadsheet manually or imported from another spreadsheet format, such as Excel.

Once a statistical routine's window has been opened in SPSS, the window will include a list of variables. **Figure 1.7** shows the **Descriptives** window, showing the Price variable in the spreadsheet. Price can be moved into the **Variable(s)** list by selecting it as shown and then clicking on the arrow between the two lists. Once this has been done, a list of descriptive statistics available to you is opened by clicking the **Options** tab shown at the bottom right of Figure 1.7.

Figure 1.7 SPSS Descriptives Window

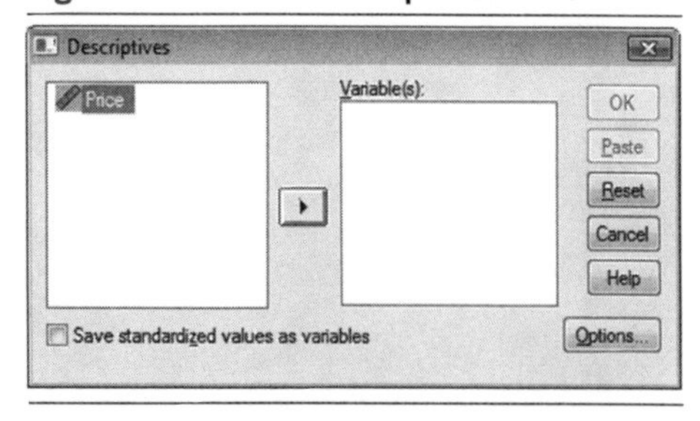

If you are planning to use SPSS to work the examples in the book, now would be a good time to open the program, type in a few variable names, and enter some fictitious data. Open the **Descriptives** window, select a variable or two, and calculate some of the descriptive statistics available in SPSS. **Figure 1.8** shows the **Descriptive Options** window and the routines available. Check and uncheck the routines to select the ones you want to run. Note that at the bottom of the window you have four options for **Display Order**, i.e., the order in which the output information is listed in the **Output** window that opens after you click **OK** in the **Descriptives** window (Figure 1.7).

Figure 1.8 SPSS Descriptives Options Window

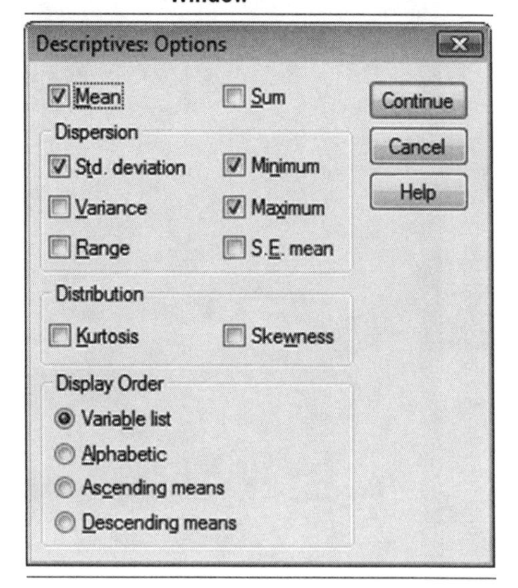

Minitab

Minitab opens with two windows running horizontally across the screen as shown in **Figure 1.9**. A menu bar also appears at the top of the screen. The menu bar contains the **Stat** and **Graph** drop-down menus used to run analyses. The upper window, immediately below the menu bar, is called the "Session" window. All of the statistical output will print there as routines are being run. Note also that the

Figure 1.9 **Minitab Program Session and Worksheet Windows**

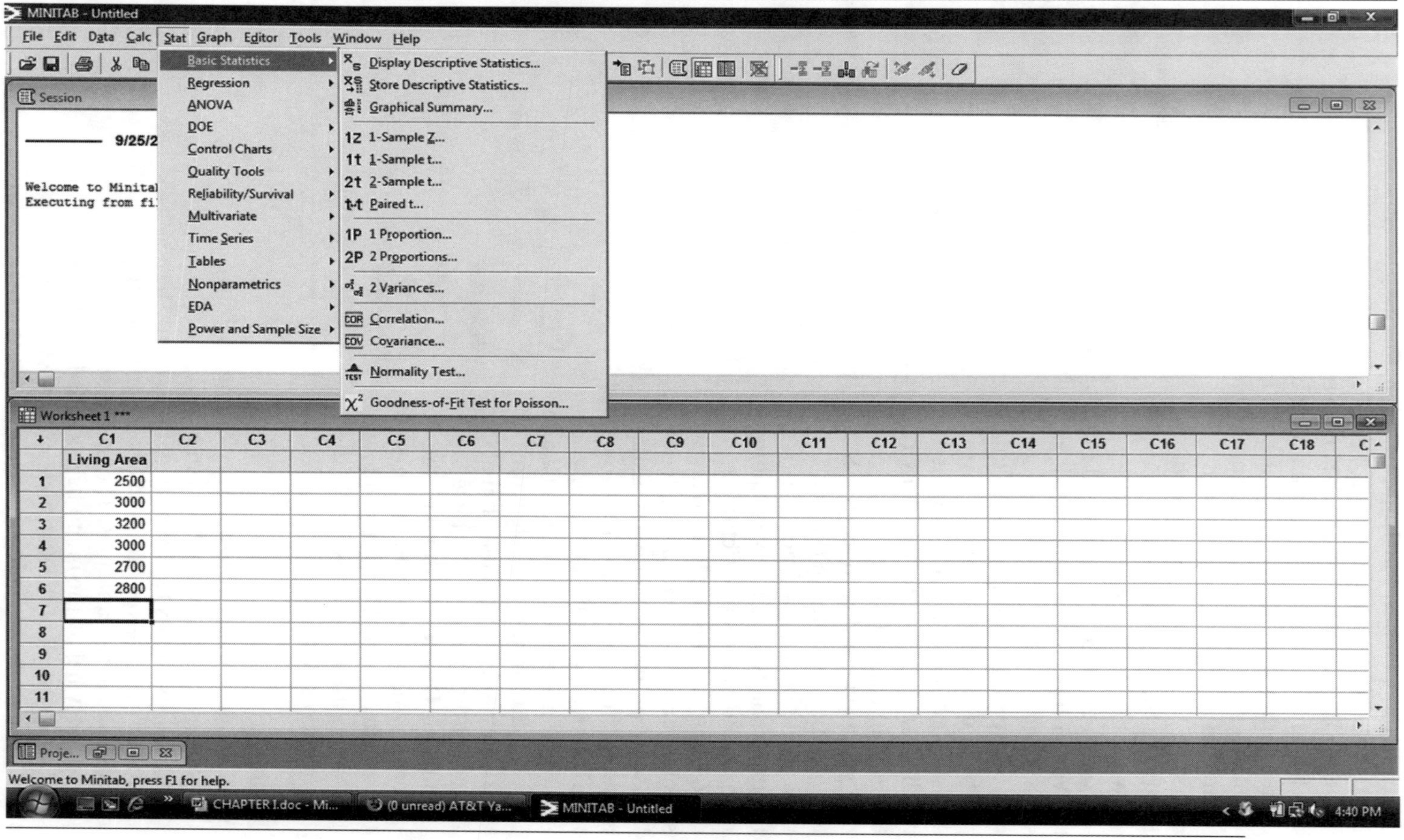

MINITAB - Untitled

File Edit Data Calc Stat Graph Editor Tools Window Help

Session

9/25/2

Welcome to Minital
Executing from fi

Basic Statistics
Regression
ANOVA
DOE
Control Charts
Quality Tools
Reliability/Survival
Multivariate
Time Series
Tables
Nonparametrics
EDA
Power and Sample Size

Display Descriptive Statistics...
Store Descriptive Statistics...
Graphical Summary...
1Z 1-Sample Z...
1t 1-Sample t...
2t 2-Sample t...
t-t Paired t...
1P 1 Proportion...
2P 2 Proportions...
2 Variances...
COR Correlation...
COV Covariance...
Normality Test...
χ^2 Goodness-of-Fit Test for Poisson...

Worksheet 1 ***

↓	C1	C2	C3	C4	C5	C6	C7	C8	C9	C10	C11	C12	C13	C14	C15	C16	C17	C18	C
	Living Area																		
1	2500																		
2	3000																		
3	3200																		
4	3000																		
5	2700																		
6	2800																		
7																			
8																			
9																			
10																			
11																			

Proje...

Welcome to Minitab, press F1 for help.

CHAPTER 1.doc - Mi... (0 unread) AT&T Ya... MINITAB - Untitled 4:40 PM

Session window includes a time and date indicating when the session was conducted–a potentially useful feature for an appraiser. Any graphs generated by Minitab will appear in new and separate windows.

The worksheet in the lower window contains the data to be analyzed. The headings C1, C2, C3, and so on identify the columns. Column names can be typed directly into the shaded spreadsheet box immediately below the column number. The contents of the two windows, along with any graphs that may have been generated, can be saved together as a "Project." You also have the option of saving the worksheet alone as a Minitab worksheet file. As with SPSS, data can be either typed directly into a Minitab worksheet or imported from a different source (e.g., an Excel spreadsheet).

Figure 1.9 shows some "Living Area" data, with the **Stat** menu open and the **Basic Statistics** window selected. Fourteen statistical routines are listed under Basic Statistics grouped into seven subcategories. **Figure 1.10** shows the window that opens when **Display Descriptive Statistics** is selected. Routines available from this window are similar to the SPSS routines listed in Figure 1.8. Variables are selected for analysis by clicking on the entries in the lefthand window and then clicking the **Select** button at the bottom of the variable list in Figure 1.10.

After selecting a variable for analysis (the only variable choice here is Living Area), click the **Statistics** button to open a list of available descriptive statistics. **Figure 1.11** shows the list of statistical measures available as descriptive statistics in Minitab. Check and uncheck the boxes to create the list of descriptive statistics you want to run, and then click **OK** to go back to the window shown in Figure 1.10. Click **OK** again to run the analysis. The results will print in the **Session** window.

Figure 1.10 Minitab Display Descriptive Statistics Window

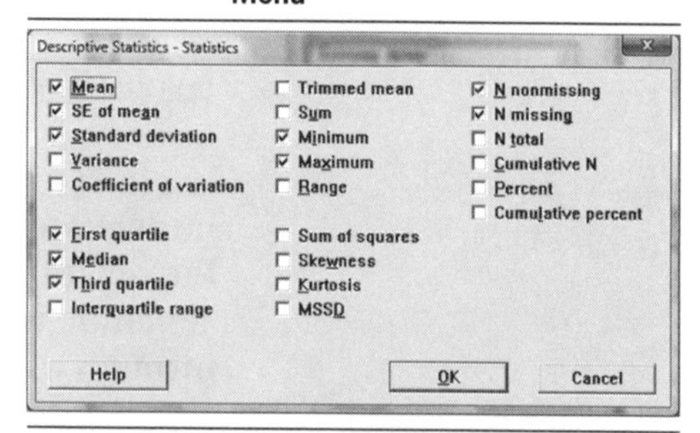

Figure 1.11 Minitab Descriptive Statistics Menu

If you will be using Minitab to work the problems in the book, this would be a good time to enter some fictitious data (or import data from an Excel spreadsheet or some other source) and attempt to generate some descriptive statistics using the procedures described above.

Data

With the exception of the case study townhouse data, all of the data needed for analysis can be found within the appropriate section of each chapter. The townhouse data set is included as an appendix to the book. In addition, the townhouse data set is available as a downloadable file on the Appraisal Institute Web site. Go to www.appraisalinstitute.org/statisticsdownloads for links to the townhouse data in both raw and analysis-ready formats. The data set is in Excel spreadsheet format, which can be easily read by either SPSS or Minitab. You can also download some of the data in the book when the data sets are so large that manual data entry would be too time-consuming. The files are identified by chapter and topic. It is probably a good idea to download all the files and save them for future use and reference.

These data sets are provided for educational use. Most of the data have been altered for pedagogical purposes, so do not rely on the data as, or represent the data as, being "market data." Use of these data for any purpose other than your own learning is prohibited, and any representation of the data as being indicative of any market phenomenon is a violation of professional standards.

Using This Book

Keep in mind that reading a page of mathematics requires much more time than reading a similarly sized page of text. This is because math is a condensed language, capturing a lot of meaning in symbolic (i.e., equation) form. The key to understanding math is to slow down and digest the meaning of each equation or symbolic presentation prior to moving forward. If you work through Chapter 2 methodically, you will discover that math is not very difficult. The payoff is being prepared to successfully tackle statistics.

The book is designed to be a self-contained learning environment. Nevertheless, learning by reading alone can be difficult. Three activities will help you reach your learning goals more easily:

1. Work all of the problems in the book as they arise, a chapter section at a time. Refer back to the example problems as you work the practice problems at the end of each section. Compare what you did to what was done in the example problems and check your answers to the practice problems against the answers provided at the end of the book. This methodical process will greatly improve your understanding and retention.
2. Enroll in an Appraisal Institute statistics class when it becomes available to you. Use the book as a reference for the class, and use the book's examples and practice problems to assist you in deciding how to solve the problems you encounter in class.
3. Enroll in an introductory general business statistics course at a university or community college, and use this book to supplement the course lectures, readings, and problems. This book should assist you in placing what you learn in a general business statistics course into a real estate context.

Finally, it would be a good idea to file this book and printouts or handwritten copies of your solutions to the problems together so that you have them available as a reference when you need to apply them in practice. Everything in this book is intended to be applicable to the practice of real property appraisal and analysis. Become competent in the use of what you will learn here and improve yourself professionally.

Mathematics Review

The language of statistics is mathematics. Like any language, mathematics has its own special rules that must be understood in order to communicate effectively. Most of us studied mathematics fairly intensely in our youth, but mathematical knowledge tends to become rusty with lack of use. The material in this chapter should refresh your math skills by reviewing a few basic rules and procedures.

> Keep in mind that reading a page of mathematics requires much more time than reading a similarly sized page of text. This is because math is a condensed language, capturing a lot of meaning in symbolic (i.e., equation) form. The key to understanding math is to slow down and digest the meaning of each equation or symbolic presentation prior to moving forward. If you work through this chapter methodically, you will discover that math is not very difficult. The payoff is being prepared to successfully tackle statistics.

Almost everyone can benefit from a review of the fundamentals, but if you are confident in your mathematical skills, you might want to skim over the chapter headings and subheadings, study only the topics you need to review, and move on. However, do not skip the section of statistics-specific information titled “Types of Data” at the end of the chapter.

The following discussion is by no means a comprehensive math review. It is tailored to the skills necessary to understand this book. All of the book’s content is presented at a pre-calculus level. No advanced math is needed.

Working with Positive and Negative Numbers

Number Sign

Numbers can be positive or negative. Negative numbers are always preceded by a minus sign.

Positive numbers may be accompanied by a plus sign. If a mathematical symbol or number is presented without a plus sign, it is positive.

$4 = +4$	$a = +a$	$xy = +xy$
$-4 = -4$	$-a = -a$	$-xy = -xy$
$4 \neq -4$	$a \neq -a$	$xy \neq -xy$

Addition and Number Sign

A three-step process provides one way to add up a group of numbers with mixed signs:

1. Group the positive numbers and negative numbers together, and sum each group as if they were all positive numbers.
2. Subtract the smaller subtotal from the larger subtotal.
3. Retain the sign associated with the larger subtotal.

The computation string examples below contain subtraction as well as addition. Remember that subtraction can be viewed as adding a negative number. For instance, the calculation string in the example below could be rewritten as $6 + (-4) + 2 + 5 + (-10) + 7$, which looks more like a pure addition problem.

Problem:			$6 - 4 + 2 + 5 - 10 + 7 = \ ?$
Step 1:	(positive group)		$6 + 2 + 5 + 7 = 20$
	(negative group)		$4 + 10 = 14$
Step 2:	20 (larger subtotal)	– 14 (smaller subtotal)	= 6
Step 3:	answer (larger subtotal is positive)		= +6

Another way to accomplish the same result is by adding successive pairs on the left of the string as you work your way through intermediate calculations.

Problem:	$2 - 10 + 2 + 3 - 9 + 7 - 5 = ?$
Step 1:	$(2 - 10) + 2 + 3 - 9 + 7 - 5 = -8 + 2 + 3 - 9 + 7 - 5$
Step 2:	$(-8 + 2) + 3 - 9 + 7 - 5 = -6 + 3 - 9 + 7 - 5$
Step 3:	$(-6 + 3) - 9 + 7 - 5 = -3 - 9 + 7 - 5$
Step 4:	$(-3 - 9) + 7 - 5 = -12 + 7 - 5$
Step 5:	$(-12 + 7) - 5 = -5 - 5$
Step 6:	$(-5 - 5) = -10$

Although it appears to be a time-consuming process, addition strings done on a calculator are similar to this procedure. Keep in mind that signs must be handled correctly to get the correct answer in the two example problems above.

Subtraction and Number Sign

Keep the following three rules in mind when subtracting numbers:

Rule 1: Any number subtracted from itself equals zero.

$$6 - 6 = 0$$

$$-3 - (-3) = 0$$

Rule 2: Subtracting a positive number is the same as adding a negative number. (We applied this rule in the preceding discussion of addition and number sign.)

$$7 - 3 = 7 + (-3) = 4$$

The inverse of this rule is adding a negative number is the same as subtracting a positive number.

$$12 + (-9) = 12 - 9 = 3$$

Rule 3: Subtracting a negative number is the same as adding a positive number.

$$5 - (-2) = 5 + 2 = 7$$

This rule is not as intuitively obvious as the other two, but it has to be true for two reasons. First, we know that

$$7 - 2 = 5$$

$$7 - 2 - (-2) = 5 - (-2)$$ (Subtracting -2 from both sides preserves the equality.)

$$7 + 0 = 5 - (-2)$$ (See Rule 1, a number subtracted from itself is zero.)

$$7 = 5 - (-2)$$

We also know that

$$7 = 5 + 2$$

and

$$5 - (-2) = 5 + 2$$

because they both equal 7. Additionally, using Rule 1 again plus a little logic:

a) $-2 - (-2) = 0$ [Rule 1]

b) $-2 + 2 = 0$

For a) and b) to both be true -(-2) must equal +2.

Multiplication and Number Sign

Multiplication of two numbers having the same sign results in a positive answer. Multiplication of two numbers having opposite signs results in a negative answer. Multiplication of any number by zero results in an answer of zero.

$$6 \times 4 = 24 \qquad (-6) \times (-4) = 24$$
$$6 \times (-4) = -24 \qquad 6 \times 0 = 0$$

Several symbols are used to indicate "multiplication" in a mathematical expression. They include the traditional "times" sign (×), a dot (·), parentheses, no symbol at all, and–as a result of the advent of spreadsheet programs–an asterisk (*).

$$a \times b = a \cdot b = a(b) = ab = a{*}b$$

Division and Number Sign

Dividing two numbers having the same sign results in a positive number. Dividing two numbers having opposite signs results in a negative number. Zero divided by any number is zero. Division by zero is undefined.

$$4 \div 2 = 2 \qquad -4 \div -2 = 2 \qquad -4 \div 2 = -2$$

$$4 \div -2 = -2 \qquad 0 \div 2 = 0 \qquad 0 \div -2 = 0$$

$$4 \div 0 = \text{undefined}$$

Symbols used to indicate division include the traditional "divide" symbol (÷), a horizontal line between two numbers, and a slash (the last is used in spreadsheet programs).

$$7 \div 3 = \frac{7}{3} = 7/3$$

String Multiplication and Division

When more than two numbers are multiplied together (i.e., a "string" of numbers), it is easiest to work from left to right a pair at a time, remembering the sign rules for multiplication and division.

$$\begin{aligned} 2 \times -6 \times 3 \times 4 \times -1 &= -12 \times 3 \times 4 \times -1 \\ &= -36 \times 4 \times -1 \\ &= -144 \times -1 = 144 \end{aligned}$$

$$\begin{aligned} 100 \div 5 \div -4 \div 5 &= 20 \div -4 \div 5 \\ &= -5 \div 5 \\ &= -1 \end{aligned}$$

$$\begin{aligned} 100 \div 5 \times -2 \div -4 \times 3 &= 20 \times -2 \div -4 \times 3 \\ &= -40 \div -4 \times 3 \\ &= 10 \times 3 \\ &= 30 \end{aligned}$$

Alternatively, do the string calculations ignoring the signs. Then count the number of negative numbers. If there is an odd number of negative numbers, the answer will be negative; otherwise the answer will be positive.

Practice Problems 2.1

1. 5 + -7 = ?	**2.** 2 + 3 + 8 – 12 + 7 = ?	**3.** -6 – (-4) = ?
4. 2 × -8 = ?	**5.** 2 · -8 = ?	**6.** 2(-8) = ?
7. 2 * (-8)/4 = ?	**8.** 4 · 5 · -2 · -3 = ?	**9.** 6 ÷ -3 ÷ -2 = ?
10. $(t)(-y)(z)(t)$ = ?	**11.** 16 × 3 × -1 × 0 = ?	**12.** 0 ÷ 27 = ?
13. 27 ÷ 0 = ?	**14.** 50 × 3 ÷ 75 × -4 ÷ -2 = ?	

Order of Operations

Parentheses

Calculations inside of parentheses should be done first, reducing each set of operations inside a parenthesis to a single number or expression. If parentheses are nested (parentheses inside of parentheses), then start from the inside and work outward.

$$\begin{aligned} 9(3+3) &= 9(6) \\ &= 54 \end{aligned}$$

$$\begin{aligned} -2(4+6) &= -2(10) \\ &= -20 \end{aligned}$$

$$(4+5+2)x = 11x$$

$$\begin{aligned} 21 \div (-247+240) &= 21 \div -7 \\ &= -3 \end{aligned}$$

$$\begin{aligned} 6 \times (3(2+1)) &= 6 \times (3(3)) \\ &= 6 \times 9 \\ &= 54 \end{aligned}$$

$$\begin{aligned} 20 \div (40 \div ((64 \div 16) + 6)) &= 20 \div (40 \div (4+6)) \\ &= 20 \div (40 \div 10) \\ &= 20 \div 4 \\ &= 5 \end{aligned}$$

No Parentheses

If there are no parentheses to govern the order of operations, then multiplication and division are done prior to addition and subtraction.

$$\begin{aligned} 2 \times 3 \div 2 + 8 + 6 \times -2 - 8 \div -2 &= (2 \times 3 \div 2) + 8 + \\ &\quad (6 \times -2) - (8 \div -2) \\ &= 3 + 8 - 12 - (-4) \\ &= 3 \end{aligned}$$

In the example above, all of the multiplication and division strings are grouped by parentheses to illustrate the multiplication and division operations that should be done prior to addition and subtraction.

Practice Problems 2.2

1. $1 + 5 \times 2 + 4 \times -5 = ?$

2. $(4 + 5) \cdot 2 + 6 = ?$

3. $(6 \div 2 - 3 + 5) \times 2 - (-7) = ?$

4. $(6 - (-4) + 10) / (4 + 4 \div 4) = ?$

5. $\dfrac{\frac{40}{4} - \frac{-6}{3}}{\frac{60}{10}} = ?$

Fractions, Decimals, and Percentages

A fraction is simply notation that denotes division. The top number is called the "numerator" and the bottom number is called the "denominator." The denominator value expresses the number of equal parts the numerator is to be separated into. For example, a fraction *X*/*Y* asks, "If *X* (the numerator) is divided into *Y* (the denominator) number of equal parts, what size will the parts be?" The fraction $8/4 asks, "If I divide $8 into four equal parts, what amount will each of the four parts be worth?" Each part will be $2 because 8 ÷ 4 = 2.

$$\frac{10}{4} = 2.5 \qquad \frac{9}{3} = 3 \qquad \frac{4}{6} = 0.667 \qquad \frac{-12}{24} = -0.5$$

$$\frac{(6 - 2)(3 - (-1))}{-7 + 2 + 1} = \frac{4 \times 4}{-4} = \frac{16}{-4} = -4$$

Notice that the sign rules for addition, subtraction, multiplication, and division apply when solving fractions. That is, it is entirely possible for a fraction to have a negative solution.

One handy thing to realize is that any number can be turned into a fraction by expressing its value as a numerator over a denominator of 1. This knowledge is useful when multiplying fractions.

$$16 = \frac{16}{1} \quad -37 = \frac{-37}{1} \quad 2 \times 5 = \frac{2 \times 5}{1} \quad 10xy = \frac{10xy}{1}$$

Multiplying Fractions

Multiplying fractions is accomplished by multiplying the numerator of one fraction by the numerator of the other fraction, repeating the process for the denominators of each fraction, and writing the products as the numerator and denominator of the answer.

$$\frac{3}{4} \times \frac{1}{2} = \frac{3 \times 1}{4 \times 2} = \frac{3}{8}$$

$$\frac{7}{8} \times (3+2) = \frac{7}{8} \times \frac{(3+2)}{1} = \frac{7 \times (3+2)}{8 \times 1} = \frac{7 \times 5}{8 \times 1} = \frac{35}{8} = 4\tfrac{3}{8}$$

$$\frac{a}{b} \times \frac{c}{d} = \frac{ac}{bd}$$

$$2ac \times \frac{3+2}{d} = \frac{2ac}{1} \times \frac{3+2}{d} = \frac{2ac(3+2)}{d \times 1} = \frac{10ac}{d}$$

For future reference, note that multiplication of the numerator and denominator by the same number does not change a fraction's value. Multiplying the numerator and denominator by the same value is the same as multiplying the entire fraction by 1, and any number multiplied by 1 is the same number.

This knowledge will be useful for addition and subtraction of fractions with different denominators.

Dividing by Fractions

The easiest way to divide by a fraction is to invert the fraction and multiply the numerator by the inverted fraction.

$$\frac{\frac{3}{4}}{\frac{1}{2}} = \frac{3}{4} \times \frac{2}{1} = \frac{6}{4} = 1.5$$

$$18 \div \frac{1}{3} = \frac{18}{1} \times \frac{3}{1} = \frac{54}{1} = 54$$

$$\frac{-50}{\frac{1}{5}} = \frac{-50}{1} \times \frac{5}{1} = \frac{-250}{1} = -250$$

$$18 \div \frac{2}{3} = \frac{18}{1} \times \frac{3}{2} = \frac{54}{2} = 27$$

Adding and Subtracting Fractions

In order to add or subtract fractions, the fractions must have a common denominator, i.e., the denominators of the fractions being added or subtracted must all have the same value. Finding a common denominator is easier when you remember that multiplying the numerator and denominator by the same amount does not change the original fraction's value. Ideally the smallest possible common denominator is preferred. This is often referred to as the "lowest common denominator" or "least common denominator." For example, the lowest common denominator of 1/5 and 1/2 is 10.

$$\frac{1}{5}+\frac{1}{2}=\left(\frac{1}{5}\right)\cdot\left(\frac{2}{2}\right)+\left(\frac{1}{2}\right)\cdot\left(\frac{5}{5}\right)=\frac{2}{10}+\frac{5}{10}=\frac{7}{10}$$

$$\frac{1}{5}+\frac{1}{2}+\frac{1}{10}=\left(\frac{1}{5}\right)\cdot\left(\frac{2}{2}\right)+\left(\frac{1}{2}\right)\cdot\left(\frac{5}{5}\right)+\left(\frac{1}{10}\right)\cdot\left(\frac{1}{1}\right)=\frac{2}{10}+\frac{5}{10}+\frac{1}{10}=\frac{8}{10}=\frac{4}{5}$$

$$\frac{1}{2}+\frac{5}{7}+\frac{3}{4}=\left(\frac{1}{2}\right)\cdot\left(\frac{14}{14}\right)+\left(\frac{5}{7}\right)\cdot\left(\frac{4}{4}\right)+\left(\frac{3}{4}\right)\cdot\left(\frac{7}{7}\right)=\frac{14}{28}+\frac{20}{28}+\frac{21}{28}=\frac{55}{28}=1\frac{27}{28}$$

Decimals

Decimals provide a shorthand means of writing fractions. The denominators of fractional equivalents of decimals are limited to the number 10 and 10 multiplied by itself any number of times. That is, denominators of fractional equivalents of decimals are limited to 10, 100 (10 × 10), 1,000 (10 × 10 × 10), 10,000 (10 × 10 × 10 × 10), and so on.

$0.1 = \frac{1}{10}$ $5.2 = 5\frac{2}{10}$ $-6.24 = -6\frac{24}{100}$ $2.001 = 2\frac{1}{1{,}000}$

The number to the right of the decimal place is the numerator of the decimal's fractional equivalent (1, 2, 24, and 1 in the examples above). The number of digits to the right of the decimal place determines the denominator. In the first example above, there is one digit to the right of the decimal place, so the denominator is 10. The second example's denominator is also 10. The third example has two digits to the right of the decimal, so its denominator is 100 (10 × 10). The fourth example has three digits to the right of the decimal and the denominator is 1,000 (10 × 10 × 10).

If a decimal has five digits to the right of the decimal, what is the denominator in its fractional equivalent? It is 100,000 (10 × 10 × 10 × 10 × 10). Therefore,

$$9.12345 = 9\frac{12{,}345}{100{,}000}$$

As you have probably figured out by now, the number of digits to the right of the decimal (referred to as the "number of decimal places") dictates the number of times 10 is multiplied by itself in the fractional equivalent denominator.

Some fractions yield decimal equivalents that literally go on forever. For example, 1/3 = 1.33333333333... . This sort of outcome is symbolized by placing a bar over the last digit when writing a decimal equivalent. Therefore, we can write 1/3 this way:

$$\frac{1}{3} = 1.3\overline{3}$$

The bar over the last 3 indicates that the 3s repeat infinitely. In this case the fraction could be seen as shorthand for the decimal equivalent, which would continue indefinitely if written out.

Rounding is another way of handling decimals that go on forever and decimals that are simply too long to be of practical use. The common rules for rounding decimals are

- If the number to the right of the last digit you want to retain is less than 5, then merely drop everything to the right of the digit you want to retain.
- If the number to the right of the last digit you want to retain is 5 or greater, then increase the amount of the last digit you want to retain by 1.

For example, 4.1324628 rounded to three decimal places is 4.132, and 4.1324628 rounded to four decimal places is 4.1325.

Percentages

The word "percent" is just another way of saying "hundredths." For example, "fifteen percent" is the same as "fifteen hundredths."

$$75\% = \frac{75}{100} = 0.75$$

$$225\% = \frac{225}{100} = \frac{200}{100} + \frac{25}{100} = 2 + \frac{25}{100} = 2\frac{25}{100} = 2.25$$

$$0.15\% = \frac{0.15}{100} = \frac{\frac{15}{100}}{\frac{10{,}000}{100}} = \frac{15}{10{,}000} = 0.0015$$

$$0.15 = \frac{15}{100} = 15\%$$

$$3.25 = 3\frac{25}{100} = \left(\frac{3}{1}\right)\left(\frac{100}{100}\right) + \left(\frac{25}{100}\right) = \frac{300}{100} + \frac{25}{100} = \frac{325}{100} = 325\%$$

$$0.0024 = \frac{24}{10{,}000} = \frac{\frac{24}{100}}{\frac{10{,}000}{100}} = \frac{0.24}{100} = 0.24\%$$

The shortcut for moving back and forth between decimal and percentage notation is to shift the decimal place two spaces and either drop or add the percent sign as appropriate. For example, to convert 0.15 to a percentage equivalent, move the decimal point in 0.15 two places to the right and add the percent sign, which yields the expression "15%." To get from 75% to 0.75, move the decimal point in 75% two places to the left and drop the percent sign. The general rule is

- Decimals to percentages–shift the decimal point two places to the right and add the percent sign. (This is the same as multiplying the decimal by 100.)
- Percentages to decimals–shift the decimal point two places to the left and drop the percent sign. (This is the same as dividing the percent amount by 100.)

If this is confusing, work the example problems provided longhand until you are more comfortable with the process of switching back and forth using the shortcut.

Practice Problems 2.3	
1. $\frac{5}{6} = ?$ **2.** $\frac{120}{8} = ?$	**3.** $\frac{-13}{4} = ?$
4. $\frac{(5+2)(7-(-2))}{9+-(-3)+1} = ?$	**5.** $\frac{2}{3} \times \frac{5}{6} = ?$
6. $\frac{4}{3} \cdot (2+6) = ?$	**7.** $\frac{2x}{5} \cdot \left(\frac{1}{2}\right) \cdot \left(\frac{3+2}{3}\right) = ?$
8. $40 \div \frac{4}{5} = ?$	**9.** $\frac{1}{2} \div \frac{2}{3} = ?$
10. $\frac{2}{x} \div \frac{-y}{10} = ?$	**11.** Decimal equivalent of $9^{26}/_{100} = ?$
12. Express $7\frac{2}{3}$ as a decimal.	**13.** Round 7.233782 to 3 decimal places.
14. Express 0.134 as a percentage.	**15.** Express 256% as a decimal.
16. Express 5 ÷ 6 as a percentage.	**17.** Express 0.00032 as a percentage.
18. If 35% of home buyers in Cleveland opted for an adjustable rate mortgage last year and 3,000 total mortgages were recorded during the year, how many of them were adjustable rate mortgages?	

Exponents and Roots

An exponent indicates how many times a number should be multiplied by itself. For example, 2^3 indicates that 2 will be multiplied by itself three times (3 is the exponent). Doing the math, $2^3 = 2 \times 2 \times 2 = 8$.

$$4^2 = 16 \qquad 10^3 = 1{,}000$$

$$\left(\frac{1}{3}\right)^2 = \frac{1}{3} \cdot \frac{1}{3} = \frac{1}{9} \qquad (xy)^2 = (xy)(xy) = x^2y^2$$

A root asks which number multiplied by itself this many times equals the given amount? (Exponents and roots are opposites.) A square root asks what number squared equals the given amount? A cube root asks what number cubed (i.e., multiplied by itself three times) equals the given amount? Roots can be expressed two ways:

1. Using the root symbol (√)
2. Using a fractional exponent

The root sign by itself means square root. The square root of 25 (which can also be expressed as $25^{1/2}$) is 5 because $5 \times 5 = 25$.

$$\sqrt{25} = 25^{1/2} = 5$$

In the expression $\sqrt[3]{27}$, the small three in front of the root sign means cube root. The cube root of 27 is 3 because $3 \times 3 \times 3 = 27$.

$$\sqrt[3]{27} = 27^{1/3} = 3$$

The easiest way to solve roots is to use a calculator. (The longhand arithmetic is about as cumbersome as it gets.) If your calculator has a y^x key, enter the amount x as the decimal equivalent of the fractional exponent that provides the root you are solving for (e.g., to solve for the fourth root of y, enter 0.25, which equals 1/4, as x). Most calculators also have a $\sqrt{x}$ key, which solves for a square root. We will limit ourselves to square roots later in this book.

Two additional aspects of exponents and roots are order of operations and multiplication (or division). When exponents and roots appear in a string of calculations, these operations should be completed first. For example, $3^3 + 2 + 3 \times \sqrt{4}$ should be expressed as $27 + 2 + 3 \times 2$ prior to considering the order of operations for addition and multiplication.

Regarding multiplication and division, exponents are added when multiplying expressions with the same base and subtracted when dividing expressions with the same base. For example, $(x^2)(x^2) = x^{(2+2)} = x^4$. It is easy to see why this is so because $(x^2)(x^2) = (xx)(xx) = xxxx = x^4$. Also, $(x^4)/(x^2) = x^{(4-2)} = x^2$. This is also easy to understand when you remember that $1/x^2 = x^{-2}$, so the preceding expression is the same as $(x^4)(x^{-2}) = x^{(4-2)} = x^2$. Another way to visualize this is to recognize that$(x^4)/(x^2) = xxxx/xx = xx = x^2$.

Practice Problems 2.4

1. $7^2 = ?$ **2.** $5^3 = ?$ **3.** $\sqrt{144} = ?$

4. $\sqrt{x^2} = ?$ **5.** $0.8^2 = ?$ **6.** $\sqrt{0.95} = ?$

7. $\sqrt{1{,}482.25} = ?$ **8.** $3(5)^2 = ?$ **9.** $(3x)^2 = ?$

10. $(3 + 5)^2 = ?$ **11.** $(3^2 + 5^2) = ?$ **12.** $2^3 \cdot 2^3 = ?$

13. $y^3 \cdot y^2 = ?$

Logarithms

Logarithms and exponents are related. Consider the equation $x = a^y$. A logarithm expresses the value of y, where $y = \log_a x$ with a couple of constraints: $a > 0$ and $a \neq 1$. The letter a in this equation is referred to as the logarithmic base. Two commonly used logarithmic bases are 10 and e. These two bases have special names. A base 10 logarithm is called a *common logarithm*. A base e logarithm is called a *natural logarithm*.

Natural logs are often used in regression models when applied to real property valuation. The symbol for a natural logarithm is "ln." In other words, $\ln x$ is a shorthand way of writing $\log_e x$. Like π, the base number e has an infinite number of decimal places, and the "approximate value" of e is 2.718281828459.[1] Fortunately, the value of e is programmed into most calculators. For instance, the HP-12C financial calculator popular with appraisers has two blue keys labeled "e^x" and "LN." These keys calculate natural logs (LN) and allow the user to reverse the process (e^x), which is called an *antilogarithm*.

Six basic properties apply to working with logarithms, which are handy to know if you find yourself working with natural logs in the future (or logs to any other base). The basic properties of logarithms are

The discussion of logarithms here is limited to what you will need in order to use logarithms in regression modeling. Plenty of math books and online resources are available if you want to study this topic more deeply.

Property 1:	$\ln xy = \ln x + \ln y$
Property 2:	$\ln(x \div y) = \ln x - \ln y$
Property 3:	$\ln x^a = a(\ln x)$
Property 4:	$\ln e = 1$
Property 5:	$\ln e^x = x$
Property 6:	$\ln 1 = 0$

Proofs of these properties can be found in most math books that have a section dealing with logarithms.[2]

1. Base e has convenient properties for financial calculations. It is derived from the compound interest equation and is the amount $1 will grow to in a year at an interest rate of 100% compounded continuously.
2. See, for instance, Marvin L. Bittinger, David J. Ellenbogen, and Barbara L. Johnson, *Intermediate Algebra: Concepts and Applications*, 8th ed. (Reading, Mass.: Addison-Wesley, 2009).

Using these properties, some algebraic manipulation, and the facts that $\ln 2 = 0.69315$ and $\ln 4 = 1.38629$ (please feel free to check these solutions) we can perform a variety of calculations without using the natural log key of a calculator:

- $\ln 8 = \ln (2 \cdot 4)$
  $= \ln 2 + \ln 4$
  $= 0.69315 + 1.38629$
  $= 2.07944$
- $\ln e^4 = 4$
- $\ln (½) = \ln (2 \div 4)$
  $= \ln 2 - \ln 4$
  $= 0.69315 - 1.38629$
  $= -0.69314$
- $\ln (½) = \ln (1 \div 2)$
  $= \ln 1 - \ln 2$
  $= 0 - 0.69315$
  $= -0.69315$ (the difference between this solution and the one above is due to rounding of ln 2 and ln 4)
- $\ln 64 = \ln (4^3)$
  $= 3 (\ln 4)$
  $= 3 \cdot 1.38629$
  $= 4.15887$
- $\ln 4e = \ln 4 + \ln e$
  $= 1.38629 + 1$
  $= 2.38629$
- $\ln y = 1.38629$
  $y = e^{1.38629}$
  $= 4$

Practice Problems 2.5

Given $\ln 3 = 1.09861$ and $\ln 5 = 1.60944$:

1. $\ln\left(\frac{3}{5}\right) = ?$

2. $\ln (3 \cdot 5) = ?$

3. $\ln 15 = ?$

4. $\ln 125 = ?$ (note that $125 = 5^3$)

5. $\ln e = ?$

6. $\ln 5e = ?$

7. $\ln y = 1.60944$, $y = ?$

8. $\ln\left(\frac{1}{5}\right) = ?$

9. $\ln e^5 = ?$

Summation Notation

The summation sign is the uppercase Greek letter sigma (Σ). The summation sign means, "Add up what follows."

$$\Sigma(2, 3, 4) = 2 + 3 + 4 = 9$$

$$\Sigma(x_1, x_2, x_3) = x_1 + x_2 + x_3$$

More generally,

$$\sum_{i=1}^{i=3} x_i = x_1 + x_2 + x_3$$

The i in the equation above is a place-holding integer. In this instance i runs from 1 to 3, meaning there will be three separate values of x to sum together (x_1, x_2, and x_3). Sometimes the equation above is abbreviated to

$$\sum_{i=1}^{3} x_i = x_1 + x_2 + x_3$$

Or as

$$\sum_{1}^{3} x_i = x_1 + x_2 + x_3$$

An even more general expression is

$$\sum_{1}^{n} x_i = x_1 + x_2 + x_3 + \ldots + x_n$$

The expression above is sometimes written in more abbreviated form as

$$\Sigma x_i = x_1 + x_2 + x_3 + \ldots + x_n$$

Or sometimes as

$$\Sigma x = x_1 + x_2 + x_3 + \ldots + x_n$$

All of the following numerical examples are based on the table below:

x values	y values
$x_1 = 2$	$y_1 = 4$
$x_2 = 3$	$y_2 = 1$
$x_3 = 1$	$y_3 = 2$

$$\Sigma x = 2 + 3 + 1 = 6$$

$$\Sigma y = 4 + 1 + 2 = 7$$

$$\begin{aligned}\Sigma xy = \Sigma(xy) &= (2 \cdot 4) + (3 \cdot 1) + (1 \cdot 2) \\ &= 13\end{aligned}$$

$$\begin{aligned}\Sigma(x + y) &= (2 + 4) + (3 + 1) + (1 + 2) \\ &= 6 + 4 + 3 \\ &= 13 \text{ (being equal to the previous answer is a coincidence)}\end{aligned}$$

$$\begin{aligned}\Sigma x^2 &= 2^2 + 3^2 + 1^2 \\ &= 4 + 9 + 1 \\ &= 14\end{aligned}$$

$$\begin{aligned}(\Sigma x)^2 &= (2 + 3 + 1)^2 \\ &= 6^2 \\ &= 36\end{aligned}$$

$$\begin{aligned}\Sigma(x^2 + y^2) &= (4 + 16) + (9 + 1) + (1 + 4) \\ &= 20 + 10 + 5 \\ &= 35\end{aligned}$$

$$\begin{aligned}\Sigma(x + y)^2 &= (2 + 4)^2 + (3 + 1)^2 + (1 + 2)^2 \\ &= 6^2 + 4^2 + 3^2 \\ &= 36 + 16 + 9 \\ &= 61\end{aligned}$$

$$\begin{aligned}(\Sigma(x + y))^2 &= ((2 + 4) + (3 + 1) + (1 + 2))^2 \\ &= 13^2 \\ &= 169\end{aligned}$$

As shown in the numerical examples above, $\Sigma(xy)$ is commonly written as Σxy. However, it is important for the meaning of a mathematical expression to be clear to the reader. For example, writing $y\Sigma x$, $y(\Sigma x)$, or $(\Sigma x)y$ is preferable to $\Sigma(x)y$. The first three expressions clearly mean "multiply y by the sum of the xs." The fourth expression could be interpreted in the same manner as the first three, or it could mean "calculate the sum of $(x)y$." When in doubt, the unambiguous choice is the best choice.

Practice Problems 2.6

Use the following table when solving the practice problems:

A values	*B* values
$A_1 = 3$	$B_1 = 3$
$A_2 = 5$	$B_2 = 1$
$A_3 = 2$	$B_3 = 4$

1. $\Sigma A = ?$

2. $\Sigma B = ?$

3. $\Sigma AB = ?$

4. $\Sigma A^2 = ?$

5. $\Sigma B^2 = ?$

6. $(\Sigma A)^2 = ?$

7. $\Sigma(A^2 + B^2) = ?$

8. $\Sigma(A + B)^2 = ?$

9. $\Sigma A^2 + \Sigma B^2 = ?$

10. $(\Sigma(A + B))^2 = ?$

Equations

An equation is two mathematical expressions joined by an equal sign. For example, $2x = 6$ is an equation. This means "2 times x is precisely equal to 6." The expressions on each side of the equal signs below are equal:

$$2 + 1 = 5 - 2$$
$$3 = 3$$

$$(2 + 1) - 6 = (5 - 2) - 6$$
$$-3 = -3$$

$$(2 + 1) + 5 = (5 - 2) + 5$$
$$8 = 8$$

$$\frac{2 + 1}{3} = \frac{5 - 2}{3}$$
$$1 = 1$$

$$3(2 + 1) = 3(5 - 2)$$
$$9 = 9$$

Three things to keep in mind when working with equations are

1. If you add or subtract something to one side of an equation, you must add or subtract exactly the same thing to the other side. Otherwise, the two sides will no longer be equal.
2. If you multiply or divide one side of an equation by something, you must multiply or divide the other side by the same thing. Otherwise, the two sides will no longer be equal.

3. You can add, subtract, divide, or multiply one side of an equation as long as you do the same thing to the other side.

Knowing these three rules about equations facilitates solving for an unknown value. For example:

- If $4x - 5 = 15$, then $4x - 5 + 5 = 15 + 5$ and $4x = 20$.
- If $4x = 20$, then $4x \div 4 = 20 \div 4$ and $x = 5$.

Check your answer by plugging the value of x into the original equation $4(x) - 5 = 15$ and solving to be certain that substituting 5 for x yields the correct value on the righthand side of 15 (i.e., $4(5) - 5 = 20 - 5 = 15$).

Practice Problems 2.7

Solve for x:

1. $3x = 12$

2. $3x + 6 = 27 - 9$

3. $x + 25 = -50$

4. $(9 \div 3) = 2x - 7$

5. $x \div 9 = 7 + 4$

Factorials and Combinations

The symbol for factorial is an exclamation point (!). In other words, you would read 5! as "five factorial." The exclamation point denotes that the series of integers counting down to 1 are multiplied together; that is, $5! = 5 \cdot 4 \cdot 3 \cdot 2 \cdot 1 = 120$.

A few special rules apply to factorials:

- Factorials are defined only for zero and positive integers.
- $0! = 1$

Recall that in the logarithm discussion e was stated to be approximately equal to 2.718281828459. The calculation of the value of e involves factorials where

$$e = 1 + \frac{1}{1!} + \frac{1}{2!} + \frac{1}{3!} + \frac{1}{4!} + \ldots$$

which is an infinite series allowing you to calculate e as precisely as you desire. Because e cannot be stated exactly, it is what is called an *irrational number.*

In addition to computing e, factorials are useful for calculating combinations and permutations, which will be useful later in the discussion of probability.

Combinations are the number of ways you can select *X* objects out of a set of *n* objects without regard to the order in which they are chosen. Sometimes this is written as $\binom{n}{X}$, which is read as "*n* choose *X*" and is calculated as follows:

$$\binom{n}{X} = \frac{n!}{X!(n-X)!}$$

For example, how many ways can you select three floor coverings out of four possible choices?

$$\binom{4}{3} = \frac{4!}{3!(4-3)!} = \frac{4 \cdot 3!}{3! \cdot 1!} = \frac{4}{1} = 4$$

If the four floor covering options were carpet, vinyl, tile, or wood, the four possible combinations of three choices are

- Wood, tile, vinyl
- Wood, tile, carpet
- Wood, carpet, vinyl
- Tile, vinyl, carpet

Permutations are similar to combinations, but the order matters. Permutations are usually written as **n_P_X**. This notation asks how many ways can you arrange *X* objects chosen from n objects and is calculated as follows:

$$n\_P\_X = \frac{n!}{(n-X)!}$$

Because order matters with permutations, "wood, tile, vinyl" and "tile, vinyl, wood" are different and are counted separately. As a result, there are more permutations than combinations. For example, in the "choose 3 floor coverings out of 4" problem there are 24 permutations:

$$4\_P\_3 = \frac{4!}{(4-3)!} = \frac{4 \cdot 3 \cdot 2 \cdot 1}{1} = 24$$

All of this matters for three reasons:

1. Combinations are important for calculating probabilities for the binomial distribution (Chapter 5).
2. Combinations are important in and of themselves (e.g., figuring out how many different floor plans and exterior elevations can occur in a tract-built, single-builder subdivision).
3. Permutations are different from combinations, which is nice to know so you don't get them confused.

Practice Problems 2.8

1. Calculate 8!, 6!, and 3!.

2. Calculate $\frac{5!}{3!}$, $\frac{5! \cdot 3!}{6!}$, $\frac{7!}{5!(7-5)!}$.

3. How many combinations of three colors can be chosen out of six possible colors?

Types of Variables

In statistics, variables are characteristics that differ from one observation to another. Using housing as an example, differing characteristics can include size, ceiling height, exterior finish material, and exterior finish color. These characteristics are all variables that can be measured in a number of ways. Collections of measurements associated with each observation are called data. Variable type falls into one of two general categories: categorical variables and numerical variables.

Categorical Variables

Categorical variables yield "yes" or "no" answers to the question, "Does the observation fit this description?" The observation is either in the category or it is not. Consider a residential subdivision where homes have either masonry, stucco, or lap siding exteriors. Assume that these three categories exhaust all possibilities of exterior finish for this subdivision. Each type of finish can be described by yes or no responses. Four homes are categorized in **Table 2.1** as an example. Rather than set up a yes/no matrix, we could describe each house as having a certain type of exterior

Table 2.1 Categorical Exterior Finish Variable, Yes/No Form

	Masonry	Stucco	Lap Siding
302 Oak Street	yes	no	no
315 Oak Street	no	no	yes
316 Oak Street	no	yes	no
319 Oak Street	yes	no	no

finish. However, creation of a yes/no matrix is advantageous because it can be coded numerically later on and entered into a statistical program for analysis.

If the cell entries for the categorical exterior finish variable were coded so that "yes" = 1 and "no" = 0, then the yes/no matrix above would look like **Table 2.2**.

Notice that in Table 2.2 we can count how many homes have each type of exterior finish by summing the columns. In addition, we can count how many total homes are in the data set by summing the "Total" row. There are 4 homes (2 + 1 + 1). Also, when coded this way the categorical variable "exterior finish" can be entered numerically into a statistical model that could be employed to analyze whether or not exterior finish is significant in price differentiation. (Note that the values 1 and 0 only signify membership–or not–in a category. They have no numerical significance as either counts or measures.) A word of caution: categorical variables must be collectively exhaustive and mutually exclusive in order to be useful for statistical analysis. This means that every observation must fit into one of the categories (collectively exhaustive) and into no more than one category (mutually exclusive).

Nominal and Ordinal Categorical Variables

There are two subcategories of categorical variable: nominal variables and ordinal variables. Nominal variables classify characteristics by nomenclature (i.e., "name"). Table 2.2 is an example

Table 2.2 Categorical Exterior Finish Variable, Recoded to 1/0 Form

	Masonry	Stucco	Lap Siding
302 Oak Street	1	0	0
315 Oak Street	0	0	1
316 Oak Street	0	1	0
319 Oak Street	1	0	0
Total	2	1	1

of a nominal variable that could be called "exterior finish." No ranking of the exterior finish characteristics is implied.

Ordinal variables facilitate ranking (i.e., "order"). For example, tax assessors often create an ordinal variable called "condition." Categories might include poor, fair, good, and excellent. Each home in a market area could then be classified in the same yes/no fashion we used for exterior finish. Ordinal data categories can also be coded "yes = 1" and "no = 0" just as we did for exterior finish and analyzed mathematically.

The difference between a nominal and ordinal categorical variable is that there is an a priori expectation implicit in an ordinal variable. All else being equal, we expect a property on the assessor's roll with the entry "fair" for the ordinal variable condition to have a higher sale price in the market than a property characterized as in "poor" condition. Ordinal categorical variables can be difficult to classify because they generally involve more judgment than is required to assess nominal categories. For consistency's sake (and to ensure better analytical results) decision rules should be written and followed for categorical variable classification. Decision rules should provide clear guidance on what constitutes membership in one category and exclusion from another. (See Chapter 6 for more about data collection and classification protocols.)

Numerical Variables

Numerical variables provide numerical responses to questions like, "How many bedrooms and bathrooms does this house have?" Numerical variables can be either discrete or continuous. Discrete variables are created by counting. Numbers of bedrooms and baths are examples of discrete variables. In contrast, continuous variables are the result of measurement. House size, measured in square feet, is a continuous variable. Other examples might include lot width, ceiling height, and distance to a transportation node.

Later in this book we will see how numerical variables and categorical variables can be analyzed simultaneously in multiple regression models when categorical variables have been coded numerically, such as yes = 1 and no = 0. Using statistical tools, it is possible to develop housing price models that simultaneously include nominal categorical, ordinal categorical, discrete numerical, and continuous numerical variables.

Practice Problems 2.9

Classify each of the variables below as nominal, ordinal, discrete, or continuous.

1. Number of garage stalls:	nominal	ordinal	discreet	continuous
2. Ceiling height in an entry foyer:	nominal	ordinal	discreet	continuous
3. Distance from a regional mall:	nominal	ordinal	discreet	continuous
4. Number of three-bedroom apartments in a large apartment complex:	nominal	ordinal	discreet	continuous
5. Type of floor covering:	nominal	ordinal	discreet	continuous
6. Construction cost category (low, medium, or high):	nominal	ordinal	discreet	continuous
7. Type of windows:	nominal	ordinal	discreet	continuous
8. Linear feet of lake frontage:	nominal	ordinal	discreet	continuous
9. Mortgage interest rate:	nominal	ordinal	discreet	continuous

Working with Tables and Charts

As we discovered in Chapter 2, variables are characteristics that differ from one observation to another. In the analysis of real property markets, differing characteristics do not necessarily affect a property's value. For example, a second-level office suite might be otherwise identical to a different, third-level office suite. Although the suites differ on the variable "floor level," this difference may not affect market rent (although there are conceivable situations in which this difference could be material). Other variables, such as differences in leaseable area, interior finish allowance, or building location, are highly likely to affect market rent and property value. Variables that affect market rent or market price are referred to as *elements of comparison* by the appraisal profession.

This chapter addresses ways of organizing variables and describing them in tables and charts. The chapter has two overarching goals:

1. To develop data presentation skills necessary to enhance written and visual communication
2. To provide a foundation for material that will be presented in future chapters of this book

As you now know, data elements can be divided into two general categories: numerical variables and categorical variables. We will discuss each in turn, beginning with numerical data.

Numerical Data

Organizing Numerical Data: The Ordered Array

An ordered array is a listing of the values associated with a variable in numerical order beginning with the smallest value and ending with the largest value. **Table 3.1** shows right-of-way easement lengths in miles for 24 properties. As you can see, in an ordered array the minimum value, maximum value, and proportions within some range of value are easily identified. The shortest right-of-way easement is 0.00227 miles long and the longest is 3.6553 miles. A large proportion of the easements are one-quarter mile or less in length (15 out of 24, or 62.5%), and relatively few exceed a mile in length (five out of 24, or 20.83%).

Ordered arrays facilitate the creation of graphics such as stem and leaf displays and histograms. Ordered arrays also support derivation of some measures of central tendency and dispersion (such as the median and interquartile range) which will be developed fully in the next chapter.

Table 3.1 Ordered Array of Right-of-Way Easement Lengths in Miles

Group	Length	Note
one-quarter mile or less	0.00227	← shortest
one-quarter mile or less	0.01515	
one-quarter mile or less	0.01780	
one-quarter mile or less	0.01970	
one-quarter mile or less	0.02235	
one-quarter mile or less	0.02367	
one-quarter mile or less	0.03220	
one-quarter mile or less	0.04830	
one-quarter mile or less	0.05682	
one-quarter mile or less	0.07330	
one-quarter mile or less	0.07727	
one-quarter mile or less	0.11364	
one-quarter mile or less	0.18939	
one-quarter mile or less	0.21591	
one-quarter mile or less	0.25000	
	0.38352	
	0.51326	
	0.77652	
	0.89924	
one mile or longer	1.11705	
one mile or longer	1.38731	
one mile or longer	1.51515	
one mile or longer	2.08333	
one mile or longer	3.65530	← longest

Sorting in Excel, Minitab, and SPSS

Ordered arrays are easily produced in **Excel** by using the **Sort** command. Under the **Data** tab, select **Sort**, and then select the column or variable heading **Sort By**, and finally choose **Ascending** or **Descending**. SPSS and Minitab also include similar sort routines. In SPSS select the **Data** pull-down menu and then select **Sort Cases.** The Minitab procedure is similar. Select the **Data** pull-down menu and then select **Sort**. As with Excel, both SPSS and Minitab will sort in either ascending or descending order.

Organizing Numerical Data: Stem and Leaf Display

A stem and leaf display is a convenient way of organizing numerical data to see how data cluster and how the data points are distributed over the full range of values. Consider the following ordered array consisting of realized compound annual rates of return for a sample of sixteen large apartment complexes located in a major U.S. city, rounded to the nearest whole percentage:

-3%	-1%	0%	3%
5%	6%	10%	10%
12%	14%	16%	17%
17%	18%	21%	23%

A stem and leaf display can show how the rates of return are distributed and illustrate how the rate of return percentages cluster in the teens. To construct the display in **Figure 3.1**, let the stem be the tens column (e.g., -3% is -03% with a -0 in the tens column, 3% is 03% with 0 in the tens column, 10% has a 1 in the tens column, and 21% has a 2 in the tens column). The leaves are the numbers in the ones column.

One advantage of the stem and leaf display is an ability to numerically reconstruct the raw data from the display, which is generally not possible with other pictorial representations of data. The disadvantage is that stem and leaf displays can sometimes be difficult to comprehend if the reader is unfamiliar with them. This can be a problem for an appraiser if the intended user needs a wordy explanation of how to read the display and how it was constructed. When it comes to choosing graphic illustrations, it is important to consider which type of graphic is suitable for the needs of the client (e.g., will the illustration be understood by the client and other intended users and be appropriate for the intended use?).

Figure 3.1 **Stem and Leaf Display of a Sample of Annual Returns on Apartment Investments**

Tens %	Ones %
-0	3 1
0	0 3 5 6
1	0 0 2 4 6 7 7 8
2	1 3

Stem and Leaf Displays in Minitab

Minitab has stem and leaf creation capability, but Excel and SPSS do not. To construct a stem and leaf display in Minitab, select the **Graph** pull-down menu and then select **Stem-and-Leaf**. A **Stem-and-Leaf** window will open that allows you to select the variable (or variables) you want to display. The **Stem-and-Leaf** window also includes a **By Variable** feature, allowing comparison of a variable's stem and leaf displays by subcategory. For example, an apartment rent variable could be subcategorized by each apartment unit's number of bedrooms. Stem and leaf displays could then be compared for differing bedroom counts to illustrate how rents cluster around larger amounts as bedrooms per unit increase. Stem and leaf diagrams can also be typed directly into a word processing document. For example, Figure 3.1 was initially typed into a Microsoft Word document using tabs prior to being typeset for this book.

Practice Problems 3.1

1. Place the lot size data shown in the stem and leaf display below into an ordered array.

Stem and Leaf Display of a Sample of 1 Acre or Larger Residential Lot Sales within the Past Year

Acres	Tenths of an Acre
1	1 1 2 3 3 4
1	5 6 7 7 9
2	0 0 0 0 1 2 4
2	5 5 6 7 8 8 8 9
3	0 1 3 4
3	5 7
4	3
4	5 8
5	0
5	6
6	2

2. What proportion of the acre-or-larger lots are greater than 4 acres in area? Less than 1.5 acres? Greater than or equal to 2 acres? Less than 3.5 acres?

3. Based on the lot acreage sample above, what is the most likely size category for 1 acre or larger residential lots?

4. The following data consists of a sample of discount rates (in percentages net of developer's profit) used in DCF models of subdivision development cash flow from appraisals submitted to a regional lending institution over the past year.

12.3	12.4	11.7	12.8	9.1
11.5	9.4	10.2	10.2	10.3
10.8	11.1	11.5	11.6	11.6
11.7	11.7	12.2	12.2	12.9
13.0	11.2	13.2	9.7	10.0

a. Place these data into an ordered array.

b. Construct a stem and leaf display for these data.

c. What discount rate category is most likely?

Frequency and Percentage Distributions

Often it is more practical to illustrate a data range and clustering in a more condensed and easy-to-interpret format. Frequency and percentage distributions are means to accomplish these goals in tabular form. The procedure essentially converts a large amount of numerical information into a small number of categories. That is, these tables convert numerical variables into ordinal categories (i.e., categorical data). While detail is lost, the creation of concise, easy-to-interpret information can be beneficial.

Frequency Distribution

The ordinal categories in frequency and percentage distributions are usually called "classes." As an example, let's look at the lot size data from the stem and leaf diagram in the first problem in Practice Problems 3.1. Eleven of the lots were 1 to 1.9 acres in area. Fifteen were 2 to 2.9 acres. Six were 3 to 3.9 acres. Three were 4 to 4.9 acres, two were 5 to 5.9 acres, and one was 6 to 6.9 acres. Viewed this way, six classes can be constructed from the data. The number of lots in each class is the frequency of occurrence (i.e., frequency = count of observations in a class). This information can be captured in tabular form (**Table 3.2**), and the resulting table is called a *frequency distribution.*

Table 3.2 Frequency Distribution of 1 Acre or Larger Residential Lot Sales within the Past Year

Lot Size (Acres)	Number of Lots
1 to 1.9	11
2 to 2.9	15
3 to 3.9	6
4 to 4.9	3
5 to 5.9	2
6 to 6.9	1
Total	38

Guidelines for good-quality frequency distribution presentation include the following:

- The frequency distribution should have 5 to 15 classes. Too few or too many classes detract from the effectiveness of the presentation.
- Each class should have the same width. Otherwise, the frequency distribution may be misleading.
- Class width is approximately determined by the range of the data (largest value minus smallest value) and the number of classes:

$$\textbf{Class Width} \cong \frac{\textbf{Range}}{\textbf{Number of Classes}}$$

For the lot-size example above, the class width is 1 acre, derived as follows:

$$\textbf{Class Width} \cong \frac{5.9}{6} \cong 0.98$$

- Class boundaries should be collectively exhaustive and mutually exclusive. This means that the classes should accommodate all of the observations in the data, but the class boundaries should not overlap.
- Do not choose class widths or boundaries that mask or misrepresent the data. Too many or too few classes can distort the distribution of the data, especially when applied to small data sets. Decisions concerning the number of classes and class width can be somewhat subjective, so it is a good practice to experiment with varying class widths to determine which best illustrates the distribution of the underlying data.

Percentage Distribution

When frequency distribution information is altered to express class counts in relative rather than absolute terms, the resulting table is referred to as either a *relative frequency distribution* or a *percentage distribution.* When class counts are converted to proportions, the result is a relative frequency distribution. When class counts are converted to percentages, the result is a percentage distribution.

Table 3.3 presents the frequency information from Table 3.2 in relative form as a percentage distribution. The form is "relative" because the counts in the lot size classes are now expressed relative to the total number of lots in the sample. For example, the relative percentage of lots in the 1- to 1.9-acre class is calculated by dividing the 11 lots in the class by the total of 38 lots in the sample ($11 \div 38 = 0.289$, or 28.9%).

Cumulative percentage distributions are variations on the basic percentage distribution. The cumulative percentage assists the reader in determining the value range that captures the bulk of the observations. It also helps illustrate how the data cluster.

In **Table 3.4**, the information presented

Table 3.3 Percentage Distribution of 1 Acre or Larger Residential Lot Sales within the Past Year

Lot Size (Acres)	Percentage of Total
1 to 1.9	28.9%
2 to 2.9	39.5%
3 to 3.9	15.8%
4 to 4.9	7.9%
5 to 5.9	5.3%
6 to 6.9	2.6%
Total	100%

Note: The percentages are rounded to the nearest tenth of a percent. Although the sum of the percentages in this exhibit is 100%, sometimes rounding error will cause the total to differ slightly from 100%. When this occurs, inclusion of a simple statement saying "The sum of the percentages differs slightly from 100% due to rounding" is sufficient.

in Table 3.3 is expanded to include a cumulative percentage column. This table could be used to illustrate and support a statement such as, "Lots are typically sized from one to four acres, but the developer has included some larger sized lots when site topography required more land to accommodate an adequately level building site or drainage easement."

Table 3.4 Percentage and Cumulative Percentage Distributions of 1 Acre or Larger Residential Lot Sales within the Past Year

Lot Size (Acres)	Percentage of Total	Cumulative Percentage
1 to 1.9	28.9%	28.9%
2 to 2.9	39.5%	68.4%
3 to 3.9	15.8%	84.2%
4 to 4.9	7.9%	92.1%
5 to 5.9	5.3%	97.4%
6 to 6.9	2.6%	100.0%
Total	100%	

Histograms

Readers of appraisal reports usually find that graphic representations of data are easier to understand than tables containing the same information. A histogram turns a frequency or percentage distribution into an easy-to-read chart.

When plotting a histogram, the class labels are plotted along the horizontal axis, keeping in mind that they represent an ordinal scale. Frequencies or percentages associated with each class are plotted along the vertical axis. The histogram in **Figure 3.2** was created by charting the frequency distribution information found in Table 3.2.

A similar histogram results when the percentage distribution presented in Table 3.3 is plotted as a histogram. As **Figure 3.3** shows, the shape is identical to the frequency histogram. The only difference is the scale and labels applied to the vertical axis (i.e., percentages replace class frequencies).

Tables 3.3 and 3.4 were referred to earlier as support for the statement, "Lots are typically sized from one to four acres, but the developer has included some larger sized lots when site topography required more land to accommodate an adequately level building site or drainage

Histograms in Excel, Minitab, and SPSS

Histograms are easily created in Excel and, in contrast to SPSS or Minitab, you have total control of the appearance of the resulting chart. When using Excel to plot a histogram, create a frequency or percentage distribution on a worksheet and then use the **Chart** feature to create a "Column" chart. The width of the resulting histogram's vertical bars can be altered by selecting the bars, right clicking on them, choosing **Format Data Series**, clicking on the **Options** tab, and then adjusting gap width. In SPSS, select the **Graph** menu, click **Chart Builder**, and then select **Histogram**. In Minitab, choose **Graph** and then **Histogram**.

Figure 3.2 Histogram of 1 Acre or Larger Residential Lot Sales Frequency Distribution

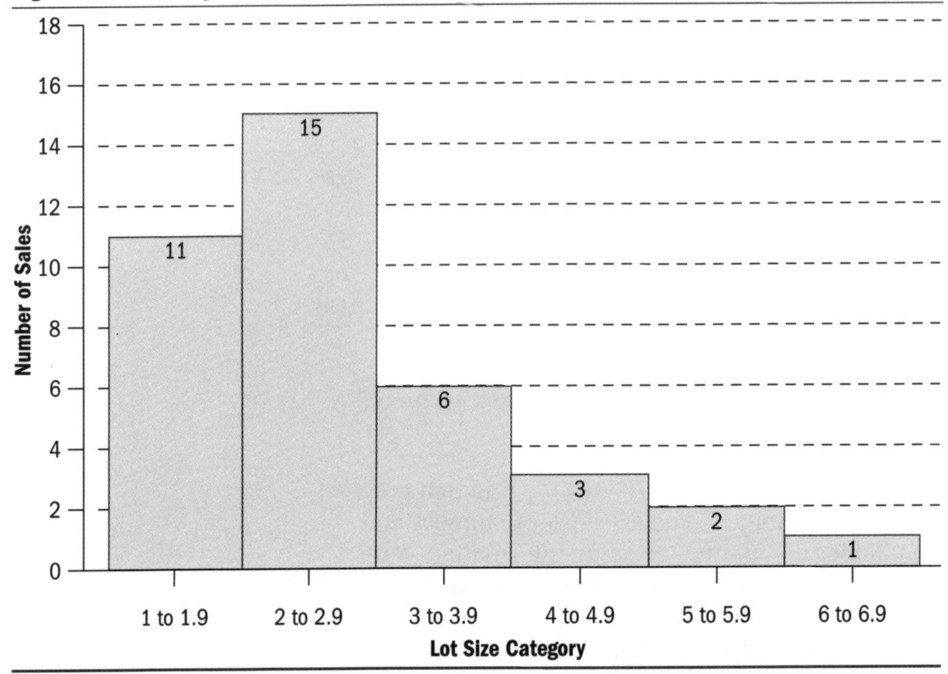

Figure 3.3 Histogram of 1 Acre or Larger Residential Lot Sales Percentage Distribution

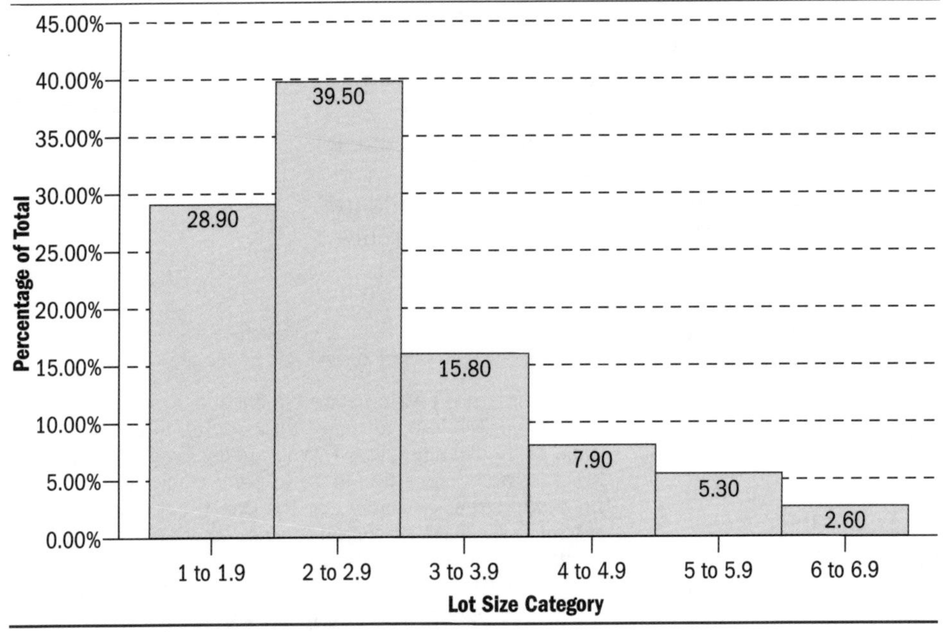

easement." The histograms shown in Figures 3.2 and 3.3 capture this idea graphically, and they give the reader a way to visually interpret the development's proportion of standard sized lots of less than four acres versus larger lots in the development.

Practice Problems 3.2

The data below represent average monthly electric bills for a random sample of two-bedroom apartment units over the past year.

89	94	110	103	94	112	87	115	97	89	109
118	82	98	98	93	99	104	106	101	102	97
105	98	106	114	97	93	108	98	102	87	112
81	98	103	92	106	95	115	97	102	88	97
93	92	94	97	99	98	103	105	102	112	109

1. Sort these data into an ordered array.
2. Create a frequency distribution with an appropriate number of classes and class intervals.
3. Convert the frequency distribution into a percentage distribution.
4. Create a cumulative percentage distribution.
5. Plot a frequency histogram based on your answers to Problem 2 above.
6. Write a brief interpretive summary of the data consistent with the distributions you have created.

Polygons

A comparison of more than one numerical data distribution on a single chart can often be meaningful. Two stem and leaf diagrams can be constructed on the same stem scale and compared side-by-side, but they retain their inherent interpretative difficulty. Histograms can be superimposed, but this format can be misleading at worst and confusing to the reader at best. By comparison, polygons provide a less confusing means of comparing two or more data distributions.

Table 3.5 compares the percentage distributions of lot sales

Table 3.5 Percentage Distribution of 1 Acre or Larger Residential Lot Sales in Phases 1 and 2.

Lot Size (Acres)	Phase 1 Percentage of Total	Phase 2 Percentage of Total
1 to 1.9	28.9%	16.2%
2 to 2.9	39.5%	59.5%
3 to 3.9	15.8%	13.5%
4 to 4.9	7.9%	5.4%
5 to 5.9	5.3%	2.7%
6 to 6.9	2.6%	2.7%
Total	100%	100%

Polygons in Excel

Polygons are easily created in Excel. To replicate the chart shown in Figure 3.4, first enter the data as shown in the table below. Leave the class midpoint blank for the 0.5 and 7.5 entries, and enter 0% in each phase for these midpoints. This keeps the horizontal axis labels consistent with the classes shown in the related histogram. Then select **Chart, Chart Type**, and **Line Chart** from the list of standard types of charts. The default "lines with markers" line chart format works well. Click **Next** and then select the **Series** tab in the chart source data window. This allows you to enter the two Phase 1 and Phase 2 data series manually, using the class midpoint data as the horizontal axis for both lot size series.

Spreadsheet Entries Required to Create the Figure 3.4 Polygons

Class Midpoint	Phase 1 Percentage	Phase 2 Percentage
	0.0%	0.0%
1.5	28.9%	16.2%
2.5	39.5%	59.5%
3.5	15.8%	13.5%
4.5	7.9%	5.4%
5.5	5.3%	2.7%
6.5	2.6%	2.7%
	0.0%	0.0%

by acreage class for two phases of a residential development. These data are easily compared by creating two, lot-size-by-class polygons as shown in **Figure 3.4**.

As shown in Figure 3.4, the central tendency for lot sales remains in the 2- to 2.9-acre class. However, the polygons illustrate that the sales in Phase 2 are more concentrated around the central tendency, whereas lot areas in Phase 1 are spread more widely both above and below the central tendency.

In this instance, variation in the lot size distribution is explained by differences in topography between the two

Figure 3.4 Percentage Polygons Comparing Lot Sale Acreages in Phase 1 and Phase 2

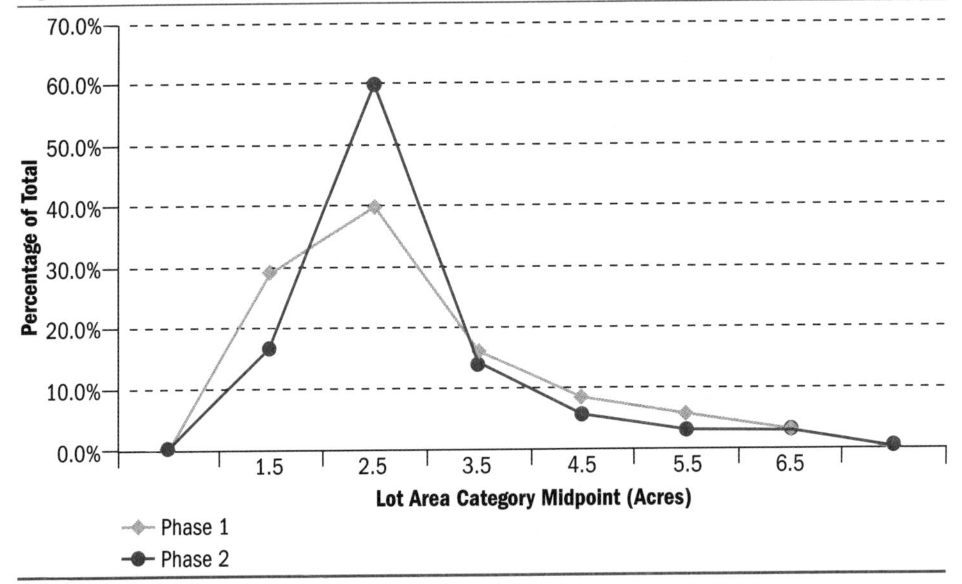

development phases. In Phase 1 more large lots were necessary to accommodate the steeper topography of the lots in that group. The resulting impact on Phase 1 lot density was mitigated by inclusion of a higher proportion of relatively smaller 1- to 1.9-acre lots in that phase. A more level topography allowed the developer to reduce the proportion of large lots in Phase 2.

Notice that the polygons in Figure 3.4 plot the lot size percentages on the vertical axis against the *midpoints* of the lot size class intervals on the horizontal axis. Two additional points were plotted on the horizontal axis corresponding with values of zero on the vertical axis. These represent the midpoints of the class intervals immediately above and below the classes shown on the equivalent histogram. As a general rule, these two extra midpoints are not labeled on the horizontal axis.

When a polygon chart is used in an analytical report, the accompanying narrative should clearly state that the horizontal axis values represent

Practice Problems 3.3

The data below represent average monthly electric bills for a random sample of one-bedroom apartment units over the past year.

79	84	92	91	83	102	77	104	87	78	98
106	74	88	87	83	89	94	96	91	92	87
95	88	96	103	86	83	98	88	92	77	102
72	88	93	82	98	87	105	87	92	78	87
83	82	84	87	89	88	93	95	93	100	98

Equivalent data on two-bedroom apartments are included below (data are taken from the previous practice problem set).

89	94	110	103	94	112	87	115	97	89	109
118	82	98	98	93	99	104	106	101	102	97
105	98	106	114	97	93	108	98	102	87	112
81	98	103	92	106	95	115	97	102	88	97
93	92	94	97	99	98	103	105	102	112	109

1. Sort these data into ordered arrays.
2. Create separate frequency and percentage distributions for the one-bedroom and two-bedroom electric bill data.
3. Plot two percentage polygons on a single chart comparing the electric bill distributions for the two different sized apartment units.

class midpoints. For example, the horizontal axis in Figure 3.4 is labeled accordingly.

Bivariate Numerical Data: Scatter Diagrams

Scatter diagrams show the strength of a relationship between paired numerical variables (i.e., bivariate numerical data). A few examples of related bivariate real estate data include

- Floor area and rent
- Bedroom count and floor area
- The quantity of insulation and annual energy consumption

The values of *directly* related variables increase together (one tends to go up when the other goes up) and decrease together. The values of *inversely* related variables move in opposite directions (one tends to go up when the other goes down). For example, floor area and rent would be directly related while insulation quantity and energy consumption would be inversely related, all else being equal.

When variables move together they are said to be *correlated.* If two paired variables are unrelated, then the movement of one provides no insight regarding the expected direction of movement of the other. Unrelated variables are said to be *uncorrelated.* Correlation of paired variables is not the same as causation. Analysis of causation is more complex than mere observation of correlation. For example, an internist might observe a direct relationship between high blood pressure and the incidence of heart disease and erroneously infer that high blood pressure causes heart disease. In reality, poor eating and exercise habits are more likely to have been root causes of the observed direct relationship between these correlated variables.

Figure 3.5 is an example of a scatter diagram, also known as a *scatter plot.* Each of the 36 dots in the plot represents a pair of observations about an apartment unit–square feet of living area and rent. Living area is measured on the horizontal axis and corresponding rent is measured along the vertical axis. The scatter plot demonstrates that these data are directly related. As living area increases, rent also tends to increase. As living area decreases, rent also tends to decrease.

Figure 3.5 Scatter Diagram of Apartment Rent and Living Area in Square Feet

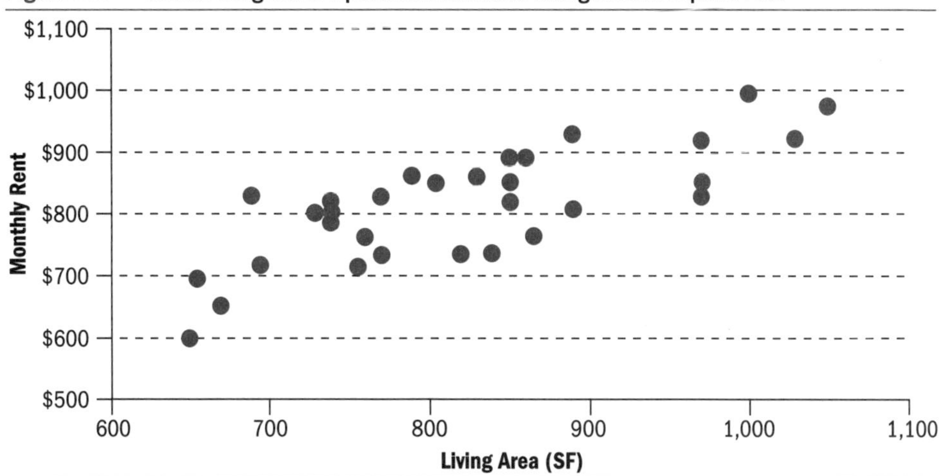

Later in the book we focus on independent and dependent variables. Traditionally, dependent variables are plotted on the vertical, or y, axis, and independent variables are plotted on the horizontal, or x, axis. The y and x designations are artifacts of the way algebraic equations are generally written with y being expressed as a function of x–that is, $y = f(x)$, where the value of y is a result of the value of x. In Figure 3.5 monthly rent is depicted as the dependent variable, which depends, in part, on an apartment's improved living area.

In a scatter plot, the strength of the relationship (i.e., extent of correlation) is illustrated by how close the markers come to forming a straight line. The extent of correlation is quantified by the correlation coefficient, which is 0.747 for the data shown in Figure 3.5. A perfect direct correlation would have a correlation coefficient of 1.00, and a perfect inverse correlation would have a correlation coefficient of -1.00. In contrast, totally uncorrelated data have a correlation coefficient of zero (see Chapter 4 for more about calculating this measure of correlation).

Scatter Plots in Excel

Scatter plots can be created in Excel by selecting **Chart** on the **Insert** tab pull-down menu (or save a step by clicking the **Chart** icon on the menu bar). Select the **XY (scatter)** chart from the list of standard types of charts. The scatter chart without lines is the default selection. Click **Next** and then select the **Series** tab in the **Chart Source Data** window. Click the **Add** button and enter the name of the series (e.g., "rent"). Select the data ranges for the horizontal axis (**X values**) and vertical axis (**Y values**) and click **Finish**. You can then right-click on various chart areas to customize their appearance.

As you can see, scatter diagrams capture the underlying mathematical relationship between two variables pictorially. As we shall see later in the book, these diagrams help visually illustrate the meaning of more complex numerical measures such as correlation coefficients and regression model results.

As an analytical tool, scatter plots can be used to assess the effect of an action or event–often called a *treatment* in statistical analysis–by either plotting data pairs before and after the event or plotting samples from affected and unaffected populations. As an example, let's analyze a rural town where almost all of the homes were built prior to adoption of modern energy conservation standards. Nearly all of the residents rely on propane to heat their homes, and many also use it for cooking and water heating. Residents were offered matching fund grants through a state program allowing them to make energy conservation renovations at half the usual cost. Qualifying renovations included adding insulation to walls and attics, installing more fuel-efficient furnaces, weather sealing doors and windows, and replacing single-pane windows with double-glazed windows. Many residents took advantage of the state program and many did not.

Figure 3.6 shows two scatter plots combined on a single chart. The data pairs are living area in square feet and annual propane use over the past year in

Figure 3.6 Scatter Plot of Renovated and Unrenovated Annual Propane Usage

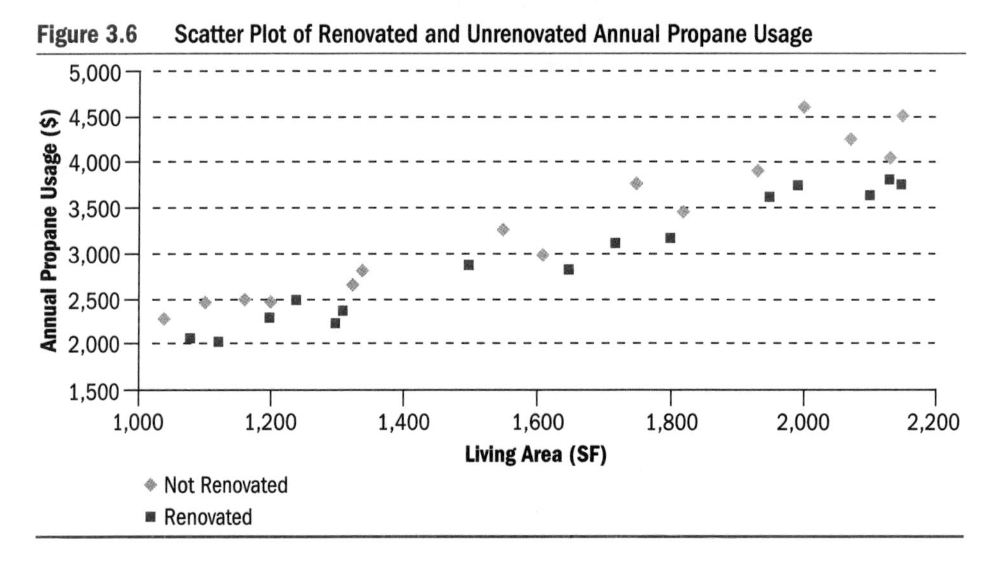

dollars for unaffected (not renovated) homes and for affected (renovated) homes. Renovated homes experienced 13% lower propane costs on average.

Bivariate Numerical Data: Time Trends

When one of the observations in a bivariate data pair is a time series (e.g., series of days, months, quarters, years), the resulting scatter plot is called a *time trend plot* or *time series plot.* The data pairs in a time trend plot are generally connected by lines to indicate that the observations occurred in sequential order. Markers may or may not be included in the plot, depending on how the illustration is being used (i.e., whether the periodic pair values are important, the overall trend is important, or both are important).

> **Time Trends in Excel**
>
> Time trends can be plotted in Excel by selecting the **Chart** icon and then choosing the **Line** option from the list of standard types of charts. **Lines with markers** is the default version of the line chart, which was used to create Figure 3.7. More than one time trend can be shown by adding a **Series** to the Excel chart. For example, average rent for one-bedroom apartment units could be added to Figure 3.7 to illustrate the extent to which the rents for the two differing apartment unit types moved together or moved in different directions over the 10-year period.

Figure 3.7 illustrates a time trend plot of average rent for two-bedroom apartment units in a defined market area. Illustrations such as this can be used to identify and emphasize historical market phenomena that may require further discussion and analysis. Questions arising from Figure 3.7 might include

- What is the historical long-term rent growth rate and what economic forces were at play in supporting this growth?

Figure 3.7 Time Trend Plot of Average Two-Bedroom Apartment Rent

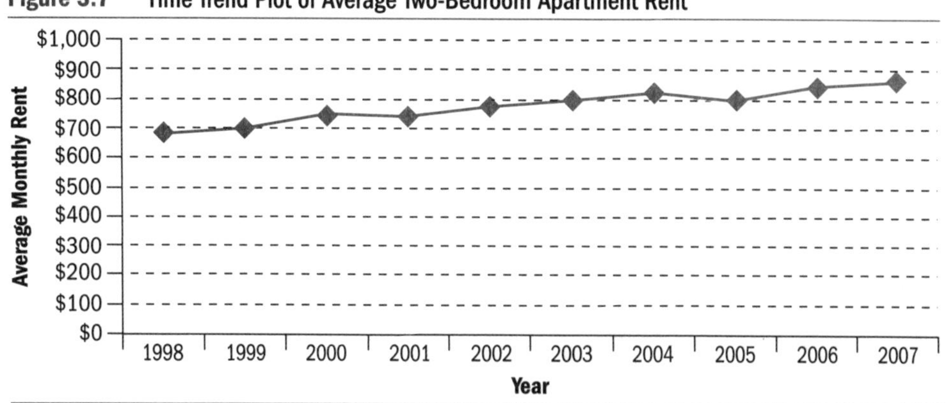

Practice Problems 3.4

1. The data below are paired samples of the improved living area and the most recent year's annual water bill for single-family homes in a suburban market area. Create a scatter plot and decide how the data are correlated (directly, inversely, or not at all).

Square Feet	1,240	1,400	1,100	1,600	1,300	1,800	2,200	1,380	1,520
Water Bill	$223	$280	$210	$319	$247	$370	$440	$285	$295

2. The data set shown here consists of city population by year. Create a population time trend plot. Does the plot indicate any particular year or years that you might want to investigate more closely?

Year	1997	1998	1999	2000	2001	2002	2003	2004	2005
Population (000)	310	315	321	332	337	343	347	347	348

- What factors were at play when average rents experienced slight reductions in 2001 and 2005?

Recognizing and analyzing market forces like these can provide support for market forecasts (in this case, a forecast of average rent for two-bedroom apartment units).

Categorical Data

Our discussion of categorical data picks up with a concept that is similar to the frequency and percentage distributions presented in the numerical data section–the summary table. Summary tables are similar to frequency and percentage distributions because they list categories and counts (frequencies) in each category. Counts can be converted to percentages and listed as well. Unlike percentage and frequency distributions, the categories in a summary table are not *necessarily* ordinal but may be ordinal if the categories are ordinal.

Table 3.6 is a summary table listing the total square footages of industrial space in an industrial district. The type of space variable is nominal, making the order of entry irrelevant. Percentages are also included in this table. Each percentage is derived from a ratio of

Table 3.6 Summary Table for Industrial Space in the ABC Industrial District

Type of Space	Floor Area (SF)	Percentage of Total Floor Area
Cold Storage	435,800	13.6%
Dock-High Warehouse	1,220,900	38.0%
Multi-Tenant Office Warehouse	675,200	21.0%
Light Industry	880,700	27.4%
Total	3,212,600	100.0%

each category's square footage to the total square footage. Percentages are often included in summary tables to assist the reader in better understanding the relative proportions in each category.

Table 3.7 is an example of a summary table for ordinal categorical data. With ordinal data like "condition," the order of entry in a summary table category list matters. The list should be in either ascending or descending order. As Table 3.7 shows, a summary table can accommodate more that one set of data, facilitating side-by-side comparisons.

Table 3.7 Summary Table of Tax Assessor Residential Condition Ratings by Market Area

	Southtown		Northtown	
Condition	**Count**	**Percentage**	**Count**	**Percentage**
Poor	27	5.3%	40	3.8%
Fair	79	15.6%	160	15.1%
Good	384	76.0%	643	60.5%
Excellent	15	3.0%	220	20.7%
Total	505		1,063	

Percentages may not total to 100% due to rounding.

Bar Charts and Pie Charts

Bar charts and pie charts provide means of representing summary table data pictorially. As with the inclusion of percentages in summary tables, charts provide a means of assessing relative proportions by category. **Figure 3.8** illustrates the square footage information found in Table 3.6 as a horizontal bar chart. In contrast, **Figure 3.9** cap-

Figure 3.8 Bar Chart of Industrial Space in the ABC Industrial District

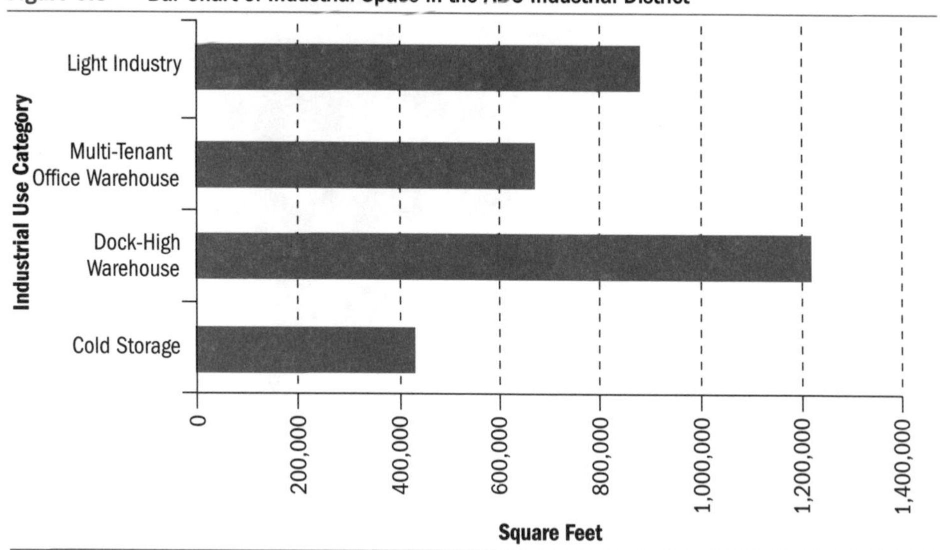

Figure 3.9 Pie Chart of Industrial Space in the ABC Industrial District

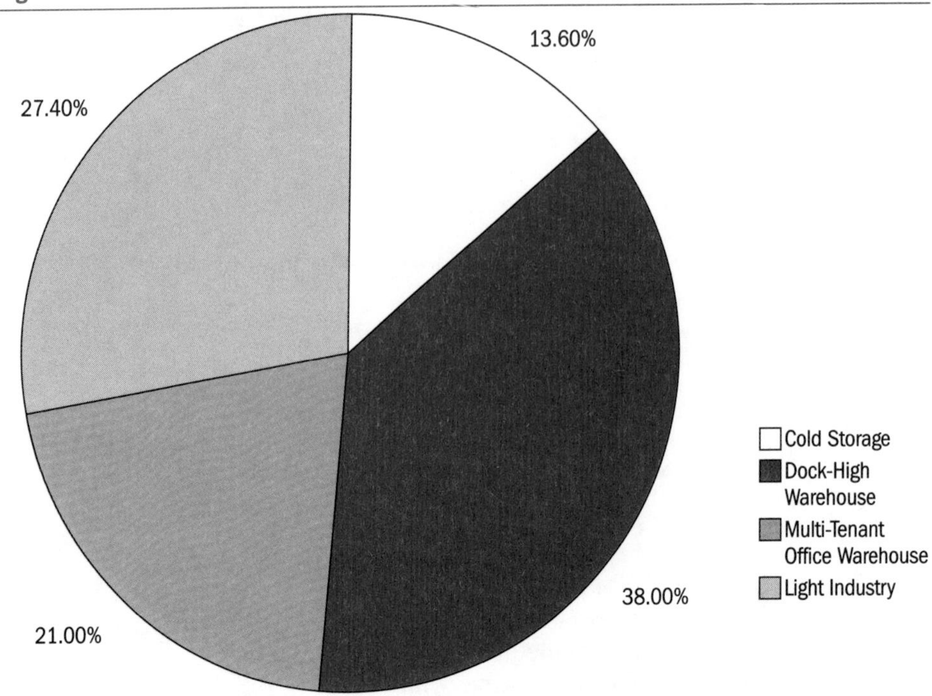

Figure 3.10 Bar Chart Comparison of Southtown and Northtown Residential Property Condition

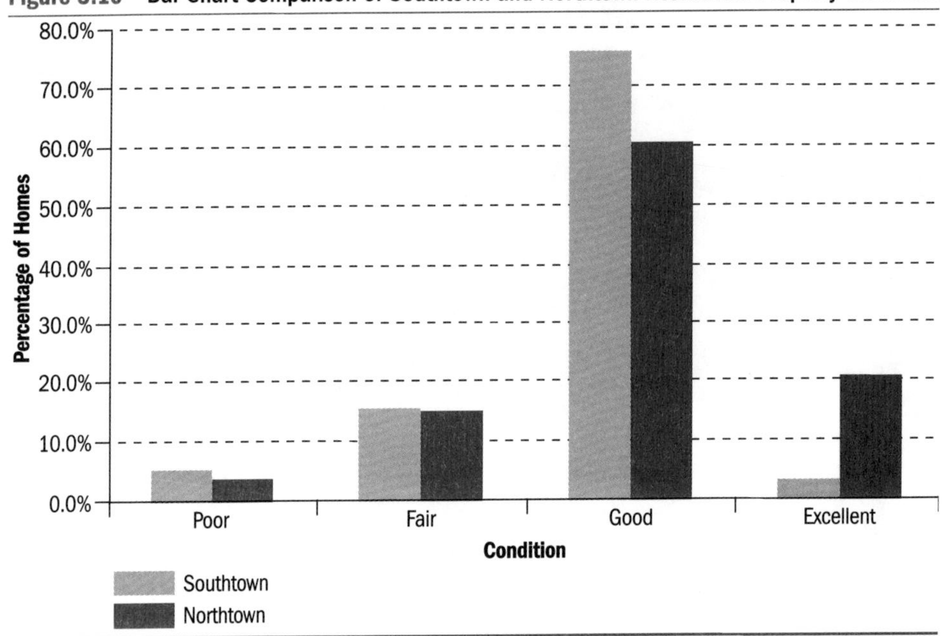

tures the same information converted to percentages and presented as a pie chart.

Bar charts are a good way to compare categorical data. For example, comparisons of the two market areas summarized in Table 3.7 are much more easily made from a bar chart than from the summary table. **Figure 3.10** illustrates one effective way to present these data. This bar chart comparison shows that the proportion of homes in excellent condition is much higher in Northtown. Northtown has proportionally fewer homes in poor and fair condition. However, the primary differences between these market areas are evident in the good and excellent categories.

Bivariate Categorical Data: The Contingency Table

Contingency tables facilitate simultaneous investigation of two categorical variables. We already did this without formally labeling it as such in Table 3.7 when we looked simultaneously at condition and market area. The comparison was charted in Figure 3.10.

Suppose you suspect some fundamental, underlying cause for the observed difference in condition by market area. If, for example, the average age of homes is greater in Southtown than in Northtown, the difference in condition ratings in the tax assessor data might be related to the age of the homes.

The contingency table in **Table 3.8** shows that 85% of homes rated "poor" are pre-1970. Similarly, about 82% of the homes rated "fair" are of this early vintage. In comparison, approximately 74% of homes rated "good" are post-1970, and about 91% of the "excellent" homes are from the post-1970 period. Based on this analysis, the association between condition and location may be related to differences in home age, with Southtown having older homes, on average. The side-by-side

Table 3.8 Contingency Table Comparing Home Age and Tax Assessor Condition Rating

	Age Category		
Condition	**Pre 1970**	**Post 1970**	**Totals**
Poor	57	10	67
Fair	196	43	239
Good	271	756	1,027
Excellent	20	215	235
Totals	457	1,111	1,568

bar chart in **Figure 3.11** illustrates the extent of the age-related condition dichotomy.

While we could include many more age categories than pre-1970 and post-1970, it is generally inadvisable to create too many comparison categories when a side-by-side bar chart will be used to facilitate data interpretation. These charts become confusing and counterproductive when too many classifications are being compared side by side. A little experimentation may be required to find a meaningful breakpoint, such as the 1970 date used in Figure 3.11. However, searching for a meaningful breakpoint may be time well spent if the resulting illustration provides a powerful communication tool and increases the depth of market knowledge.

Figure 3.11 Side-by-Side Bar Chart of Condition Rating by Age Category

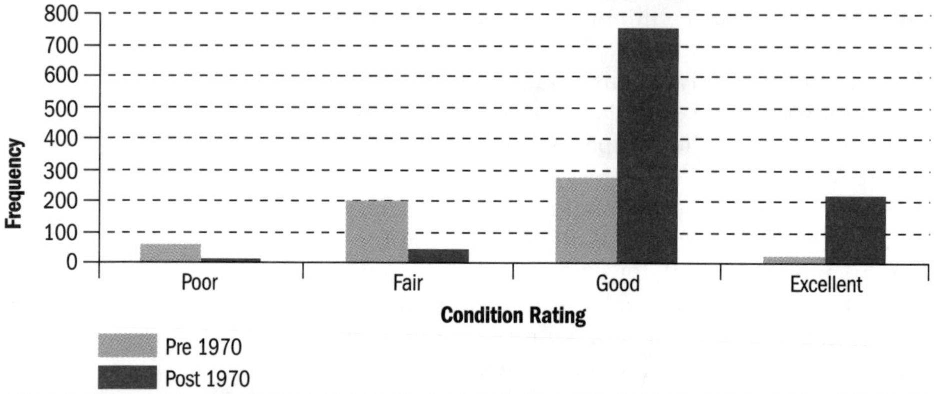

Practice Problems 3.5

1. A survey of real estate appraisers indicates that 60% of appraisers work at firms having a local practice, 28.3% work at firms with regional practices, 10.1% at national firms, and 1.6% at international firms. Organize these data into a summary table and create the related bar and pie charts.

2. An appraiser gathered a random sample of unit types in large apartment complexes located in two different market areas of a major city. One market area is influenced by proximity to a major university with an enrollment of 47,000 students. The other market area is primarily populated by middle-income families, many of whom work at one of two large employers making aircraft components and drive-train components for large trucks. The university market area sample includes 1,360 apartment units consisting of 320 studios, 428 one-bedroom units, 560 two-bedroom units, and 52 three-bedroom units. The manufacturing market area sample includes 1,580 units consisting of 38 studios, 264 one-bedroom units, 780 two-bedroom units, and 498 three-bedroom units. Create a contingency table comparing unit type and market area. Construct a side-by-side bar chart to illustrate unit mix differences between the two market areas.

Excellence and Ethics in Charting

Proper graphical data presentation should focus on two overarching themes:

1. Enhancing reader comprehension
2. Avoiding distortion

Many books and statistics textbook chapters are devoted to this topic. Ideas from several authors are summarized here.[1]

Ideally a chart should

- Include all of the relevant data
- Focus on the substance of the data presentation
- Not distort the data
- Serve a clear purpose
- Be verbally integrated into textual material
- Encourage data comparison and facilitate depth of understanding
- Communicate the information using the least ink possible

Ethical and excellent chart presentations should provide substance rather than style and communicate ideas clearly. Avoid the "*USA Today*" syndrome –that is, the inclusion of iconic symbols as markers or bars. Graphs using trees, cars, soda bottles, airplanes, and the like as markers are distracting and do not contribute to concise communication.

Most importantly, a graphic illustration should not misrepresent the data or mislead the reader. Distortion occurs when the visual presentation is inconsistent with underlying numerical values. In particular, avoid pictorial representations that distort volatility or rates of change.

As an example, the first chart in **Figure 3.12** shows the time series chart of two-bedroom apartment rents presented earlier as Figure 3.7. The second chart is a representation of the same data

1. Mark L. Berenson, David M. Levine, and T. C. Krehbiel, *Basic Business Statistics: Concepts and Applications*, 9th ed. (Upper Saddle River, N.J.: Pearson/Prentice Hall, 2004); Edward R. Tufte, *The Visual Display of Quantitative Information*, 2nd ed. (Cheshire, Conn.: Graphics Press, 2001); Edward R. Tufte, *Visual Explanations* (Cheshire, Conn.: Graphics Press, 1997); Edward R. Tufte, *Envisioning Information* (Cheshire, Conn.: Graphics Press, 1990).

Figure 3.12 Time Trend Plots of Average Two-Bedroom Apartment Rent on Two Different Scales

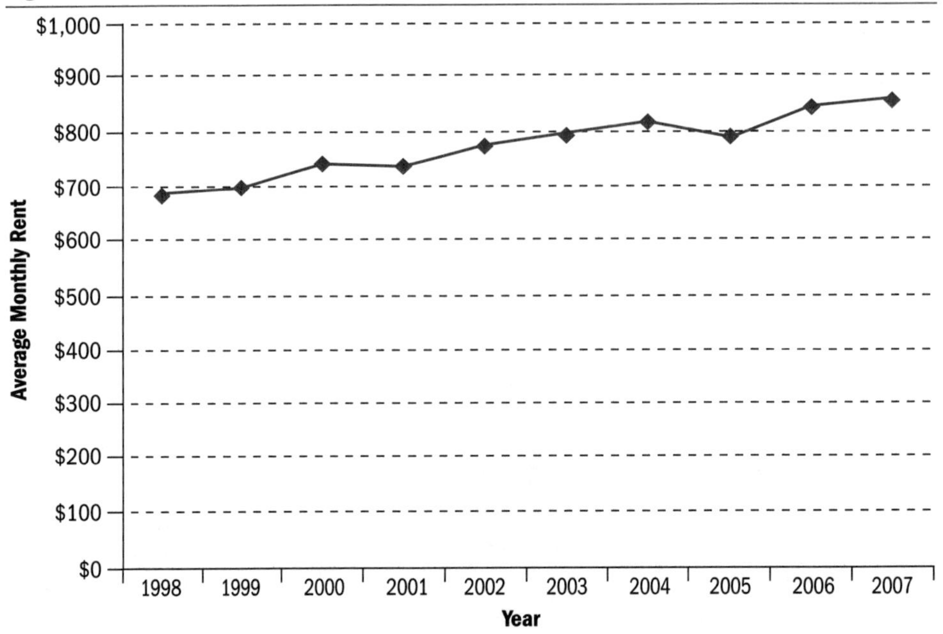

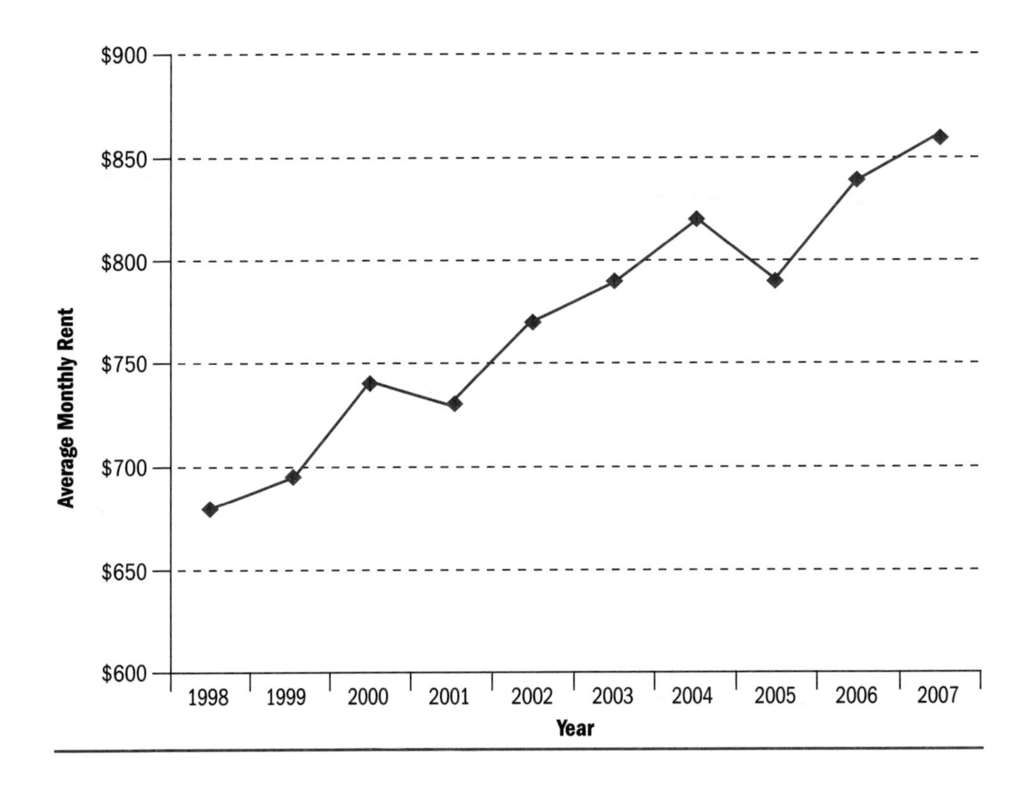

using a different vertical axis scale. In addition, the vertical and horizontal frame proportions of the second chart were altered to make the chart look taller and slimmer.

The second chart distorts the information in two ways. First, altering the vertical axis scale makes the rent change over time appear to have been more volatile than it actually was. The small decreases in 2001 and 2005 appear to be severe declines. Second, changing the chart's vertical and horizontal proportions distorts the appearance of the overall rate of increase in rent. A reader who does not "run the numbers" could be given a false impression of the historical strength of this apartment market.

A good rule-of-thumb to remember is, "If exclusion of the zero point on a scale distorts the presentation, it is usually best to not alter the scale." This said, truncation of the vertical axis sometimes renders important trends or patterns more visible. It is not truncation of an axis or alteration of chart proportions itself that is unethical. Ethics becomes an issue when a chart is altered to distort, rather than illuminate, a relationship.

Real-World Case Study: Part 1

Appendix G contains data on 129 townhouse sales located in zip code 89015, which sold during a period beginning in early 2001 through the first half of 2004. The data include

- Sale price
- Sale date
- Number of bathrooms
- Number of bedrooms
- Living area in square feet
- Number of fireplaces
- Existence of a garage
- The year the townhouse was built
- Number of days the property was on the market
- Occupant information

This portion of the case study applies several charting tools to the data provided in Appendix G. You are encouraged to replicate the charts shown here and to work with the data to discover other ways to describe it and effectively communicate salient aspects of the data.

Slightly more than 54% of the townhomes were vacant at the time of sale (54.26%), 41.86% were owner occupied, and 3.88% were tenant occupied. The **pie chart (Townhouse Occupancy at Time of Sale)** shows occupancy by type at the time of sale.

Sale price and living area were organized into ordered arrays, categorized, and plotted in the **histograms (Histogram of Sale Price and Histogram of Living Area).**

Dotplots provide another way of illustrating the distribution of sale price. Each dot represents a sale, and they are stacked up over the corresponding sale price category. This type of chart is an option in Minitab. If you are using Minitab, try to replicate the **dotplot (Dotplot of Sale Price).**

Generally speaking, sale prices are expected to increase as living area increases. The **scatter plot (Scatter Plot of Sale Price and Living Area)** illustrates this expected relationship. Although the price/size relationship seems to hold in general, note that some of the smaller townhomes sold for higher prices than some of the larger ones. What

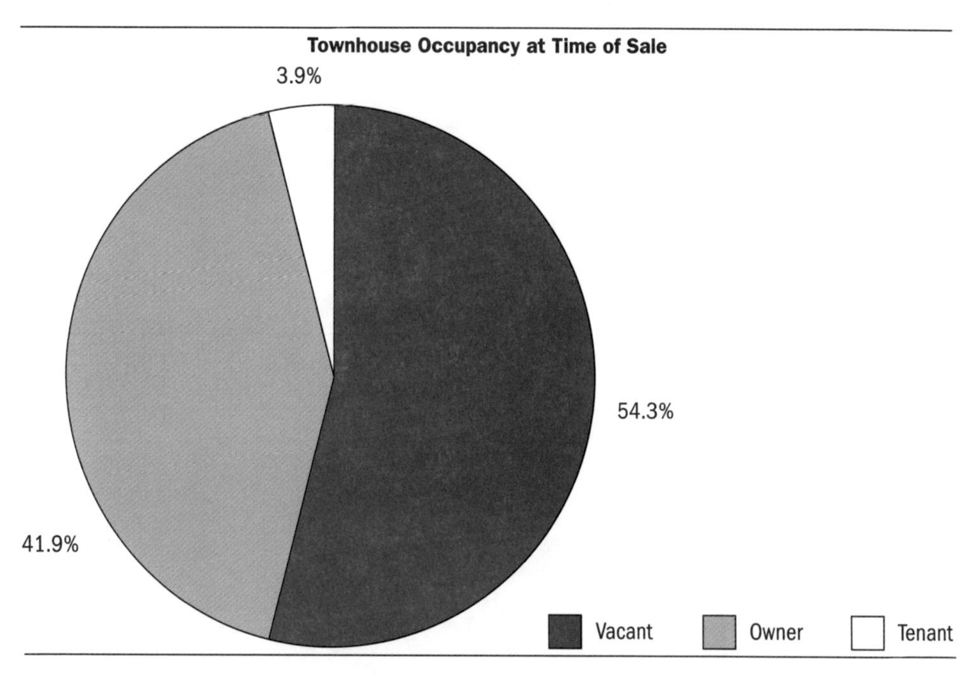
Townhouse Occupancy at Time of Sale
3.9%
54.3%
41.9%
Vacant
Owner
Tenant

Histogram of Sale Price
Frequency
35
30
25
20
15
10
5
0
$60,000
$80,000
$100,000
$120,000
$140,000
$160,000
$180,000
$200,000
Sale Price

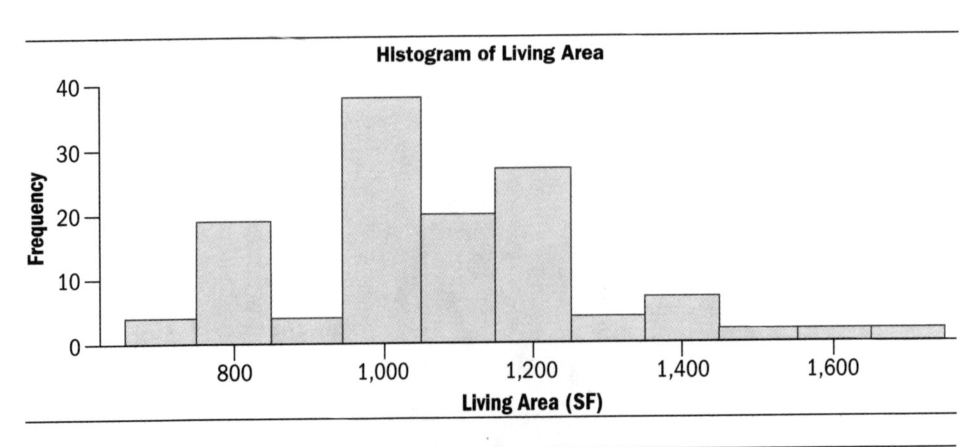
Histogram of Living Area
40
30
20
10
0
Frequency
800
1,000
1,200
1,400
1,600
Living Area (SF)

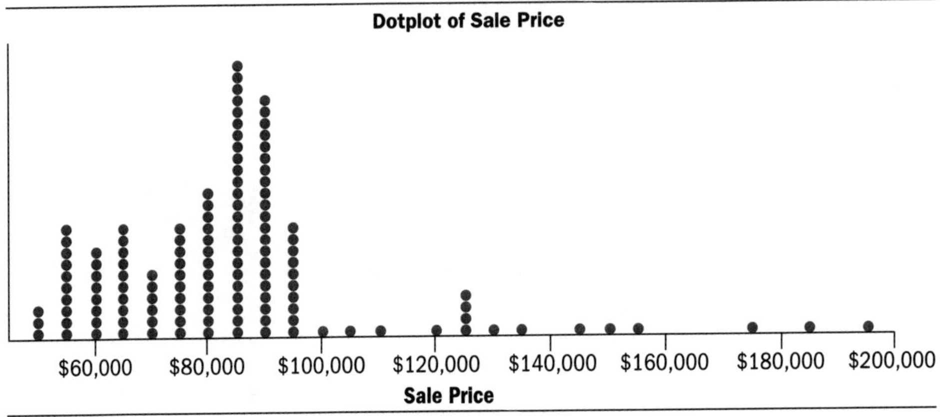
Dotplot of Sale Price
$60,000
$80,000
$100,000
$120,000
$140,000
$160,000
$180,000
$200,000
Sale Price

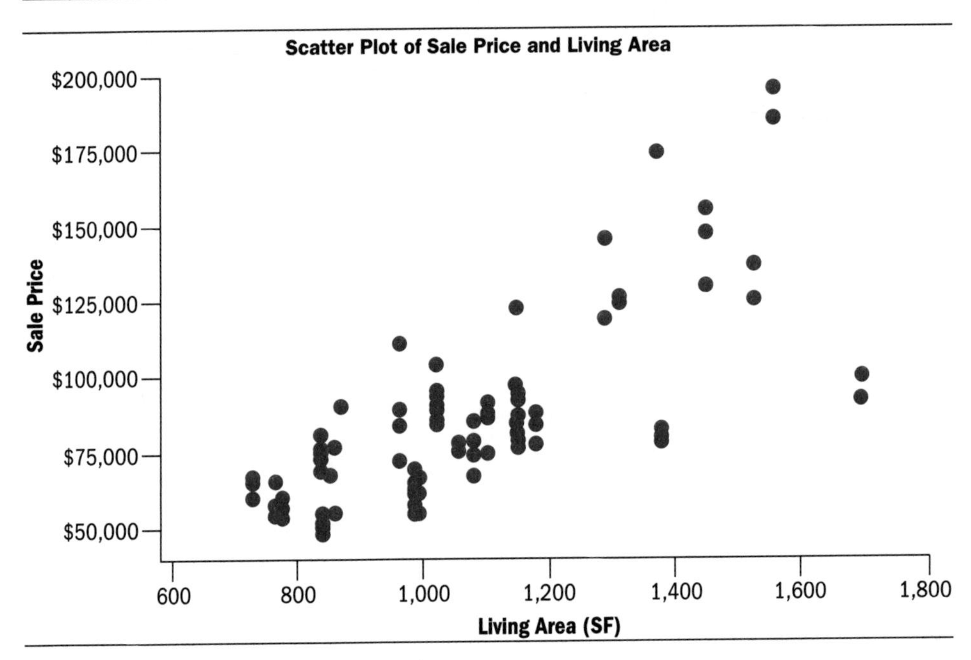
Scatter Plot of Sale Price and Living Area
$200,000
$175,000
$150,000
$125,000
$100,000
$75,000
$50,000
Sale Price
600
800
1,000
1,200
1,400
1,600
1,800
Living Area (SF)

could account for the relationship between price and size being less than perfect? (Hint: All of the data are not equivalent in terms of the elements of comparison. Also, market prices are not always exactly equal to market value.) Later in the book we will explore ways to account for significant elements of comparison in order to be able to account for variation in price more precisely than shown on the Scatter Plot of Sale Price and Living Area on page 66.

How would you illustrate differences in several elements of comparison for these data? Devise some tables and charts that will effectively capture and communicate variation in bedroom and bathroom counts, garages, fireplaces, year of sale, age of the improvements at the time of sale, lot area, and days on market.

Describing Numerical Data

Chapter 3 demonstrated that histograms, frequency distributions, and stem and leaf diagrams are useful for presenting variable values pictorially. Nevertheless, they do not provide all of the information we might need or want to know about a variable. It is often desirable to find a single number that captures and quantifies some aspect of a variable, such as the central tendency, amount of variability, or the shape of a variable's distribution.

For example, we might ask questions such as

- What is the average floor area of a Subway sandwich shop when it occupies an in-line suite in a neighborhood shopping center?
- How does the average floor area of an in-line suite compare to the average floor area of a Subway sandwich shop located in a freestanding building?

What we want is a single number that "represents" or describes the central tendency of a variable–in this case, Subway sandwich shop floor area. We call this number an "average."

The word *average* on its own is not descriptive enough for use as a technical term because there is more than one way to calculate an average. We will explore four ways of computing an average in this chapter–the arithmetic mean (or simply the "mean"), the median, the mode, and the geometric mean. Each is an average in the sense that each is a measure of central tendency. In Chapter 4 we will also explore measures of dispersion such as range, interquartile range, standard deviation, variance, and coefficient of variation. The shape of a variable's distribution is also an important characteristic, which is often described in terms of its symmetry or asymmetry.

The final section of this chapter will introduce the concept of comparing two variables. We will examine a simple but effective variable analysis and comparison tool called a *box and whisker plot* (or simply a "box plot") and a measure of the strength of the relationship between two variables called a *correlation coefficient.*

Many of the measurements found in this chapter are so commonly used to describe data that most statistics programs and packages include a preprogrammed routine labeled **Descriptive Statistics** or something similar. Also, nearly any analytical study employing sample data will report the data's descriptive statistics. These statistics reduce descriptions of central tendency and dispersion to simple, universally understood numbers that are consistent with how informed readers are accustomed to assessing variable characteristics.

Measures of Central Tendency

Measures of central tendency, or what are commonly called *averages,* answer the question, "Is there a single number that best *represents* the variable in question?" However, there is no universally preferable answer to the related question, "Which measure of central tendency is best?" Selection of a "best" measure depends on the data.

The Arithmetic Mean

The arithmetic mean is the most commonly used measure of central tendency. Its use is so prevalent that the "arithmetic mean" is often referred to simply as the "mean" or the "average." The arithmetic mean is calculated by adding up all of a variable's numerical values and dividing the sum by the number of observations in the sample (n).

When a symbol is used to refer to a sample variable, we put a bar over the symbol to indicate the sample variable's arithmetic mean. For example, the arithmetic mean of the variable x would be symbolized as $\overline{x}$, and the arithmetic mean of the variable y would be expressed as $\overline{y}$.

$$\overline{x} = \frac{\Sigma x}{n} = \frac{x_1 + x_2 + \ldots + x_n}{n}$$

Assume an appraiser draws a random sample consisting of six one-bedroom apartment rents from a market area. The sample rents are

\$625 \$600 \$650 \$630 \$640 \$630

We can calculate the sample's arithmetic mean (i.e., the sample mean) by substituting *RENT* for x in the sample mean equation and setting $n = 6$:

$$\overline{RENT} = \frac{\Sigma RENT_i}{n} = \frac{625 + 600 + 650 + 630 + 640 + 630}{6} = 629.17$$

The sample mean rent is \$629.17, which would probably be rounded to \$629 for appraisal purposes.

We refer to the global or general group from which a sample has been drawn as a "population." In the prior example, the population consisted of all of the one-bedroom apartments located in the market area. Statisticians use a different symbol, the lowercase Greek letter "mu" (μ), to refer to a population mean. The population mean is calculated similarly to the sample mean by adding up the numerical values associated with all the members of the population and dividing this sum by the number of observations in the population, symbolized as N:

$$\mu_x = \frac{\sum_i x_i}{N} = \frac{x_1 + x_2 + \ldots + x_N}{N}$$

Notice the subtle difference here: the lowercase n from the sample mean calculation has been replaced by an uppercase N. This small difference in notation has a significant effect on application because it denotes that you must know a variable's numerical value for *each* element of the entire population in order to calculate a population mean. In some situations this amount of information is nearly impossible, or too costly, to obtain, which is why researchers and analysts often rely on sampling (see Chapter 6).

The Median

The discussion of ordered arrays in Chapter 3 mentioned that the creation of an ordered array facilitates the derivation of some measures of central tendency and dispersion. In particular, the median and the mode can be derived easily by looking at an ordered array. The median is simply the ordered array's middle value.

Two ordered arrays are shown in **Table 4.1**, with the middle value (or values) circled. The median of x is 20. This is the middle value of x in the five-observation ordered array of x values. The six-observation ordered array of y values has two "middle" values, 7 and 10. When a data set has an even number of observations, the median is expressed as the arithmetic mean of the two middle values. So, the median of y is $8.5 = (10 + 7) \div 2$.

A simple counting rule is useful for finding the median of a large sample. The median observation in an ordered array is found by counting to the $(n + 1) \div 2$ ordered observation. For x in Table 4.1 the median is the third ordered observation because $(5 + 1) \div 2 = 3$. For y the median is the midpoint between the third and fourth ordered observation because $(6 + 1) \div 2 = 3.5$:

Table 4.1 Two Simple Ordered Arrays

x	y
12	3
15	6
(20)	(7)
26	(10)
27	11
	14

$$\textbf{Median} = \frac{(\mathbf{n} + \mathbf{1})}{\mathbf{2}} \textbf{ ordered observation}$$

The Mode

The mode, which is also a measure of central tendency, is defined as the most frequently occurring value in a sample. If the sample values are

3 5 7 7 8 9 9 9 10 12 13

then the mode is 9, which occurs three times in the sample set. The mode becomes less useful as a measure of central tendency when the sample is multimodal (i.e., has more than one mode). Consider the following sample values:

2 2 3 3 3 4 5 6 6 7 8 8 8 9 10

This sample is bimodal with modes at 3 and 8.

When a sample is multimodal, it may not be easy to determine which of the modes best represents the data. In addition, it could be possible that the sample was mistakenly drawn from two different populations, one centered on 3 and the other centered on 8. In the latter case, the multiple modes may be a signal to look more closely at where the data came from.

Which Central Tendency Measure is Preferred?

The answer to the question, "Which measure of central tendency is preferred?" is, "It depends." As an extreme example, suppose a small but successful local retail business has five locations in a major metropolitan area. The arithmetic mean of the floor area for the retail outlets is 3,568 square feet. Suppose that you have a vacant 3,500-sq.-ft. space in a location you believe would serve perfectly as a sixth store. You would be in error, however, if you assumed the company would be interested in renting your 3,500-sq.-ft. space if the floor areas of the other stores were distributed as shown in **Table 4.2**.

Table 4.2 **Floor Area of Five Retail Outlets**

Floor Area
1,200 square feet
1,320 square feet
1,320 square feet
1,500 square feet
12,500 square feet

The fifth location is the company's "flagship" store. The other four stores have no inventory in storage on site, being served daily from the flagship location, and their sizes are more indicative of the amount of space that would be required for a sixth outlet. In this case the median and mode of 1,320 square feet both do a better job of representing the typical floor area than the mean of 3,568 square feet.

The example from Table 4.2 illustrates the weakness of the mean–it is unduly influenced by extreme values in the sample. When extreme values exert too much influence, the mean may not be the best choice to represent central tendency. Also, note the variation in space use within the sample. Sometimes extreme values exist for a discernable reason (such as in

Mean, Median, and Mode in Excel

Excel includes built-in functions for computing the mean, median, and mode. If data are entered into Excel as depicted below, the mean, median, and mode can be calculated using the **=AVERAGE**, **=MEDIAN**, and **=MODE** function operators as shown in the table.

Microsoft Excel - Book1

File Edit View Insert Format T

F13 fx

	A	B	C	D
1		2.0		
2		3.0		
3		4.0		
4		4.0		
5		8.0		
6	Mean	4.2		
7	Median	4.0		
8	Mode	4.0		
9				
10				

=AVERAGE(B1:B5)

=MEDIAN(B1:B5)

=MODE(B1:B5)

this example), and a more detailed investigation might reveal that some of the data points are not relevant to the analysis. When this is so, it may be appropriate to focus solely on the most relevant observations. While the existence of extreme values in a data set warrants further investigation, it is inappropriate to alter or remove observations having extreme values without cause (e.g., data entry errors or lack of relevance).

When a sample distribution is symmetrical, the mean and median are equal. The mean is the preferred measure for symmetrical (or nearly symmetrical) distributions because many statistical inference tools are designed specifically for analyzing means. And, as we will see later in the book, the distribution of the sample mean is sufficiently symmetrical, regardless of the distribution of the underlying population, when the sample is large enough.

The Geometric Mean

The geometric mean is another measure of central tendency that is applicable to varying rates of change over time. This measure is especially useful to appraisers because it relates directly to the geometric mean rate of return, which we often refer to as the *effective rate of return.*

In its general form, the geometric mean $\overline{X}_G$ is the tth root of a product of changes over time (t), which can be expressed mathematically as

$$\overline{X}_G = \sqrt[t]{(x_1 \cdot x_2 \cdot \ldots \cdot x_t)}$$

This expression takes on special meaning in appraisal and finance when the xs in the equation above are replaced with $(1 + R)$, where R is a periodic rate of return. Making this substitution into the geometric mean equation results in the following relationship:

$$\overline{R}_G = \sqrt[t]{(1 + R_1) \times (1 + R_2) \times \ldots \times (1 + R_t)} - 1$$

The geometric mean rate of return is an accurate measure of an effective rate of return whereas an arithmetic mean is not. To illustrate this point, consider a three-year investment that returns 10% in Year 1, minus 10% in Year 2, and 30% in Year 3.

The arithmetic mean of the annual returns is 10%. In contrast, the geometric mean return is 8.77%:

$$\begin{aligned}\overline{R}_G &= \sqrt[3]{(1+0.10)\times(1-0.10)\times(1+0.30)}-1\\ &= (1.287)^{1/3}-1\\ &= 1.0877427-1\\ &= 0.0877427\end{aligned}$$

The accuracy of the effective annual rate of return solution of approximately 8.77% can be checked by comparing two simple multiplicative products, as follows:

$$\begin{aligned}(1+\overline{R}_G)^3 &= 1.0877427\times1.0877427\times1.0877427\\ &= 1.287\end{aligned}$$

and

$$\begin{aligned}(1+R_1)(1+R_2)(1+R_3) &= 1.10\times0.90\times1.30\\ &= 1.287\end{aligned}$$

Therefore, an 8.77427% annually compounded rate of return for three years yields the same future value as the combined effects of the three actual annual rates of return.

Practice Problems 4.1

1. A sample of branch bank locations provides the following site areas in acres.

1.3	0.9	1.2	0.8	1.5	1.2	0.9	1.3	1.1	1.2	1.4	0.8
1.2	1.6	0.6	1.1	1.2	1.4	1.2	0.9	1.1	1.0	1.0	1.2

Calculate the mean, median, and mode for these data.

2. Investment returns over a five-year holding period were 5% in Year 1, 5% in Year 2, -7% in Year 3, 10% in Year 4, and 15% in Year 5. What is the effective annual rate of return?

Measures of Dispersion

Measures of dispersion facilitate comparisons of variability between two or more data sets. Consider the lot samples from Subdivisions A and B shown in **Table 4.3**. Although both samples have the same mean lot size, the sample from Subdivision B appears to be much more variable. Calculating the measures of dispersion for these data sets will allow us to quantify the differences in variability that we observe by looking at the data in tabular form.

Table 4.3 Subdivision Lot Size Comparison

Subdivision A Lot Size (SF)	Subdivision B Lot Size (SF)
7,000	6,000
7,250	9,950
7,800	6,700
7,700	8,200
7,300	7,400
7,500	6,300
ΣA = 44,550	ΣB = 44,550
$\bar{A}$ = 7,425	$\bar{B}$ = 7,425

Range

The range is the simplest way to measure variability. The range is the maximum value in the sample minus the minimum value:

Range = Maximum Value – Minimum Value

Subdivision A
Sample Range = 7,800 – 7,000 = 800 square feet

Subdivision B
Sample Range = 9,950 – 6,000 = 3,950 square feet

The comparison of ranges confirms the earlier casual observation that the Subdivision B lot size sample is more variable. Due to Subdivision B's greater lot size variability, we might reasonably expect property prices to vary more there than in Subdivision A, all else being equal.

Notice that the range is based solely on a sample's extreme values. As a result, range comparisons can be misleading when a sample is actually not very dispersed but includes an extreme "outlier." In other words, as we discovered with the mean, the range is also sensitive to extreme values.

Interquartile Range

The interquartile range, also sometimes referred to as the "middle range" or "middle spread," is less sensitive than the range to extreme values. In addition, computation of what is known as a "five-number summary," including sample quartiles along with a sample's minimum, maximum, and median values, facilitates construction of box and whisker plots. These plots, discussed a bit later in the chapter, provide a handy means of visually comparing central tendency, variability, and symmetry.

Quartiles divide an ordered array into quarters by finding, in addition to the median, the middle value for the half of the data set below the median and the middle value for the half above the median. The first quartile position (*Q1*) in the ordered array shown in **Figure 4.1** is located between the third and fourth ordered observations. The third quartile (*Q3*) is located between the ninth and tenth ordered observations. More generally

$$Q1 = \frac{(n+1)}{4} \quad \text{ordered observation}$$

and

$$Q3 = \frac{3(n+1)}{4} \quad \text{ordered observation}$$

One way to derive a quartile value by hand (suggested by Levine, et al.[1]) when the ordered observation count is not a whole number is as follows:

- When $(n + 1)/4$ results in a decimal of 0.5, *Q1* is the mean of the two sample values immediately below and above the calculated *Q1* position.
- When the *Q1* position decimal is less than 0.5, *Q1* is the value immediately below the calculated *Q1* position
- When the *Q1* position decimal is greater than 0.5, it is the value immediately above the calculated *Q1* position.

When $n = 12$ as in Figure 4.1, then $(n + 1)/4 = 13/4 = 3.25$, and $Q1 = 3$, which is the value occupying the third position because $0.25 < 0.5$.

Similarly, when the value of $[3(n + 1)]/4$ is not a whole number and its decimal is equal to 0.5, *Q3* is the mean of the two observations immediately below and above the calculated *Q3* position. *Q3* is the value immediately below

Figure 4.1 Quartiles of a 12-Observation Ordered Array

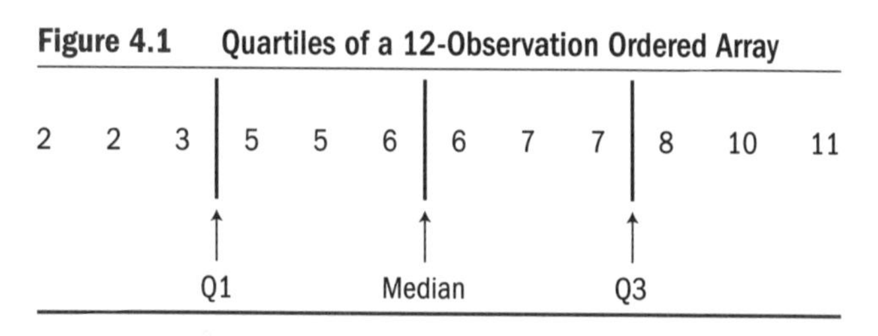

1. David M. Levine, Timothy C. Krehbiel, and Mark L. Berenson, *Business Statistics: A First Course*, 3rd ed. (Upper Saddle River, N.J.: Prentice Hall, 2003).

the calculated *Q3* position when the *Q3* position decimal is less than 0.5, and it is the value immediately above the calculated *Q3* position when the *Q3* position decimal is greater than 0.5. When n = 12, as in Figure 4.1, then $[3(n + 1)]/4 = 39/4 = 9.75$, and *Q3* = 8, which is the value occupying the tenth position because 0.75 > 0.5.

Once *Q1* and *Q3* have been identified, the interquartile range is easily calculated as follows:

Interquartile Range = Q3 – Q1

The interquartile, or middle, range for the data arrayed in Figure 4.1 is *Q3* – *Q1* = 8 – 3 = 5. The range, however, is 11 – 2 = 9. Although the range can be affected by extreme values in the data, the interquartile range is more resistant to the effects of extreme values (i.e., outliers). For example, if the maximum value in the ordered array shown in Figure 4.1 were 30 instead of 11, the range would increase to 28 while the interquartile range would remain unchanged.

The protocol above is not the only way to calculate *Q1* and *Q3*. SPSS and Minitab calculate *Q1* for the data in Figure 4.1 as 3.5 and calculate *Q3* as 7.75. In both SPSS and Minitab the 3.25th position for *Q1* and 9.75th position for *Q3* are taken literally. The *Q1* result of 3.5 is 3 plus one-quarter of the distance between 3 and 5, or 3 + 0.25(5 – 3) = 3.50. Likewise, the *Q3* result of 7.75 is three-quarters of the distance between 7 and 8, or 7 + 0.75(8 – 7) = 7.75.

Excel also uses an interpolation algorithm, but the Excel algorithm is different from the one used by SPSS and Minitab. When our simple example is solved using the "=QUARTILE" macro in Excel, the software's proposed solution is *Q1* = 4.5 and *Q3* = 7.25. Let's examine the Excel solutions for *Q3* and *Q1*. Excel uses a truncation command "TRUNC" in its calculation, which means "truncate at the decimal point by dropping everything to the right of it." This routine is used to identify the position of the lower boundary of the location containing the quartile.

In our example, where n = 12 the position for the lower boundary of the *Q3* location, symbolized as k in Excel, is

k = TRUNC(3/4*(12 – 1)) + 1 = 8 + 1 = 9

The $k + 1$ position is $9 + 1 = 10$, which identifies the upper boundary of the *Q3* location. The Excel interpolation factor *f* is

$$
\begin{aligned}
f &= 3/4*(12 - 1) - \text{TRUNC}(3/4*(12 - 1) \\
&= 8.25 - 8 \\
&= 0.25
\end{aligned}
$$

The ninth position is occupied by a numerical value of 7 in our example and the 10th position is occupied by a numerical value of 8, so the Excel *Q3* interpolation is

$$
\begin{aligned}
Q3 &= 7 + (f*(8 - 7) \\
&= 7 + (0.25*(8 - 7) \\
&= 7 + 0.25 \\
&= 7.25
\end{aligned}
$$

The position for the lower boundary of the *Q1* location is

$$
\begin{aligned}
k &= \text{TRUNC}(1/4*(12 - 1)) + 1 \\
&= 2 + 1 \\
&= 3
\end{aligned}
$$

The $k + 1$ position is $3 + 1 = 4$, which identifies the upper boundary of the *Q1* location. The Excel interpolation factor *f* is

$$
\begin{aligned}
f &= 1/4*(12 - 1) - \text{TRUNC}(1/4*(12 - 1) \\
&= 2.75 - 2 \\
&= 0.75
\end{aligned}
$$

The third position is occupied by a numerical value of 3 in our example and the fourth position is occupied by a numerical value of 5, so the Excel *Q1* interpolation is

$$
\begin{aligned}
Q1 &= 3 + (f*(5 - 3) \\
&= 3 + (0.75*(5 - 3) \\
&= 3 + 1.5 \\
&= 4.5
\end{aligned}
$$

Let's now compare the results of calculating the interquartile range (IQR) using the three differing methods discussed here:

- IQR (hand-calculated) = 8 – 3 = 5
- IQR (SPSS and Minitab) = 7.75 – 3.5 = 4.25
- IQR (Excel algorithm) = 7.25 – 4.5 = 2.75

This range of results underscores two important observations. First, it is good practice to disclose how quartiles were calculated and perhaps to state that numerous algorithms exist and results will vary by the method or software used for the analysis. Second, never compare two or more data sets based on quartiles derived by different methods.

Variance and Standard Deviation

Variance and standard deviation measures reflect how the data are clustered around the mean by accounting for the distribution of all of the data. The population standard deviation is the mean of the deviations of all elements of the population from the population mean. Likewise, the sample standard deviation can be thought of as the mean (approximately) of the deviations of a sample's observations from the sample mean. Variance is simply the square of the standard deviation.

The lowercase Greek letter sigma, σ, is used to symbolize population standard deviation (σ) and population variance (σ^2). The equations for calculating σ and σ^2 are

$$\sigma_X = \sqrt{\frac{\Sigma(x_i - \mu)^2}{N}}$$

and

$$\sigma_X^2 = \frac{\Sigma(x_i - \mu)^2}{N}$$

Notice that deviations from the mean ($x_i - \mu$) are squared prior to summing. Squaring of the terms removes the offsetting effects of positive and negative deviations from μ while preserving information on deviation from the mean.

Samples are often used to represent and make inferences about populations. When samples are used, sample standard deviations serve as estimates of underlying population standard deviations. The uppercase letter S symbolizes the sample standard deviation and S^2 symbolizes sample variance. The corresponding equations for calculating S and S^2 are

$$S_X = \sqrt{\frac{\Sigma(x_i - \bar{x})^2}{n-1}}$$

and

$$S_X^{\ 2} = \frac{\Sigma(x_i - \bar{x})^2}{n-1}$$

Two subtle differences between the prior sample measures and population measures are that $\bar{x}$ replaces μ and $n - 1$ replaces N in the sample standard deviation and sample variance equations. When a sample is used to represent a population, μ is typically not known with certainty, and S is an unbiased estimator of σ when $n - 1$ is used as a divisor instead of sample size n.

Table 4.4 includes further analysis of the Subdivision A data in Table 4.3. The second column of this table shows deviation from the mean lot area ($A - \bar{A}$) for each lot in the six-lot sample ($n = 6$). The deviations are squared $(A - \bar{A})^2$ and summed $\Sigma(A - \bar{A})^2$ in the last column in preparation for calculation of the sample standard deviation S. Substituting A for x in the equation for sample standard deviation, the sample deviation is

Table 4.4 Subdivision A Lot-Size Deviation

Subdivision A Lot Size (SF)	Deviation $(A - \bar{A})$	Squared Deviation $(A - \bar{A})^2$
7,000	7,000 – 7,425 = -425	180,625
7,250	7,250 – 7,425 = -175	30,625
7,800	7,800 – 7,425 = 375	140,625
7,700	7,700 – 7,425 = 275	75,625
7,300	7,300 – 7,425 = -125	15,625
7,500	7,500 – 7,425 = 75	5,625
$\bar{A} = 7,425$		$\Sigma(A - \bar{A})^2 = 448,750$

$$S_A = \sqrt{\frac{\Sigma(A - \bar{A})^2}{n-1}}$$

$$= \sqrt{\frac{448,750}{6-1}}$$

$$= \sqrt{89,750}$$

$$= 299.58 \text{ square feet}$$

and the sample variance is

$$S_A^2 = 89{,}750 \text{ square feet squared}$$

Sample variance provides an estimate of how much the population's average *squared* deviation differs from the mean. For this reason, the units associated with variance estimates are not very intuitive (in this case the variance is measured in square feet squared). In contrast, the sample standard deviation is an estimate of how much the underlying population's average deviation differs from the mean, which has much more intuitive meaning

Standard Deviations and Variances in Excel, Minitab, and SPSS

Standard deviations and variances are easily calculated in Excel using one of the four functions shown **below.** If the lot size sample data for Subdivisions A and B shown in Table 4.3 were to be entered into an Excel spreadsheet in cells A1 through B6 (Subdivision A in A1:A6 and Subdivision B in B1:B6), the commands for calculation of standard deviations and variances for these samples would be

S_A: =STDEV(A1:A6)	answer =	299.58
S_A^2: =VAR(A1:A6)	answer =	89,750
S_B: =STDEV(B1:B6)	answer =	1,469.61
S_B^2: =VAR(B1:B6)	answer =	2,159,750

These results are consistent with the prior observation based on the ranges for the two subdivisions—lot size for the Subdivision B sample is considerably more dispersed than for the Subdivision A sample.

Excel Functions for Calculation of Standard Deviation and Variance

Calculation	**Function**
Sample Standard Deviation	=STDEV
Sample Variance	=VAR
Population Standard Deviation	=STDEVP
Population Variance	=VARP

If you have no experience using Excel to calculate standard deviations and variances, take some time to enter the lot size data from Table 4.3 into an Excel spreadsheet and replicate the answers shown above. Note also that the Excel answers for Subdivision A are exactly the same as the answers we obtained doing the calculations by hand in Table 4.4, which illustrates the benefit of doing these calculations electronically. Standard deviation and variance are also included in Excel's descriptive statistics routine. Select **Tools**, then **Data Analysis.** A window will open containing a menu of statistical analysis tools (**see below**).

Highlight **Descriptive Statistics** and click **OK**. The **Descriptive Statistics** window will open allowing you to identify the input and output ranges for your analysis (**see below**). Also, be sure to check the **Summary statistics** box in the **Output options** area. If you are because standard deviation units are the same as the original data units (in this case square feet).

Coefficient of Variation

The mean values for Subdivisions A and B in the Table 4.3 example were equal, which made the analysis of dispersion simply a matter of comparing sample ranges and sample standard deviations. Sometimes, however, it is more difficult to determine which data set is more dispersed. Consider the floor area samples from the two different subdivisions illustrated in **Table 4.5**. Based on nominal measures of range and standard de-

including labels for your data columns, check the **Labels in first row** box in the **Input** area.

In SPSS, sample standard deviation and sample variance are included in the descriptive statistics options. Select **Analyze, Descriptive Statistics**, and then **Descriptives.** Select the variable or variables to include and then click **Options** to select which descriptive statistics to include. Once this has been done, click **OK**. Note that the SPSS descriptives procedure does not calculate population variance or population standard deviation. SPSS assumes you are working with a sample.

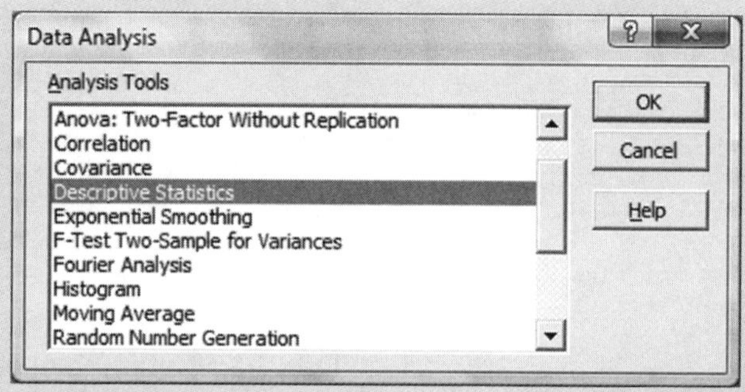

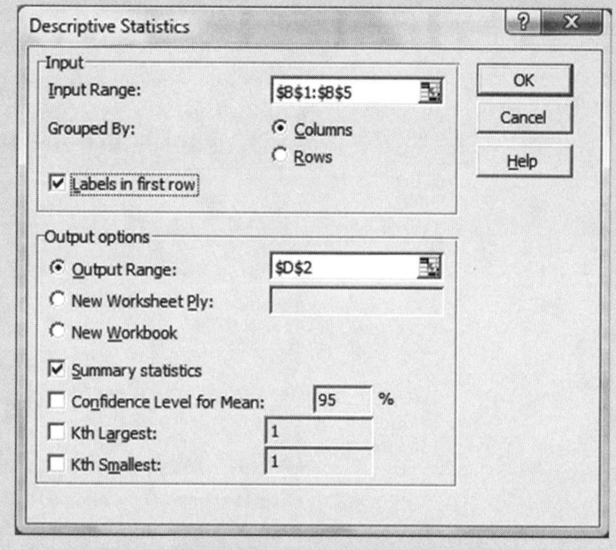

In Minitab, you can select the **Calc** menu, choose **Column Statistics,** and then select **Standard Deviation**, in which case you must square the answer to compute variance. Alternatively, you can select the **Stat** menu and then select **Basic Statistics** and then **Display Descriptive Statistics**. Select the variables to analyze and click **Statistics** to select the statistics you want to calculate and then click **OK** to obtain the result. As with SPSS, these procedures do not calculate population variance or population standard deviation.

Table 4.5 Two Samples of Single-Family Residential Floor Area

Grand Vista (SF)	Valley View (SF)
2,200	2,600
2,650	2,950
2,400	3,100
3,220	3,650
2,800	4,080
3,150	3,380
2,340	2,880
2,400	3,800
3,300	3,020
2,500	3,560
$\Sigma X = 26{,}960$	$\Sigma X = 33{,}020$
$\bar{X} = 2{,}696$	$\bar{X} = 3{,}302$
Range = 1,100	Range = 1,480
S = 400.7	S = 466.9

viation, the residential floor areas of homes in the Valley View subdivision appear to be more dispersed. The range is larger (1,480 compared to 1,100) and the sample standard deviation is greater (466.9 compared to 400.7).

In situations such as this, where the means of the two samples differ, use of the coefficient of variation allows us to draw conclusions about relative variability. When viewed from this perspective, we can see that floor areas in Valley View are actually slightly *less* dispersed than in Grand Vista.

The coefficient of variation expresses sample standard deviation as a percentage of the corresponding sample mean. It is calculated as

$$\mathrm{COV} = \frac{S}{\bar{X}} \bullet 100\%$$

Coefficient of variation calculations for the two subdivisions illustrated in Table 4.5 are

$$\mathrm{COV}_{GV} = \frac{400.7}{2{,}696} \bullet 100\% = 14.9\%$$

and

$$\mathrm{COV}_{VV} = \frac{466.9}{3{,}302} \bullet 100\% = 14.1\%$$

Grand Vista's standard deviation is 14.9% of its mean floor area, making that data set slightly more dispersed than Valley View's, with a floor area standard deviation of 14.1% of its mean. By this relative comparison, these two subdivisions actually exhibit essentially similar dispersion.

As this simple example demonstrates, nominal comparisons of dispersion measures can be misleading. When comparing data sets that have different central tendencies, it is good practice to base dispersion comparisons on a relative measure, such as the coefficient of variation.

Coefficient of Variation in Excel, Minitab, and SPSS

Excel does not include a coefficient of variation function. However, it can be easily calculated by dividing the result of the =STDEV function by the result of the =AVERAGE function and multiplying the answer by 100 to convert it into a percentage. Coefficient of variation is an option in the **Display Descriptive Statistics** menu in Minitab that we discussed in the previous section. As with Excel, use of SPSS requires that coefficient of variation be calculated by hand after solving for sample standard deviation and sample mean.

Practice Problems 4.2

The table below is the basis for all of the practice problems in this section. If possible, work these problems in Excel, SPSS, or Minitab.

Office Building Capitalization Rate Samples from Two Different U.S. Cities

City A	City B
8.2%	6.9%
8.0%	7.2%
7.9%	8.0%
7.4%	7.4%
7.4%	6.8%
8.1%	7.8%
8.3%	7.5%
7.2%	7.1%
7.5%	7.0%
8.0%	6.7%
7.6%	7.9%
7.9%	8.0%
7.0%	6.9%

1. Calculate the mean, median, and mode capitalization rate for each city.
2. Calculate the range and interquartile range of capitalization rates for each city.
3. Calculate the sample standard deviation of capitalization rates for each city.
4. Calculate the coefficient of variation for each city.
5. Based on your answers to Problems 1 through 4, what is the least reliable measure of central tendency and which city's capitalization rates are more dispersed by nominal and relative measures?

Coefficient of Dispersion

The coefficient of dispersion (COD) is often used as a measure of uniformity in tax assessment ratio studies. The statistic is rarely, if ever, used in any other context. Ideally, appraised values for all properties within a property tax assessment district should be reasonably close to the same ratio of market value. For example, assume that an assessor's goal is to appraise properties at 100% of market value. Assume also that assessed values have been derived using a linear regression model.[2] One way to test the uniformity of assessed values is to compare assessed values of recently sold properties to actual sale prices. If the assessment is accurate and uniform, we would expect the ratio of assessed value to actual sale price to be close to 1 for all of the sold properties. Systematic departure from a ratio central tendency of 1 is an indication of inaccuracy, and excessive dispersion in ratios from property to property is an indication of lack of uniformity. Calculation of the *COD* tests uniformity.[3]

$$\mathrm{COD} = \frac{\mathrm{AAD}}{\mathrm{MED(A/S)}} \times 100\%$$

As shown above, *COD* is the ratio of average absolute deviation (*AAD*) from the median ratio of assessed value to actual sale price divided by the median ratio of assessed value to actual sale price expressed as a percentage. The calculations involved in a simplified example are shown in **Table 4.6.**

Assessment professionals note that *COD*s of less than 5% are exceptionally rare, usually limited to situations where a developer has tight control over lot and home prices. Coefficients less than or equal to 15% are considered to be indicative of good uniformity, while higher coefficients are acceptable in extremely heterogeneous market areas.

There are some drawbacks to using the *COD* to assess uniformity. One is the ratio's sensitivity to the size of the base number–i.e., the denominator

2. Appraised value can be derived in any manner, but use of linear regression is assumed here to fit the context of this book.
3. International Association of Assessing Officers, *Assessment Administration* (Chicago: IAAO, 2003), 333-334.

in the *COD* equation, which is MED(*A/S*). For example, if the goal is to appraise properties at 50% of market value, a nominal error of $5,000 indicates twice the percentage error it would indicate if the assessment ratio goal were 100%. The 15% *COD* indication of uniformity would most likely have to be relaxed in such situations. For consistency's sake, assessed values could be converted to a 100% assessment ratio prior to calculating the *COD* and comparison with a given uniformity standard.

Table 4.6 Coefficient of Dispersion (*COD*) Calculation

Sale	Assessed Value (*A*)	Sale Price (*S*)	(*A/S*)	(*AD*) = |(*A/S*) – MED(*A/S*)|
1	$100,000	$110,000	0.909	0.091
2	$150,000	$140,000	1.071	0.071
3	$140,000	$135,000	1.037	0.037
4	$170,000	$174,000	0.977	0.023
5	$125,000	$118,000	1.059	0.059
6	$150,000	$156,000	0.962	0.038
7	$165,000	$180,000	0.917	0.083
8	$190,000	$175,000	1.085	0.085
9	$160,000	$160,000	1.000	0.000

Median (*A/S*) = 1.000

Average Absolute Deviation (*AAD*) = Σ(*AD*) ÷ 9 = 0.487 ÷ 9 = 0.054

COD = 0.054 ÷ 1.000 × 100% = 5.4%

Another drawback is the inability to support inferences about the true underlying measure of dispersion because there is no means of associating a probability statement with a *COD* calculation. For this reason, it is beneficial to also calculate mean *A/S* ratios and accompanying standard deviations to construct confidence intervals for the true population *A/S* ratio. Coefficients of variation can then be used as another means to compare uniformity across markets where typical prices are dissimilar.

Shape

Symmetry

When the distribution of a data set is symmetrical, the mean and median are equal, and the shape of the distribution on one side of the mean is a mirror image of the other side. **Figure 4.2** shows a dotplot of a frequency distribution of annual retail sales volumes for 194 affiliated stores. The distribution is symmetric. It is also bell-shaped with a unique mode, which is identical to the median and the mean. Distributions similar to this–with an obvious central tendency, statistical measures densely concentrated near the central tendency, and a symmetrical shape–are encountered often in both the natural world and the business world. They are indicative of phenomena or processes where more extreme values are less likely than less extreme values, and outcomes falling above or below the central tendency are equally likely to occur.

In statistics, two mathematical functions are often used to represent phenomena similar to the annual sales pattern shown in Figure 4.2. The functions are continuous, meaning that they are plotted as lines rather than discrete counts. **Figure 4.3** shows one of these equations, called the *normal distribution*, overlaid on Figure 4.2's annual sales information. Another very similar bell-shaped mathematical function that we will use often later in the book is called the *t distribution.*

In many instances it is necessary to assess the symmetry of a sample as a prelude to using the normal distribution or the *t* distribution. Several of the tools we discussed earlier are useful for assessing symmetry. In particular, a look at histograms and stem and leaf diagrams can help determine how closely one side of the pictorial representation's central tendency matches the other side. Also, a comparison of the median to the mean provides a quantitative test of symmetry. When the data distribution is symmetric, they will be equal.

Figure 4.2 Symmetrical Data Distribution

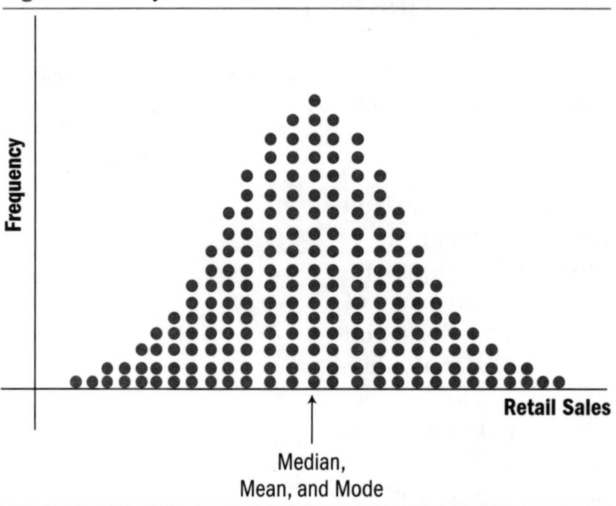

Figure 4.3 Symmetrical Data Distribution with Normal Curve

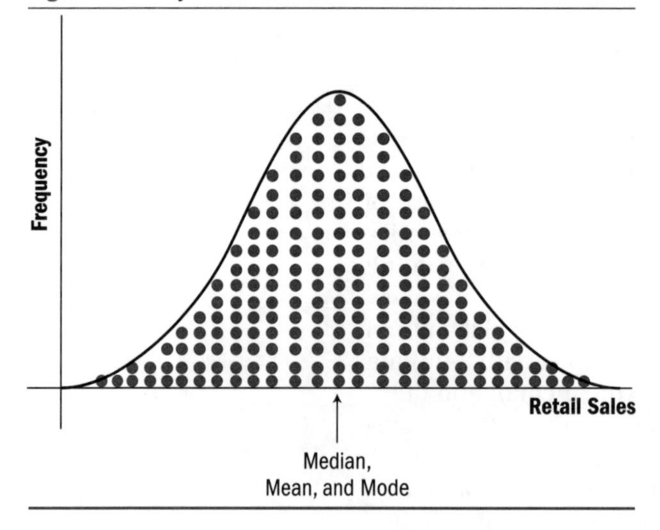

Skewness

When a data distribution is not symmetrical it is said to be *skewed*, or asymmetrical. A comparison of the median and the mean is a good way to assess skewness.

Mean = Median	⟶	Symmetrical
Mean > Median	⟶	Right-skewed
Mean < Median	⟶	Left-skewed

Figure 4.4 illustrates right and left skewness. Right-skewed data will contain a high proportion of large values, pulling the mean upward so it is greater than the median. This is also known as *positive skewness.* Left-skewed data will contain a high proportion of small values, pulling the mean downward so it is less than the median, which is also known as *negative skewness.*

Excel and many other statistics programs provide quantitative assessments of skewness based on the following equation:

$$\text{Skewness} = \frac{n}{(n-1)(n-2)} \Sigma\left(\frac{x_i - \bar{x}}{s}\right)^3$$

When a data distribution is symmetrical, the value of the expression in the parentheses following the summation sign will be equal to zero. So a value of zero on this skewness measure means the data are perfectly symmetric. The value of the expression in the parentheses will be positive when data are right-skewed. Likewise, a negative value indicates left (negative) skewness. Large values on this skewness measure indicate a greater degree

Figure 4.4 Right and Left Skewness

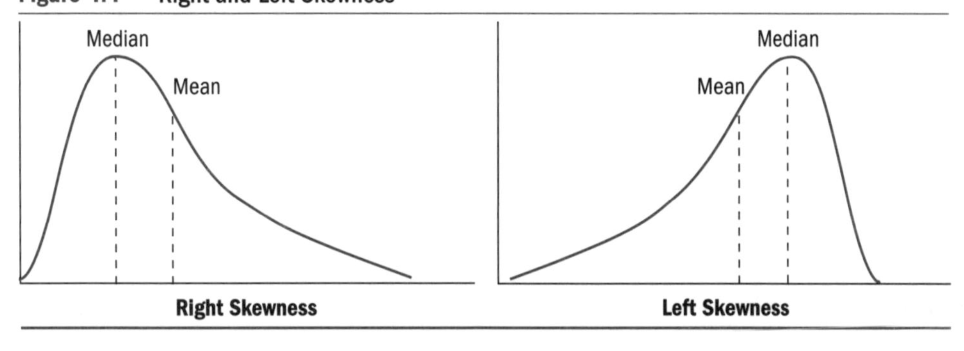

of skewness than small values. For example, a data distribution with a skewness measure of 1.50 is more right-skewed than a data set with a skewness measure of 0.50.

Capturing and Comparing Central Tendency, Dispersion, and Shape Using Box and Whisker Plots

Quantitative measures such as mean and median, standard deviation, and the skewness measure just discussed are ways of capturing and communicating information about the central tendency, dispersion, and shape of a data distribution. Multiple data sets can be compared as well using these quantitative measures. The problem with these comparisons, from an appraisal perspective, is that quantitative measures are sometimes difficult for the intended users of appraisal reports to understand and visualize.

Consider the data sets in **Table 4.7**, which list average monthly electric bills for a random sample of one- and two-bedroom apartment units over the past year. Descriptive statistics on central tendency, dispersion, and shape for the apartment electricity data are presented in **Table 4.8.**

Table 4.8 indicates that two-bedroom apartment electric bills are typically higher, variation in electricity cost is very similar for both apartment sizes, and both data sets are slightly right-skewed. While accurate and revealing, the information contained

Table 4.7 Average Monthly Electricity Bills for One- and Two-Bedroom Apartments

One-Bedroom Apartment Average Monthly Electricity Bill										
79	84	92	91	83	102	77	104	87	78	98
106	74	88	87	83	89	94	96	91	92	87
95	88	96	103	86	83	98	88	92	77	102
72	88	93	82	98	87	105	87	92	78	87
83	82	84	87	89	88	93	95	93	100	98
Two-Bedroom Apartment Average Monthly Electricity Bill										
89	94	110	103	94	112	87	115	97	89	109
118	82	98	98	93	99	104	106	101	102	97
105	98	106	114	97	93	108	98	102	87	112
81	98	103	92	106	95	115	97	102	88	97
93	92	94	97	99	98	103	105	102	112	109

These data are taken from Practice Problems 3.3 found on page 51.

in tables like these can be difficult to visualize.

Table 4.8 Descriptive Statistics on Average Monthly Electricity Bills

	One Bedroom	Two Bedroom
Mean	$89.65	$99.91
Median	$88.00	$98.00
Standard Deviation (*S*)	$8.04	$8.44
Coefficient of Variation (*COV*)	8.97%	8.45%
Skewness	0.043	0.035

Box and whisker plots (sometimes augmented with a sample mean) provide a simple way to capture this information graphically. Traditionally, box and whisker plots map the five-number summary. These five numbers–the minimum value, the value associated with the first quartile (*Q1*), the median, the value associated with the third quartile (*Q3*), and the maximum value–capture a data set's central tendency, dispersion, and shape.

Figure 4.5 is derived from the same data as Table 4.8, depicting the five-number summary in box and whisker plot format. Range and interquartile range are easily compared using these plots. As the plots reveal, most of the skewness is located within the interquartile ranges (i.e., the middle spreads are right-skewed).

Figure 4.6 presents the same data with the box and whisker plots augmented with each data set's sample mean. Including the sample mean shows that the distribution is right-skewed (i.e., the mean is greater than the median) and the extent of the difference between the median and the mean. The box and whisker plots shown here could be used to illustrate several concepts in an appraisal report:

- The extent to which expected values for monthly electricity bills differ between these two unit types.

Figure 4.5 Box and Whisker Plots of Average Monthly Electricity Expense (1 BR vs. 2 BR)

Figure 4.6 Box and Whisker Plots from Figure 4.5 Augmented with Sample Means

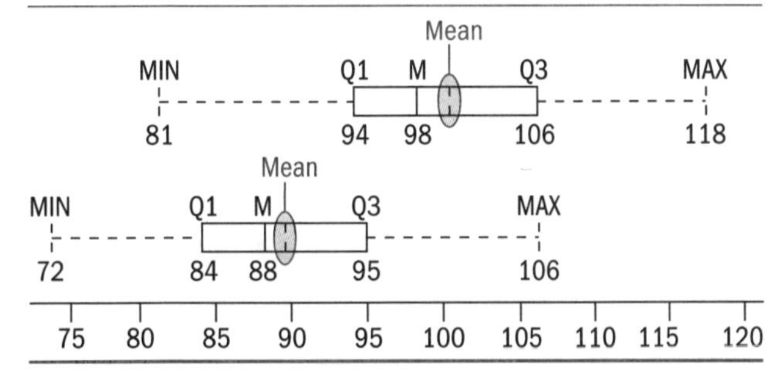

- Dispersion of monthly bills, which could be attributed to variation in unit size within each bedroom class, variation in energy efficiency, and variation in conservation behavior.
- Right skewness, which is consistent with extreme values being associated with failure to conserve energy. (That is, it is easier to waste energy than to conserve energy.)

Utility expense history at a subject property can be compared with these data, and the subject property's central tendency can be analyzed to show how its average unit size compares to the units in the sample set, how energy-efficient the subject property is, and any incentives in place for energy conservation.

While this sort of analysis may be viewed as "overkill" in many appraisal situations, it is beneficial to be able to communicate in this manner if the need arises. However, be cautious about generalizing from samples that are too small or samples that are not representative of the underlying population.

Box and Whisker Plots in Minitab and SPSS

The box and whisker plots shown in Figures 4.5 and 4.6 were created in PowerPoint. This is a time-consuming process, but it yields effective results. Excel does not include box and whisker plots in its charting options. Minitab does include a "boxplot" option. The box and whisker plot below was prepared in Minitab using the same electric bill data. It was created by selecting **Graph** and then **Boxplot** using the **Simple, Multiple Y's** option.

SPSS also includes a boxplot routine. To create a similar boxplot in SPSS, you must enter all of the electric bill data in one column and then create a second column identifying the electric bill data as either one-bedroom or two-bedroom. Once this has been done, select **Graphs** and then **Legacy Dialogs**. Choose the **Boxplot** option then select the **Simple** graph form. The data label for the electric bill column is entered into the **Variable** box, and the data label for the one-bedroom or two-bedroom identification column is entered into the **Category Axis** box.

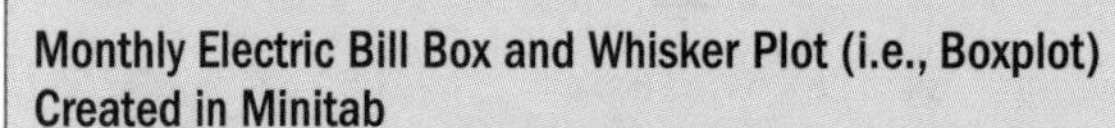

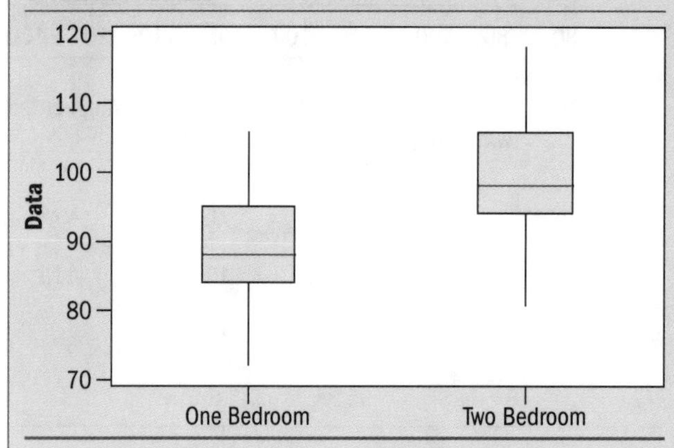

Practice Problems 4.3

The data below include gross living areas for three samples of size $n = 20$ randomly drawn from single-family subdivisions developed in a large southwestern U.S. city in the 1960s, 1980s, and 2000s.

1960 Sample	1980 Sample	2000 Sample
1,120	1,230	1,320
1,200	1,300	1,420
1,080	1,200	1,280
1,320	1,450	1,560
1,450	1,600	1,790
1,360	1,500	1,600
1,180	1,300	1,400
1,600	1,760	2,000
1,520	1,680	1,880
1,100	1,210	1,300
1,180	1,300	1,400
1,160	1,280	1,380
1,080	1,200	1,280
1,230	1,340	1,450
1,300	1,430	1,530
1,410	1,540	1,740
1,430	1,580	1,680
1,510	1,660	1,880
1,130	1,230	1,320
1,150	1,280	1,340

1. Compare means, medians, standard deviations, coefficients of variation, and skewness statistics for these three samples.
2. Create box and whisker plots comparing these three data sets.
3. Assuming these data are representative, what do they reveal about housing trends in this market?

Correlation

Important two-variable relationships in real property analysis include paired variables such as residential lot size and improved living area, hotel room rate and occupancy rate, apartment rent level and vacancy (or occupancy) rate, and trade area purchasing power and gross retail sales to name a few.

The correlation coefficient reveals the strength of the relationship between two variables such as those listed above, and a scatter plot can be created to capture the relationship pictorially. Two methods of computing a correlation coefficient will be discussed in this section–the Pearson Product-Moment Correlation Coefficient *r* and Spearman's *Rho*. The first (Pearson) assumes a linear relationship between two variables. The second (Spearman) relaxes this assumption and assumes only that the relationship is "monotonic" (i.e., the values move together in some fashion).

Figure 4.7 shows four scatter plots. Panel A is positively correlated and linear in appearance. Panel B is negatively correlated and linear in appearance. A Pearson Product-Moment Correlation Coefficient is appropriate for measuring the strength of these relationships. Panel C shows a curvilinear relationship and is not consistent with the Pearson Product-Moment assumption of linearity. Spearman's *Rho* is more appropriate for measuring the strength of this relationship. Panel D shows a relationship that is essentially uncorrelated (no pattern). The correlation coefficient for Panel D would be at or near zero.

Figure 4.7 Scatter Plots Showing Variation in Correlation

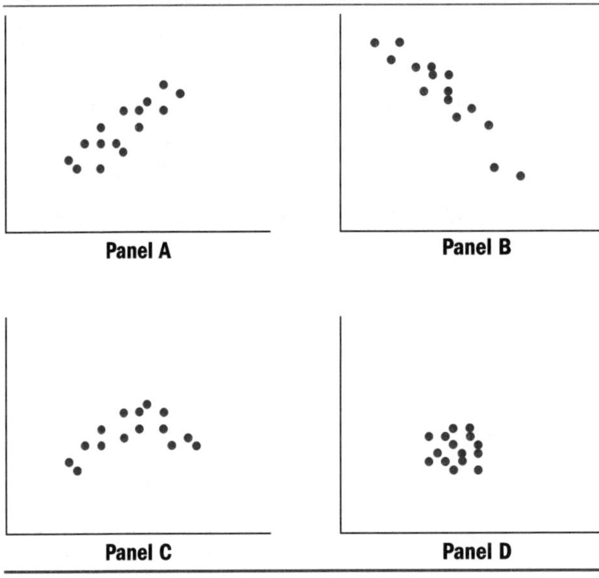

Pearson Product-Moment Correlation Coefficient

The Pearson Product-Moment Correlation Coefficient should be used when data have a linear relationship or when you want to assess the extent of a linear relationship. A lowercase *r* is used as a symbol for the Pearson Product-Moment Correlation Coefficient. For example, r_{XY} signifies the correlation between variables *X* and *Y*.

A perfect linear relationship between two variables *X* and *Y* will fit an equation of the form $Y = a + bX$, where *a* is the vertical axis intercept and *b* is the slope of the line. A perfect fit will have a correlation coefficient of 1 when *b* is positive and -1 when *b* is negative. Perfect linear correlations are said to be *deterministic* relationships because the value of one of the variables fully determines the value of the other based on the equation that captures the relationship.

Consider the deterministic relationship $Y = 2 + 3X$. When $X = 0$, $Y = 2$ (the vertical *y*-axis intercept); when $X = 1$, $Y = 5$; when $X = 2$, $Y = 8$; and so forth. The scatter plot in **Figure 4.8** captures this relationship. Notice that all of the *X* and *Y* pairs form a straight line. In each instance $Y = 2 + 3X$. The correlation coefficient r_{XY} equals 1 because the slope of the deterministic relationship is positive (slope = 3). A slope of 3 means that *Y* changes by three units whenever *X* changes by one unit.

The Pearson Product-Moment Correlation Coefficient is limited to values ranging from -1 to +1. (Correlation cannot be better than perfect.) A coefficient of 0 indicates that there is absolutely no linear relationship among two variables.

Most real-world data sets are not deterministic. In practice, knowing the value of variable *X* does not mean you know the value of the related variable *Y* with certainty. Data that is not deterministic is called *stochastic*–that is,

Figure 4.8 Deterministic Relationship where $Y = 2 + 3X$

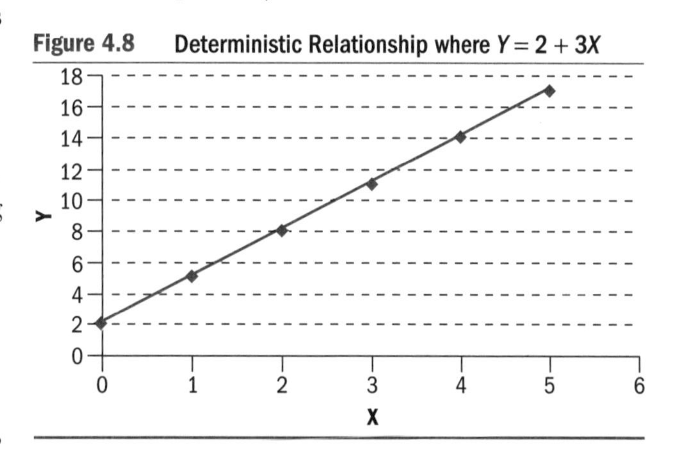

some uncertainty is involved when predicting the value of one variable based on the value of another variable. Stochastic linear relationships are written in equation form as $Y = a + bX + e$, where e in the equation indicates uncertainty, or error, in the linear relationship between X and Y.

Figure 4.9 is a scatter plot illustrating a stochastic, upward-sloping relationship between X and Y. The correlation is positive and quite high (r_{XY} = 0.988), but not perfect. It is possible (and often beneficial) to derive an equation that best fits stochastic data such as X and Y in Figure 4.8 and to quantify the error, or uncertainty, in the relationship. (Chapters 9 and 10 deal with these topics in greater detail.)

The equation for computing r_{XY} looks daunting at first, but when it is broken down into parts the calculations are relatively easy. We will calculate correlation using the equation for r_{XY} using a small data set. Once you understand how the calculation works, it is more efficient to compute r electronically in Excel, SPSS, or Minitab.

$$\mathbf{r_{XY} = \frac{n(\Sigma XY) - \Sigma X\, \Sigma Y}{\sqrt{(n\Sigma X^2 - (\Sigma X)^2)(n\Sigma Y^2 - (\Sigma Y)^2)}}}$$

Consider **Table 4.9** showing bedroom and bath counts for five residences ($n = 5$). We substitute X for bedrooms and Y for baths so the calculations are consistent with the equation above for r_{XY}. Three additional columns provide X^2, Y^2, and the product XY for each residence. The bottom row of the table includes all of the summations needed to calculate r_{XY}. Once this is accomplished, the calculation is a simple matter of plugging numbers into the formula and doing some arithmetic.

Figure 4.9 Stochastic Relationship between X and Y

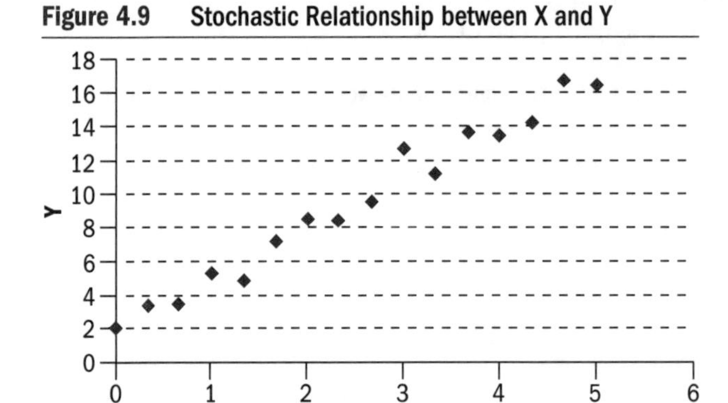

Table 4.9 Bedroom and Bath Counts for Five Residences

Residence	Bedrooms (X)	Baths (Y)	X^2	Y^2	XY
403 Ivy St.	3	2	9	4	6
312 Birch Ln.	2	1.5	4	2.25	3
400 Elm St.	3	2.5	9	6.25	7.5
408 Ivy St.	4	3	16	9	12
515 Elm St.	2	2	4	4	4
$n = 5$	$\Sigma X = 14$	$\Sigma Y = 11$	$\Sigma X^2 = 42$	$\Sigma Y^2 = 25.5$	$\Sigma XY = 32.5$

First we calculate the numerator of the r_{XY} equation:

$$\begin{aligned} n(\Sigma XY) - \Sigma X\,\Sigma Y &= 5 \cdot 32.5 - 14 \cdot 11 \\ &= 162.5 - 154 \\ &= 8.5 \end{aligned}$$

Next we calculate the denominator:

$$\begin{aligned} \sqrt{(n\Sigma X^2 - (\Sigma X)^2)(n\Sigma Y^2 - (\Sigma Y)^2)} &= \sqrt{(5 \cdot 42 - 14^2)(5 \cdot 25.5 - 11^2)} \\ &= \sqrt{(14 \cdot 6.5)} \\ &= \sqrt{91} \\ &= 9.539 \end{aligned}$$

Therefore,

$$\begin{aligned} r_{XY} &= \frac{8.5}{9.539} \\ &= 0.89 \end{aligned}$$

Not surprisingly, bedroom and bath counts are positively correlated, and the relationship is fairly strong.

A correlation coefficient can be interpreted in different ways. One method of analyzing the statistical significance of the correlation coefficient is to ask the question, "What is the probability of obtaining an r_{XY} equal to the value indicated by the sample if the actual population $r_{XY} = 0$?" This probability is known as a *p-value.* If the probability of obtaining the calculated value when the actual $r_{XY} = 0$ is small enough, then the result is considered to be statistically significant.[4] The probability of obtaining a sample correlation coefficient of 0.89 when $n = 5$ and the population is actually uncorrelated ($r_{XY} = 0$) is 0.042. v other words, 42 times out of 1,000 sampling trials you can expect to obtain $r_{XY} = 0.89$ when the actual $r_{XY} = 0$. Given this low *p*-value, there is a high probability[5] that the actual $r_{XY} \neq 0$. Because $r = 0.89$, there appears to be a systematic, but not deterministic, relationship between bedroom counts and bath counts.

Excel does not compute the *p*-values associated with the Pearson Product-Moment Correlation Coefficients it calculates. Associated probabilities are provided in SPSS and Minitab, however.

The coefficient of determination provides another way to interpret r_{XY}. The coefficient of determination is the square of the correlation coefficient, $(r_{XY})^2$, and is symbolized as R^2. This figure tells us the extent to which variation in X accounts for variation in Y and vice versa. When the correlation coefficient is 0.89, the coefficient of determination is equal to 0.89^2, or 0.79. So 79% of the variation in bath count in the sample is accounted for by variation in the sample's bedroom count. The other 21% is not accounted for by variation in bedroom

Correlation in Excel, Minitab, and SPSS

To calculate r_{XY} in Excel, select **Tools** and then **Data Analysis**. Highlight **Correlation** in the data analysis window and then click **OK**. Once the input and output information has been entered into the correlation window as shown below, click **OK** and a **correlation matrix** will appear in your spreadsheet. Note also that the correlation routine will allow you to assess the correlations among numerous pairs of variables simultaneously.

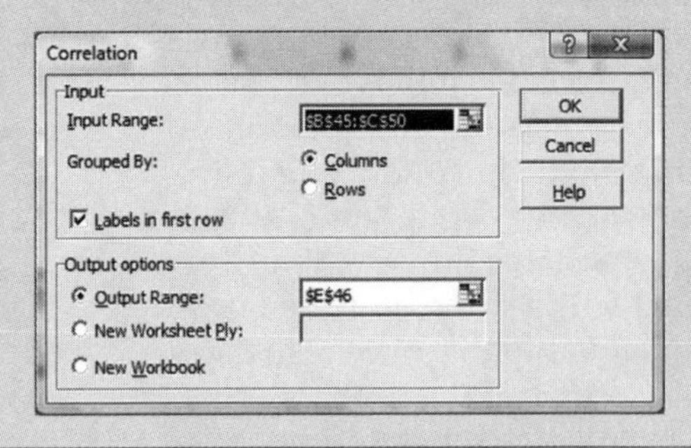

4. How small is small enough? We will cover this in detail in Chapter 7. At this stage, it is sufficient to say that the analyst determines the level of probability that is sufficiently small.
5. The probability is 0.958, i.e., 1 – 0.042.

In Minitab select **Stat, Basic Statistics,** and then **Correlation.** The **following window** will open. Select the variables you want to analyze. (Here Column 1 is BR and Column 2 is BA.) Check **Display p-values** if you want the program to compute the probability associated with the correlation coefficient. When you click **OK**, the result will appear in the **Session** window.

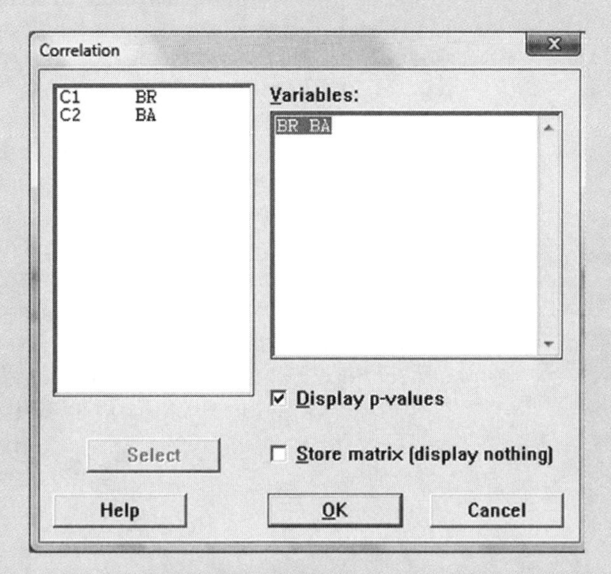

In SPSS select **Analyze, Correlate,** and then **Bivariate.** The **Bivariate Correlations window shown below** will open. Fill in the **Variables** window by selecting the variables to analyze. (Here we selected BR and BA.) Pearson will be the default as will two-tailed significance, which will provide the probability associated with the correlation coefficient. If you check **Flag significant correlations,** SPSS will place an asterisk by any correlation coefficient with a *p*-value of 5% or less. Note that SPSS gives you the option of computing Spearman's *Rho* within this same window.

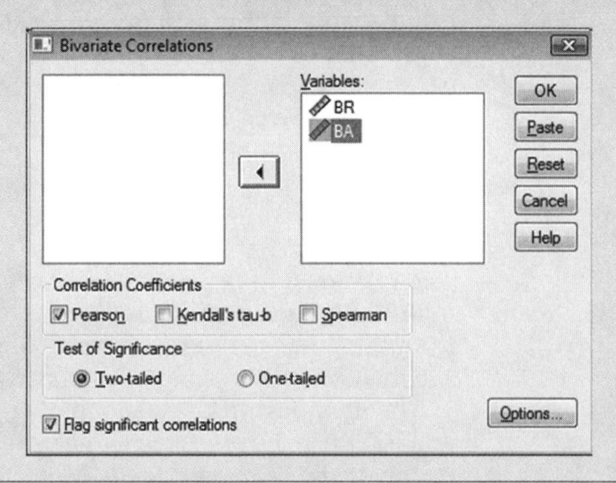

count. The other 21% represents the error inherent in attempting to predict number of baths knowing only bedroom count. Other factors may account for some of the remaining variation in number of baths (such as total living area or age of the residence), or the remaining variation might be totally random.

Spearman's Rho

As mentioned earlier, Spearman's Rank Correlation Coefficient (Spearman's *Rho*, represented by the symbol ρ) relaxes the assumption that the relationship between two variables is linear and assumes only that the relationship is monotonic (that is, they move together in some fashion). Spearman's *Rho* is also useful when at least one of the variables has not been measured on a numerical scale (e.g., ordinal). Spearman's *Rho* is derived from ranked data, computed as follows:

$$\rho_{XY} = \frac{\frac{\Sigma d_X d_Y}{n - 1}}{S_{Xr} S_{Yr}}$$

where d_X and d_Y are deviations of each observation's rank from the mean rank, n is sample size, and S_{Xr} and S_{Yr} are the standard deviations of the ranks on X and Y.

The following example demonstrates how Pearson and Spearman results differ when a relationship is nonlinear and takes us through how to calculate ρ_{XY}. **Table 4.10** presents data on average residential living area in square feet by lot-size category in acres. It also includes a building-to-land ratio derived as

$$\text{Building-to-Land Ratio} = \frac{\text{Improved Living Area}}{\text{Lot Area} \times 43{,}560}$$

The relationship between lot size category and building-to-land ratio in this example is monotonic, but not linear. The downward-sloping curvilinear relationship is pictured in **Figure 4.10**.

The relationship between building-to-land ratio and lot-size category is curvilinear rather than linear, so the Pearson Product-Moment Correlation Coefficient of -0.903 is not a reliable estimate of correlation. The Spearman's Rank Correlation

Coefficient correctly estimates the correlation to be -0.97, which is more indicative of how well the data fits a curvilinear path.

Table 4.11 provides the information required to calculate Spearman's *Rho* from the building-to-land ratio data. Notice that the building-to-land

Table 4.10 Building-to-Land Ratio by Lot-Size Category

Lot-Size Category (AC)	Mean Improved Living Area	Building-to-Land Ratio
0.2	2,500	0.286961
0.4	3,600	0.206612
0.6	3,700	0.141567
0.8	3,600	0.103306
1.0	4,500	0.103306
1.2	4,100	0.078436
1.4	4,900	0.080349
1.6	4,400	0.063131

Figure 4.10 Building-to-Land Ratio by Lot-Size Category

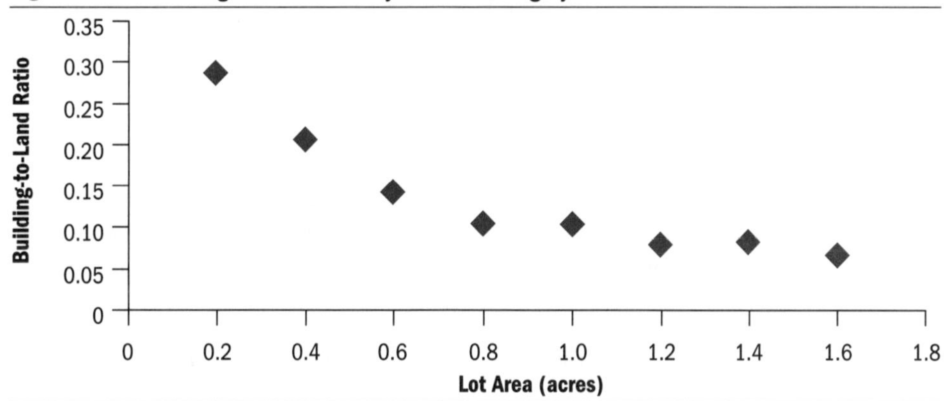

Table 4.11 Spearman's Rho Setup and Calculations

Lot Size Category (X)	Rank (X_r)	Building-to-Land Ratio (Y)	Rank (Y_r)	d_X	d_Y	$d_X d_Y$
0.2	1	0.286961	8	-3.5	3.5	-12.25
0.4	2	0.206612	7	-2.5	2.5	-6.25
0.6	3	0.141567	6	-1.5	1.5	-2.25
0.8	4	0.103306	4.5	-0.5	0.0	0.0
1.0	5	0.103306	4.5	0.5	0.0	0.0
1.2	6	0.078436	2	1.5	-2.5	-3.75
1.4	7	0.080349	3	2.5	-1.5	-3.75
1.6	8	0.063131	1	3.5	-3.5	-12.25
	$\bar{X}_r = 4.5$		$\bar{Y}_r = 4.5$			
	$S_{Xr} = 2.449$		$S_{Yr} = 2.435$			$\Sigma d_x d_y = -40.5$

$$\rho_{XY} = \frac{(\Sigma d_x d_y)/(n-1)}{S_{Xr} S_{Yr}} = \frac{-40.5/7}{2.449 \cdot 2.435} = \frac{-5.786}{5.963} = -0.9703$$

ratios for the 0.8-acre and 1.0-acre categories are the same. Because their ranks are tied, the mean rank for the two categories is entered as each observation's rank.

Spearman's *Rho* in SPSS

Spearman's *Rho* can be calculated in SPSS by choosing **Analyze** then **Correlate** then **Bivariate.** When the **Bivariate Correlation** window opens, enter the variables you want to compare and check the Spearman selection before clicking **OK.** Spearman's *Rho* in SPSS is -0.97 for the Table 4.10 data, which is identical to our hand calculations.

The picture below shows the example data in SPSS along with the **Analyze, Correlate,** and **Bivariate** selections necessary to open the **Bivariate Correlation** window.

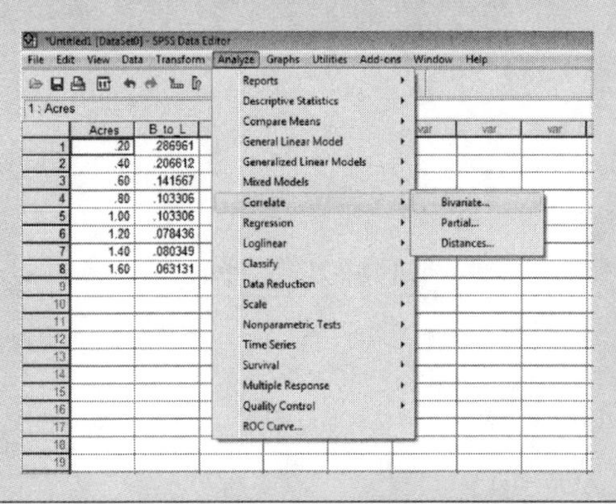

Practice Problems 4.4

Listed below are data on 25 pairs of information on sale price and improved living area in square feet randomly selected from the Las Vegas housing market in 2003.

	Price	Square Feet		Price	Square Feet
1.	$420,750	3,503	14.	$300,000	2,244
2.	$360,000	3,289	15.	$265,000	2,222
3.	$398,000	3,099	16.	$317,000	2,219
4.	$359,900	3,040	17.	$285,000	2,218
5.	$335,000	2,876	18.	$325,000	2,207
6.	$360,000	2,728	19.	$280,000	2,178
7.	$365,000	2,728	20.	$283,500	2,156
8.	$320,150	2,645	21.	$280,000	2,144
9.	$316,000	2,643	22.	$270,000	2,113
10.	$325,000	2,457	23.	$230,000	2,060
11.	$300,000	2,312	24.	$250,000	2,022
12.	$295,700	2,311	25.	$276,000	2,020
13.	$302,500	2,300			

1. Construct a scatter diagram showing the relationship between sale price and improved living area.

2. Calculate the Pearson Product-Moment Correlation Coefficient and the coefficient of determination.

3. What percentage of variability in price is accounted for by variability in improved living area?

4. The data below show price and price per square foot for a nine-property sample. Create a scatter plot and calculate Spearman's *Rho* for this data. Do you think Spearman's *Rho* is the appropriate measure of correlation for these data?

Price	Price/SF
$230,000	135
$245,000	132
$276,000	127
$280,000	128
$285,000	126
$295,000	126
$300,000	125
$320,000	124
$340,000	123

Causality

The fact that two variables are *correlated* does not mean that one causes the other. Correlation is a necessary, but not sufficient, condition for causation.

Causality is difficult to assess because three conditions must be satisfied. For example, to conclude that A caused B:

- A must have preceded B.
- A and B must be correlated.
- Any other possible cause must be ruled out.

The first two steps to assessing causality are usually easy to perform. The third step is difficult and may be impossible to perform in some situations. The point for now is not to confuse correlation with causation. When two variables are correlated, all we can conclude is that they are related.

Real-World Case Study: Part 2

The following is a continuation of the analysis of the real-world townhouse data found in Appendix G, illustrating the statistical tools found in Chapter 4. The **table below** shows the descriptive statistics derived in SPSS for central tendency and dispersion.

Note that SPSS uses a capital *N* for sample size in contrast to the convention of using *N* to represent population size. For reporting purposes it is advisable to change this to a lowercase *n* unless your data are in fact population data (in which case SPSS will miscalculate the standard deviation slightly by treating the data as a sample).

Sale price seems to be centered on $84,000, based on the sample mean and median. Typical living area is about 1,020 to 1,070 square feet located on a roughly 2,800-sq.-ft. lot. Improvement age at the date of sale and days on market appear to be highly right-skewed with each mean being much larger than the median. Central tendencies for these variables are best captured by their medians, at 9 years and 34 days respectively.

Information an analyst might want to stress depends on the scope of the assignment. Possibilities are broad, including land-to-building ratios, market analysis topics such as days on market vs. property age or price bracket, identification of proportion of market by property age, and much more given the amount of data available in this simple spreadsheet. With no particular assignment scope in mind, we will focus on a small subset of the

Descriptive Statistics for Townhouse Sales

		Sale Price	Improved Living Area (SF)	Lot Area (SF)	Age at Sale	Days on Market
N		129	129	129	129	129
Mean		$84,585	1,067	2,789	13.6	51
Median		$83,500	1,019	2,809	9	34
Std. Deviation		$25,261	200	418	8.5	58
Range		$147,000	972	2,236	32	321
Minimum		$48,000	724	1,742	1	0
Maximum		$195,000	1,696	3,978	33	321
Percentiles	25th (*Q1*)	$67,750	984	2,500	8	7.5
	75th (*Q3*)	$90,000	1,151	2,970	20	69

data. Feel free, of course, to explore these data from any perspective that interests you.

Because the days on market variable appears to be quite skewed, we will begin our investigation there. The **boxplot below**, created in Minitab, shows that the middle spread (the interquartile range) of time on market is slightly right-skewed and the upper quartile is highly right-skewed. It appears that a few extreme values for days on market, identified in Minitab by asterisks, are influencing the mean days on market calculation. This supports the conclusion that the median of 34 days is a better indicator of expected time on market than the mean of 51 days. This is confirmed by a skewness statistic value of 1.70 and the **dotplot below**.

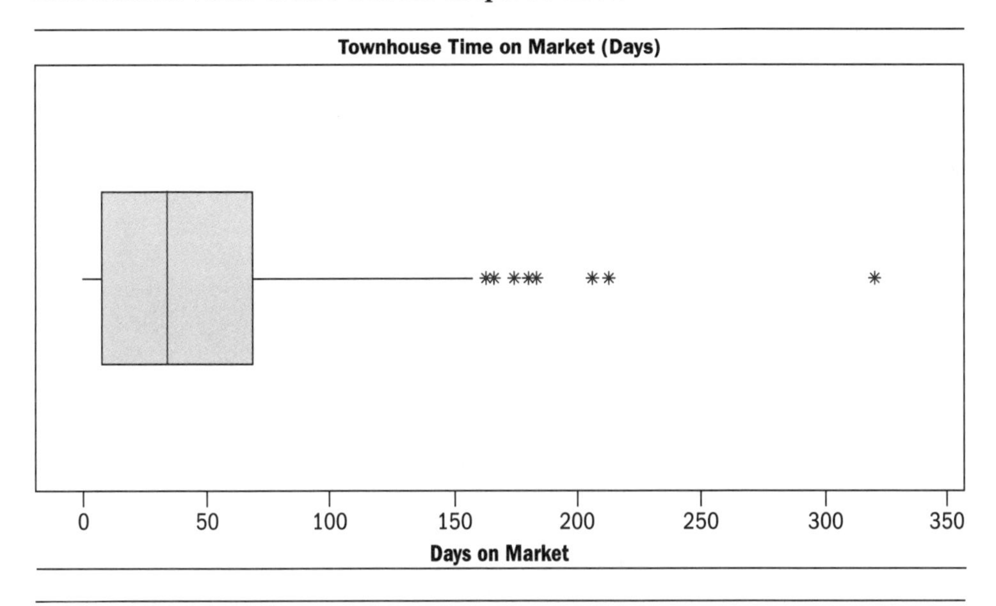

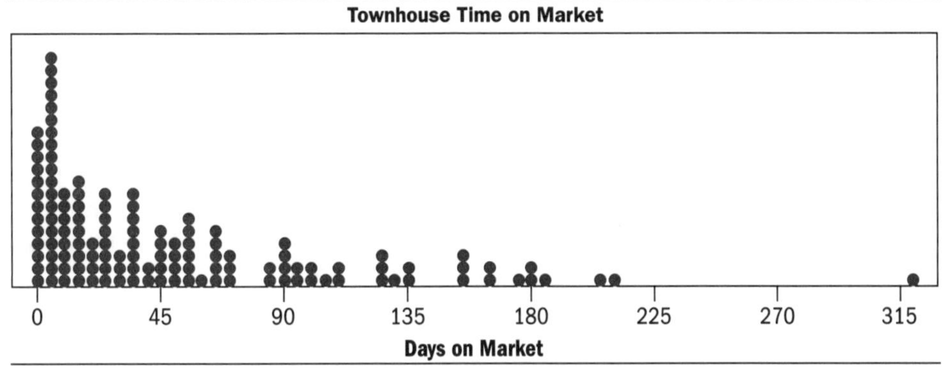

As you will recall, we looked at a **scatter plot** in Chapter 3 comparing price and living area, which is reproduced **below**. The significant extent of linear correlation between price and living area is quantified by the Pearson Correlation Coefficient, which is 0.72. The coefficient of determination for these two variables is 0.518, indicating that 51.8% of price variability is associated with living area differences. The remaining variation in price variability can be attributed to other factors.

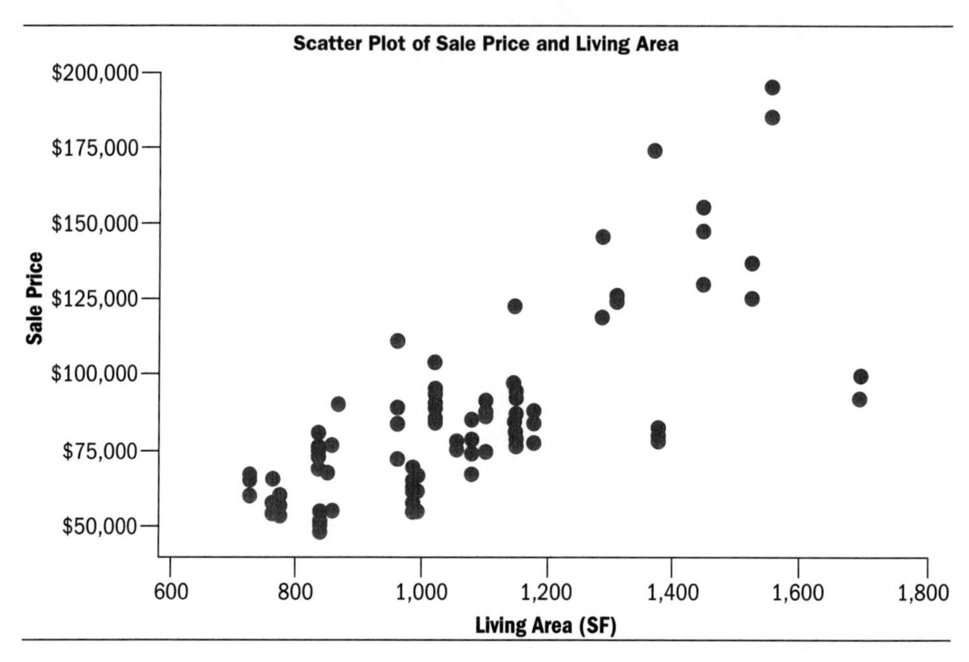

In comparison, depreciation is generally considered to be a nonlinear phenomenon, occurring rapidly initially and slowing down as short-lived items wear out and long-lived building components persist in their contribution to value. The **scatter plot on the following page** indicates this to be the case, showing sale price falling in a curvilinear fashion with the less current construction dates. In this case the Pearson Correlation Coefficient of 0.574 probably understates the correlation between price and building age. Spearman's *Rho* is 0.756 for these data, and is a better indication of correlation.

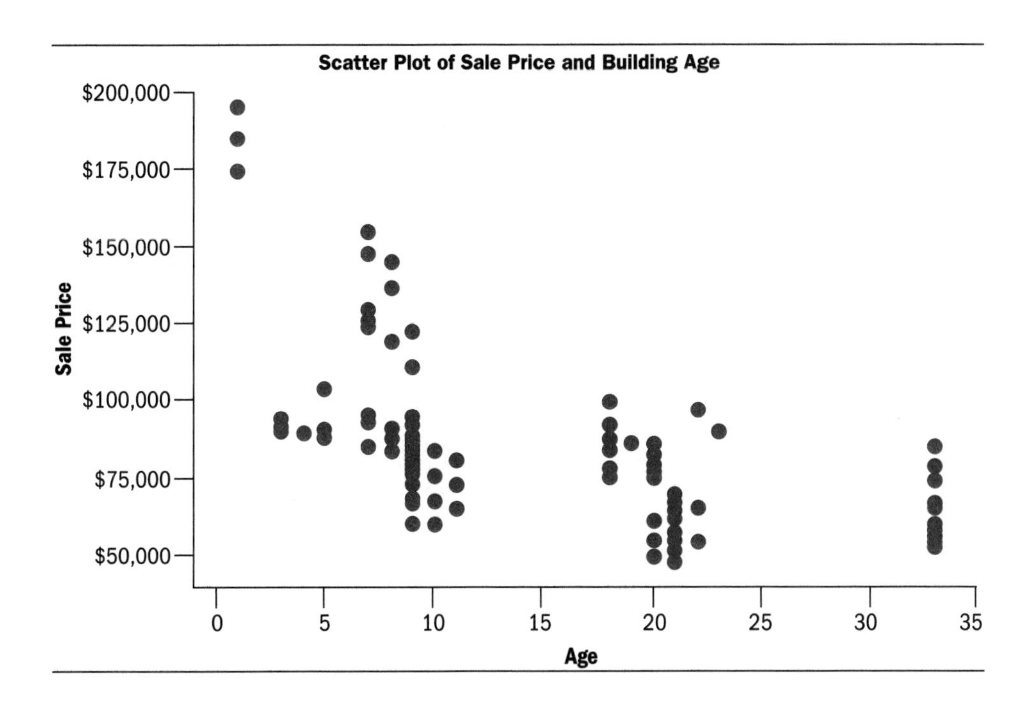
Scatter Plot of Sale Price and Building Age
$200,000
$175,000
$150,000
$125,000
$100,000
$75,000
$50,000
Sale Price
0
5
10
15
20
25
30
35
Age

5 Probability

What Is Probability?

Probability is the chance (likelihood, odds) that something will happen (an outcome). Probability can be defined as *expected relative frequency*. Viewed this way, probability is the proportion of times a given outcome is expected to occur over the long run if an event is repeated again and again.

Why is probability important? In short, because it gives meaning to sample data. Also, making credible inferences about a population based on a sample from the population is not possible without an understanding of probability. While it is not necessary to become an expert in probability to use the sophisticated statistical tools covered later in this book, you should develop an understanding of what probability is, and this chapter will help you achieve a requisite level of understanding.

Two mathematical theories of probability typically prevail–classical and frequency determined. When a probability expectation is based on a *theoretical assessment* of the proportion of all outcomes that should conform to some criterion, probability assessments are said to have been classically derived. Coin tosses and dice rolling are examples of classical probability assessment. On the other hand, *empirically based assessments* define what is called a frequency determined assessment of probability. Surveys, samples, and experiments are examples of frequency determined assessments.

Consider a coin toss (assuming a fair coin) with two possible outcomes: heads and tails. The probability of "heads" on any individual toss of the coin is the expected proportion of heads over the long run as a coin is hypothetically flipped forever. Symbolically, we can refer to this as *P*(heads), where *P* means probability and the outcome in parentheses is the outcome being referred to. So, "*P*(heads)" means the probability of tossing heads. When probability is viewed in this manner,

$$\text{Probability} = \frac{\textbf{Number of Ways an Outcome Can Occur}}{\textbf{Number of Possible Outcomes}}$$

When the number of ways an outcome can occur and the number of possible outcomes are known, then probability can be assessed classically. For example, no experimentation is needed to assess the probability of "heads" in a coin toss. The outcome can occur only one way (i.e., "heads") and the number of possible outcomes is two: 1, "heads" or 2, "tails." Therefore, *P*(heads) = 1 ÷ 2 = 0.50.

More generally, if *X* is a random outcome of an event, then *P(X)* is classically defined as the function

$$P(X) = \frac{X}{T}$$

where *X* is the number of ways in which an outcome can occur and *T* is the number of possible outcomes.

Conditional Probability

Probability can be based on a simple outcome as expressed above, or it can be based on a *condition*, meaning the outcome is conditional on some other event already having occurred. We write this symbolically as $P(X|A)$, where the vertical line means "given." The expression $P(X|A)$ would be read as "probability of *X* given *A*," which means the probability of Event *X* occurring assuming Event *A* has *already* occurred.

A comparison of simple and conditional probability is easily illustrated by the roll of two dice. First we can ask, "What is the probability of rolling a total of 4 with two dice?" The answer would be based on a classical assessment of simple probability. There are three ways to roll a 4:

1. 1 and 3
2. 2 and 2

3. 3 and 1

There are 36 total possible outcomes–6 on the first die and 6 on the second, (i.e., $6 \times 6 = 36$). Therefore, $X = 3$ and $T = 36$, and $P(4) = 3 \div 36$:

$$\mathbf{P(4) = \frac{3}{6 \cdot 6} = \frac{3}{36} = 0.083}$$

If the math is unclear, take a look at **Figure 5.1**. This matrix shows all of the possible sums from rolling two fair six-sided dice numbered 1 through 6. The top row shows the possible outcomes of one die, which are 1 through 6. The left column shows the possible outcomes of the other die, which are also 1 through 6. The intersections of the rows and columns are filled in with the sum of the die in the upper row and the die in the left column. Note that there are 36 intersections (6 × 6). Note also that only three intersections sum to 4: 1 and 3, 2 and 2, and 3 and 1. So $P(4) = 3 \div 36$. You can calculate the probability of any simple outcome of two dice using this matrix.

Figure 5.1 Roll of Two Dice

	1	**2**	**3**	**4**	**5**	**6**
1	2	3	4	5	6	7
2	3	4	5	6	7	8
3	4	5	6	7	8	9
4	5	6	7	8	9	10
5	6	7	8	9	10	11
6	7	8	9	10	11	12

Now think for a moment about conditional probability. What is the probability of rolling a 4 given that one die has been rolled and it is a 3, i.e., $P(4|3)$? This probability is much higher than the previous simple outcome. $P(4|3) = 0.167$, or 1 out of 6, because there is only one way to roll a sum of 4 on the second die (rolling a 1) and there are six possible outcomes for the second die.

More generally,

$$\mathbf{P(A|B) = \frac{P(A \ \& \ B)}{P(B)}}$$

where $P(A \ \& \ B) = P(A) \times P(B)$. In our conditional dice roll example, let Event B be rolling a 3 on the first die, and let Event A be rolling a 1 on the second die, which is required for an outcome of

4 with two dice. $P(A) = 1 \div 6$, and $P(B) = 1 \div 6$ as well. Using the general conditional probability equation above, the probability of rolling a 4 with two dice given the condition of having already rolled a 3 on the first die is

$$P(4|3) = \frac{P(4) \times P(3)}{P(3)} = \frac{\frac{1}{6} \bullet \frac{1}{6}}{\frac{1}{6}} = \frac{\frac{1}{36}}{\frac{1}{6}} = \frac{6}{36} = \frac{1}{6} = 0.167$$

Figure 5.1 provides the same result as the mathematical solution. Look at the "3" column, which has six possible outcomes consisting of the integers 4 through 9. A total of 4 occurs once, therefore there is only one way to roll a 4 after having rolled a 3. (The intersecting row is labeled "1," and 4 = 3 + 1.) There are only six possible outcomes under the "3" column, so the probability associated with each outcome is 1/6, or 0.167.

What is $P(4|5)$? This outcome requires rolling a negative 1 on the second die, which is not possible and has a probability of zero. When $P(A) = 0$ in the conditional probability equation, then $P(A \& B) = 0$ and $P(A|B) = 0$. This is also consistent with Figure 5.1. The six possible outcomes under the "5" column are the integers 6 through 11. Rolling a total of 4 is not possible.

Subjective Probability

People often assess probability based on very little in the way of empirical evidence, or based on misinterpretation of the available evidence. These sorts of probability assessments are based on subjective probability and are often incorrect, given the lack of mathematical or empirical evidence supporting them. Nevertheless, they are important.

Subjective probability assessments are important because people act on them. That is, human behavior is often determined by subjective assessments of probability. Since markets are affected by human behavior, an understanding of subjective probability assessment is often crucial to understanding a market. Consider, for example, the concept of stigma. Even if the mathematical probability of an event reoccurring is very low, subjective assessments by market participants can be much higher due to the stigma associated with some past event. A classic example of stigma is a residence that was the scene of a past murder in a market where that fact must be disclosed to potential buyers. The actual probability of this event reoccurring is not what matters to a market analyst. Subjective assessments by buyers and sellers of the effect of the prior event on the desirability of the property are more important. Potential buyers are likely to be forming a subjective assessment of concerns such as what effect this past event will have on the buyer's ability to sell the home in the future or whether the members of the household will be able to enjoy living there knowing the past event occurred in the home. Sellers may be assessing the psychological need to sell the home and move on given the bad memories associated with the property.

Appraisers may be faced with discovering how such subjective probability assessments will affect expected market price or how a subjective probability assessment affected the price of a comparable sale. Appraisers must also often examine the consequences of subjective probability assessments by market participants. Often tasks such as these are not easily completed due to a lack of empirical evidence (market data). In such cases appraisers may have to rely on other methods such as interviews and surveys.

Practice Problems 5.1

1. What is the probability of rolling a total of 6 by rolling two dice simultaneously?
2. What is the probability of rolling a total of 6 by summing two dice when one die has already been rolled and shows a 2?
3. What is the probability of drawing a spade from a deck of 52 standard playing cards?
4. What is the probability of drawing an ace from a deck of 52 standard playing cards?
5. What is the probability of drawing a spade from a standard deck of 52 cards if one card has already been drawn and set aside and is a heart?
6. What is the probability of drawing a spade from a standard deck of 52 cards if one card has already been drawn and set aside and is a spade?

Laws of Chance

The Law of Large Numbers

The *Law of Large Numbers*, which is often commonly referred to as the "law of averages," can be stated as follows:

> If P represents the probability of an event outcome occurring and the event is repeated n times, as n gets larger the frequency of the occurrence of the outcome approaches $P \times n$.

Practically speaking, this concept can be boiled down to coin flipping. If you flip a coin twice, you cannot expect "heads" to *always* follow "tails" or vice versa. When $N = 2$, the law of large numbers does not apply. When you flip "tails," half of the time this event will be followed by another "tails" because a truly random event is uninfluenced by the prior outcome. However, if you flip a coin a million times (large N), odds are that about 500,000 flips will be tails and 500,000 flips will be heads. This is the so-called "law of averages." As N grows large, the proportion of occurrences of outcome X will eventually approach $P(X)$.

This law has implications for appraisal. Have you ever heard an appraiser say, "I just proved ________________"? You can fill in the blank with any number of statements:

- "a garage is worth $5,000."
- "a pool is worth $15,000."
- "a lot adjoining a busy street is worth $8,000 less."

The "proof" is often one or two paired sales that are the same in all respects except the characteristic that is the subject of the "proof."

A paired sale or two proves nothing. Many uncontrolled circumstances may account for some or all of the observed difference in a small sample: buyer or seller ignorance, buyer or seller motivation, or buyer or seller ambivalence. Such uncontrolled circumstances are likely to fall either way and offset each other when there is a large amount of data, but when uncontrolled effects are not "averaged out" drawing an inference from a small number of sale pairs becomes difficult.

"Proof" would require many, many pairs of sales, with the actual contributory value of the characteristic being the mean difference. The Uniform Standards of Professional Appraisal Practice stresses credibility in valuation analyses, and credibility is enhanced when inference-dependent adjustments are based on an adequately sized sample. A procedure we will discuss in Chapter 10, multiple linear regression, will allow us to control for other factors and synthetically create a large enough number of pairs to support inferences such as these.

General Addition Law

The General Addition Law says

$$P(A \text{ or } B) = P(A) + P(B) - P(A \,\&\, B)$$

Consider **Table 5.1**, which is referred to as a contingency table. The table reflects a survey of home buyers in a given market. It indicates that 1,000 home buyers were surveyed (bottom right total). Of these, 300 purchased a new home, leaving 700 who purchased a resale home (these are the column totals). It also shows that 400 of the buyers had planned to purchase a new home, while 600 had not planned to purchase a new home (these are the row totals). Therefore, actual behavior differed slightly from planned behavior.

Table 5.1 Home Buyer New Home Purchasing Behavior Survey

	Actually Purchased a New Home		
Planned to Purchase a New Home	**Yes**	**No**	**Total**
Yes	200	200	400
No	100	500	600
Total	300	700	1,000

The General Addition Law can be applied here. For example, if the table probabilities are representative of the population of all home buyers in this market, you could ask what is the probability that a home buyer either actually purchased a new home [*P*(*A*)] or planned to purchase a new home [*P*(*B*)]?

$P(A) = 0.30$ (300 out of 1,000 purchased a new home)

$P(B) = 0.40$ (400 out of 1,000 planned to purchase a new home)

$P(A \,\&\, B) = 0.20$ (200 out of 1,000 planned to purchase a new home *and* actually did so)

P(A or B) =0.30 + 0.40 – 0.20 = 0.50

Therefore, half of those surveyed (500 people) must have either planned to purchase a new home *or* actually purchased one.

We can test this result with the contingency table. It turns out that 300 people actually purchased a new home and 400 planned to purchase one, making a total of 700. However, 200 people fall into both categories, so the upper left 200 cell would be double-counted if we were to add the 300 and 400 totals. The general addition law accounts for double-counting the 200 cell. Note that $300 + 400 - 200 = 500$, and $500 \div 1{,}000 = 0.50$, which is the same as the result of the application of the General Addition Law.

A special case of the General Addition Law applies when Events A and B are mutually exclusive (cannot occur together). When two Events A and B are mutually exclusive, $P(A \& B) = 0$. Therefore, in this situation, $P(A \text{ or } B) = P(A) + P(B)$.

General Multiplication Law

The General Multiplication Law says

$$P(A \& B) = P(A|B) \cdot P(B)$$

Let's take another look at Table 5.1. What is the probability of both actually purchasing a new home [$P(A)$] *and* planning to purchase one [$P(B)$]? This "Yes/Yes" category has 200 respondents out of 1,000. So, $P(A \& B) = 0.20$.

Now, let's apply the General Multiplication Law and see if we get the same answer. First, $P(A|B) = 200 \div 400 = 0.50$. That is, 200 of the 400 who planned to purchase a new home actually did so. $P(B) = 400 \div 1{,}000 = 0.40$. That is, 400 of those surveyed had planned to purchase a new home. Therefore, $P(A \& B) = 0.50 \times 0.40 = 0.20$. This analysis indicates that 200 out of 1,000 planned to purchase a new home *and* actually did so. Note that this defines the upper left cell, which has an entry of 200. Therefore, application of the General Multiplication Law is consistent with Table 5.1.

Notice also that if we rearrange the equation for the General Multiplication Law by solving for $P(A|B)$, the result is the conditional probability equation we started with when we discussed dice rolling. That is,

$$P(A|B) = \frac{P(A \& B)}{P(B)}$$

So all of this ties together mathematically.

Statistical Independence

When the outcome of an event does not affect the next outcome of the event, or the outcome of some other event, the outcomes are said to be statistically independent. The coin toss example demonstrates statistical independence because the previous flip of a coin has no effect on the outcome of the next coin toss.

When outcomes are statistically independent, special versions of the Conditional Probability Law and General Multiplication Law apply. When Event A and Event B are statistically independent,

$$P(A|B) = P(A)$$

because $P(A)$ is unaffected by the occurrence of Event B.

Also, when Event A and Event B are statistically independent,

$$P(A \& B) = P(A) \times P(B)$$

For example, the roll of a second die is unaffected by the roll of another die. Therefore, the probability of rolling a 4 with two dice by rolling a 3 on the first one and a 1 on the second one is 1 in 36 ($1/6 \times 1/6$). This is consistent with Figure 5.1 where only one cell out of the 36 cells corresponds with rolling a 3 and then a 1.

An easy way to test for statistical independence is to compare $P(A|B)$ with $P(A)$. Looking again at Table 5.1, $P(A|B) = 0.50$ when Event A was defined as purchasing a new home and Event B was defined as planning to purchase a new home. By comparison, $P(A) = 0.30$. Since $0.50 \neq 0.30$, purchasing a new home and planning to purchase a new home are not independent events. From a marketing perspective, $P(A|B) > P(A)$ indicates that attempting to sell new homes to people who are planning to purchase one is much more fruitful than attempting to sell to everyone. This is, of course, not a surprising bit of insight.

Practice Problems 5.2

1. If $P(A) = 0.20$, $P(B) = 0.60$, and $P(A \& B) = 0.20$, what is $P(A \text{ or } B)$? What is $P(A \mid B)$?
2. If $P(A \mid B) = 0.50$ and $P(B) = 0.50$ what is $P(A \text{ and } B)$?
3. Consider the following contingency table based on a survey of 2,200 homeowners who just completed the process of building a new home.

	Home Completed when Promised		
Satisfied with the Building Experience	**Yes**	**No**	**Total**
Yes	1,000	250	1,250
No	500	450	950
Total	1,500	700	2,200

a. What is P(Completed when Promised *and* Satisfied with the Experience)?

b. What is P(Satisfied with the Experience *given* Home not Completed when Promised)?

c. Are being satisfied with the experience and timely completion statistically independent?

Two Important Probability Distributions

Most of what we have done to this point has applied to discrete random variables, where outcomes are expressed as "200 buyers," "3 bedrooms," "a roll of 4 on two dice," and the like. Probability has therefore been expressed as the ratio of two discrete numbers. For example, in the dice rolling illustration

$$P(4) = \frac{3}{36}$$

Many discrete outcomes can be modeled mathematically by use of what is known as the *binomial distribution.* When discrete outcomes can be categorized into two mutually exclusive and collectively exhaustive groups, the binomial distribution can usually be applied. As we will see, most events can be classified in this manner.

By contrast, many inferential methods rely on another form of random variable called a continuous random variable. With continuous random variables, as compared to discrete random variables, the number of possible outcomes in an interval is infinite. Several important continuous random variable distributions will be investigated and employed in this book. Two of them are particularly noteworthy, the *Normal Distribution* and the *Student's t Distribution.* We will look at the Normal Distribution later in this chapter, leaving the *t* distribution for a future chapter.

Binomial Distribution

The binomial distribution relies on two mutually exclusive and collectively exhaustive categories generally referred to as "success" and "failure."[1] By definition, binomial outcomes are mutually exclusive because both cannot apply to a single event, and they are collectively exhaustive because success and failure cover all possible outcomes.

The binomial distribution applies to data that meet the following five criteria:

1. There are a fixed number of observations (n) in a sample.
2. Each observation is classified into one of two collectively exhaustive and mutually exclusive groups, typically referred to as "success" and "failure."
3. The probability of an outcome being a "success" (p) is constant from trial to trial. Therefore, the probability of an outcome being a "failure" is also constant and equal to $(1 - p)$.
4. Outcomes are independent, meaning the current outcome is not influenced by prior outcomes.
5. Because p is constant, each outcome is the result of either sampling without replacement from an infinite population or with replacement from a finite population.[2]

---

1. Success and failure are the outcome labels typically used in statistics. However, "yes" and "no" are perhaps better labels because the binomial distribution is often used to assess the probability of a negative event occurring. In this context, the negative event is perversely labeled a "success" for probability assessment purposes. The air-conditioner compressor failure example problem in this section is an example of labeling a failure as a success for probability assessment purposes.
2. If the sample were to be from a finite population without replacement, then p would change as sampling continued because the size of the remaining unselected population would become smaller as sample elements were withdrawn.

When all of the above is true, then the Binomial Distribution can be expressed by the equation

$$\mathbf{P(X)} = \frac{\mathbf{n!}}{\mathbf{X!(n - X)!}} \mathbf{p^{X} (1 - p)^{n-X}}$$

where

X = the number of successes in a sample

P(X) = the probability of **X** successes in a sample of size **n** having success probability of **p**

n = sample size

p = the probability of a success

1 – p = the probability of a failure

n – X = the number of failures in a sample

The binomial distribution is quite useful for assessing the probability of an event occurring. Take, for example, the homeowner's warranty that is often associated with the resale of a home. Assume that the probability of an air-conditioner compressor failure within the warranty period is 0.05, based on empirical data (i.e., a frequency determined probability).[3] Knowing this, a typical underwriting concern would be, "If a sample of 10 homes is selected, what is the probability that no more than one will require a new air-conditioner compressor during the homeowner's warranty period?" This can be expressed mathematically as $P(X \leq 1)$. (Assume for illustrative purposes that each home has just one air-conditioning unit.)

Two probabilities must be assessed to calculate $P(X \leq 1)$:

1. The probability that $X = 0$ (none will require a new compressor)
2. The probability that $X = 1$ (one will require a new compressor)

3. Although p is the probability of a "success," success and failure are merely labels. In this example, $p = 0.05$, and failure of a compressor is a "success" for calculation purposes.

In this situation

$$P(X = 0) = \frac{10!}{0!(10 - 0)!} 0.05^0(1 - 0.05)^{10-0}$$

$$= \frac{10!}{10!} (0.95)^{10}$$

$$= 0.5987$$

and

$$P(X = 1) = \frac{10!}{1!(10 - 1)!} 0.05^1(1 - 0.05)^{10-1}$$

$$= \frac{10!}{9!} (0.05)(0.95)^9$$

$$= 0.3151$$

Since $X = 0$ and $X = 1$ are mutually exclusive outcomes, the addition rule says that $P(X \leq 1)$, which is also $P(X = 0$ or $X = 1)$, is the sum of $P(X = 0)$ and $P(X = 1)$. Therefore, $P(X \leq 1) = 0.5987 + 0.3151 = 0.9138$.

As this example demonstrates, given reliable historical information, the Binomial Distribution is a powerful tool for pricing costs associated with expected future events. This knowledge should be useful for tasks such as pricing valuation-related services (see this section's practice problems).

Mean of the Binomial Distribution

The mean of the binomial distribution is the expected value of X, which is the number of successes in n trials. The expected number of successes is the probability of success times the number of trials. Therefore,

$$\mu_{Binomial} = E[X] = np$$

Refer again to the failed compressor example. How many compressors would you expect to fail out of each 10 homes covered by a homeowner's warranty? Because $p = 0.05$, you would expect 0.5 failures for every 10 homes, or 1 out of 20 (or 5 out of 100, 50 out of 1,000, and so forth).

Standard Deviation of the Binomial Distribution

The mathematical expression for standard deviation of the binomial distribution is

$$\sigma_{Binomial} = \sqrt{np(1 - p)}$$

Referring again to the air conditioner compressor example, the standard deviation when $n = 10$ and $p = 0.05$ is

$$\sigma_{Binomial} = \sqrt{10 \cdot 0.05(1 - 0.05)} = \sqrt{0.5(0.95)} = 0.689$$

With an expected value of 0.5 compressor failures per 10 houses and a standard deviation of 0.689, you would expect the typical compressor failure rate to be between 0 and about 1 for each 10 homes ($0.5 \pm 0.689 = -0.189$ to 1.189). This is supported by the assessment of $P(X \leq 1)$ being 0.9138.

Information such as this is can be useful to a variety of analysts. For example, assume a housing development had a highly improbable compressor failure rate of 4 out of 10 during the warranty period. Knowing that such a failure rate is highly unlikely, a warranty provider, builder, subcontractor, or manufacturer would benefit from investigating the reason for the failures–perhaps faulty installation throughout the subdivision or a run of defective compressors at the manufacturer.

Practice Problems 5.3

An appraisal firm analyzed its historical experience appraising warehouse buildings in its market and found that typically each appraisal's set of comparable sales included sales from the appraisal company's own database and new data purchased from a proprietary data source. Newly purchased data accounted for 60% of the comparable sales data, and sales derived from the company's own database accounted for 40% of the comparable sales data.

1. If using a sale from the company's own database is defined as a success, what is the probability of four sales coming from the company's own database when seven comparable sales are included in an appraisal?
2. If each comparable sale purchased from the proprietary data source costs $50 and on average six comparable sales are included in a warehouse appraisal, how much should the appraisal firm budget for purchased sales data when pricing warehouse appraisals?

Normal Distribution

The normal distribution is sometimes referred to as the most important probability distribution in the field of statistics. It is a continuous random variable that is symmetric and bell-shaped, and that theoretically extends from negative infinity ($-\infty$) to positive infinity ($+\infty$). In other words, the function applies to very large positive and negative departures from its mean. Because it is symmetric, its mean and median are equal.

The normal distribution encompasses a family of normal curves that is infinitely large due to differences in the mean and standard deviation of the normal curves. The *standard* normal distribution has a mean of zero and a standard deviation of 1. Statisticians often rely on the standard normal distribution for probability assessment purposes, and convert normal distributions having different means or standard deviations to the standard normal distribution prior to analysis.

The equation for the normal distribution is

$$Y = \frac{1}{\sqrt{2\pi\sigma}} e^{-(1/2)[(X-\mu)/\sigma]^2}$$

where π = approximately 3.14159 and e = approximately 2.71828.

Substituting 0 for the population mean μ and 1 for the population standard deviation σ, results in the following much simpler equation for the standard normal distribution:

$$Y = \frac{1}{\sqrt{2\pi}} e^{-(1/2)Z^2}$$

where

$$Z = \frac{X-\mu}{\sigma}$$

Values of Z are measured in standard deviation units. For example, $Z = 1.5$ is 1.5 standard deviations above the mean of the standard normal distribution, and $Z = -1.5$ is 1.5 standard deviations below the mean of the standard normal distribution.

The simple exercise illustrated in **Figure 5.2** provides another way of presenting the standard normal distribution. *Z* values are listed at intervals of 0.1 beginning at -3.5 and extending up to +3.5. Associated values of *Y* are then calculated for each *Z* value using the standard normal distribution equation. (You can easily do this yourself in Excel using the equation for *Y* on page 123.) The two values (*Y* and *Z*) are then plotted in Excel (see the chart in Figure 5.2), replacing the scatter plot points with a line, which illustrates the values of the standard normal distribution ranging from -3.5 standard deviations to +3.5 standard deviations.

As **Figure 5.2** shows, although the equation for the standard normal distribution is defined for very large negative and positive numbers, the solution for *Y* becomes a very small number above $Z = 3$ and below $Z = -3$. The chart approaches zero at these levels, but it is "asymptotic" to the horizontal axis. (That means it continues to approach the horizontal axis at large values of Z and negative Z but never gets completely there.) For this reason, most standard normal probability tables range from $Z = -4.0$ to $Z = +4.0$, with probabilities above and below these values being viewed as negligible.

Assessing Normality

Several procedures are very instructive for assessing how closely population data or sample data fit the normal distribution:

- First, chart the data to see if it is generally bell-shaped and symmetrical. For small data sets, construct a stem and leaf display to assess shape and symmetry or a box and whisker plot to assess symmetry. A histogram can be used for large data sets.
- Second, look at summary descriptive statistics for the data. How similar are the mean and median? Compare the interquartile range and the standard deviation. The interquartile range is about 1.33 times the standard deviation in a normal distribution. Also compare the range to the standard deviation. If the data are normally distributed, the range should be approximately equal to 6 standard deviations (i.e., from 3 stan-

Figure 5.2 Standard Normal Distribution

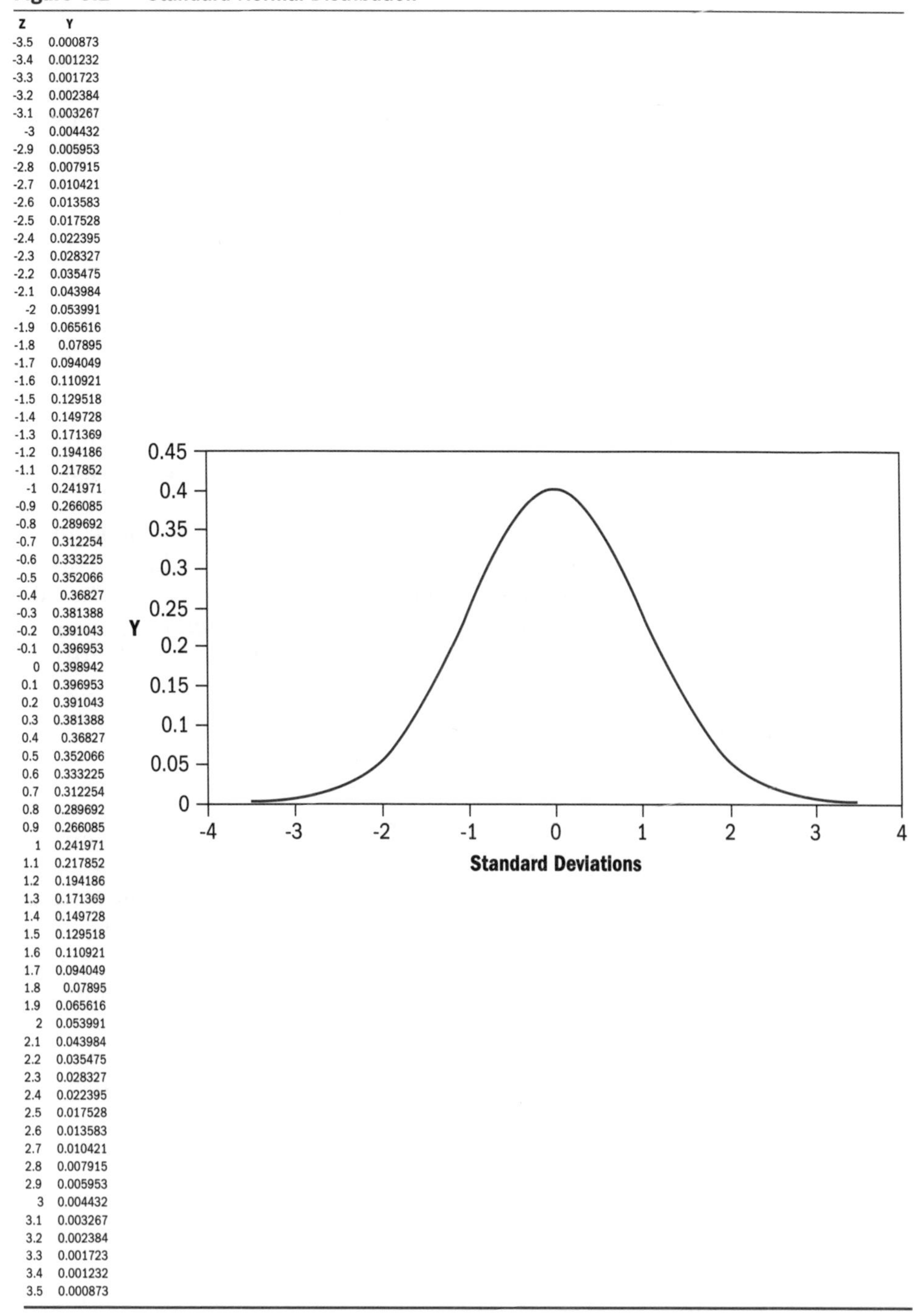

Z	Y
-3.5	0.000873
-3.4	0.001232
-3.3	0.001723
-3.2	0.002384
-3.1	0.003267
-3	0.004432
-2.9	0.005953
-2.8	0.007915
-2.7	0.010421
-2.6	0.013583
-2.5	0.017528
-2.4	0.022395
-2.3	0.028327
-2.2	0.035475
-2.1	0.043984
-2	0.053991
-1.9	0.065616
-1.8	0.07895
-1.7	0.094049
-1.6	0.110921
-1.5	0.129518
-1.4	0.149728
-1.3	0.171369
-1.2	0.194186
-1.1	0.217852
-1	0.241971
-0.9	0.266085
-0.8	0.289692
-0.7	0.312254
-0.6	0.333225
-0.5	0.352066
-0.4	0.36827
-0.3	0.381388
-0.2	0.391043
-0.1	0.396953
0	0.398942
0.1	0.396953
0.2	0.391043
0.3	0.381388
0.4	0.36827
0.5	0.352066
0.6	0.333225
0.7	0.312254
0.8	0.289692
0.9	0.266085
1	0.241971
1.1	0.217852
1.2	0.194186
1.3	0.171369
1.4	0.149728
1.5	0.129518
1.6	0.110921
1.7	0.094049
1.8	0.07895
1.9	0.065616
2	0.053991
2.1	0.043984
2.2	0.035475
2.3	0.028327
2.4	0.022395
2.5	0.017528
2.6	0.013583
2.7	0.010421
2.8	0.007915
2.9	0.005953
3	0.004432
3.1	0.003267
3.2	0.002384
3.3	0.001723
3.4	0.001232
3.5	0.000873

dard deviations above the mean to 3 standard deviations below the mean).

- Look at the distribution of the data. When data are normally distributed, about 68% of the data are within one standard deviation of the mean. Also, about 80% of the data should be between -1.28 and +1.28 standard deviations from the mean, and about 95% of the data should be between -2 and +2 standard deviations from the mean. See how close the data come to these criteria.

Many statistical programs include tests for normality such as normal probability plots and test statistics. However, once you become experienced at assessing normality, the three tests above can be more instructive than attempting to interpret a normal probability plot or relying solely on a statistical test.

Normal probability plots assess bell shape and skewness in a data distribution. As a data set approaches being normally distributed (bell-shaped and symmetrical), its normal probability plot approaches the shape of the straight line in **Figure 5.3** labeled "Normal." Left-skewed data will result in a curved line bowed upward. Right-skewed data will result in a curved line bowed downward. Some statistics programs reverse the axes (i.e., data values are on the horizontal axis and Z values are on the vertical axis). When the axes are reversed, the shapes of left- and right-skewed curves are the opposite of how they are shown in Figure 5.3. Note also that Figure 5.3 is simplified for illustrative purposes. For many data sets normal probability plots are more complex.

Figure 5.3 Exemplar Normal Probability Plots

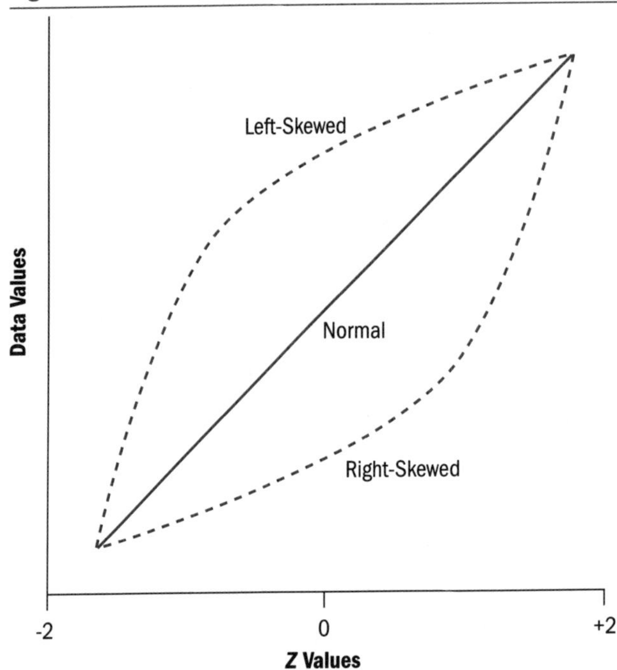

Reading and interpreting normal probability plots can be difficult due to subtle variations in plot shape. Therefore, probability plots should not be relied on exclusively but should be included as one tool along with statistical tests and the list of assessment criteria included in this section. Ultimately, assessment of normality requires judgment about how close to normal is close enough for statistical inference. Some statistical procedures, such as regression analysis, are mathematically robust and remain useful despite violations of normality assumptions.[4] Also, the sampling distribution of the mean is approximately normal regardless of the shape of the underlying distribution when n is sufficiently large. (Chapter 7 deals with this topic in detail.)

Figure 5.4, which was created using Minitab, illustrates an investigation of normality of assessment ratios for 1,054 sales, computed as taxable value divided by sale price. The mean ratio is 0.784 and the median ratio is 0.732, indicating that the central tendency is to assess properties at about 73% to 78% of market value, as estimated by actual sale price. The mean is greater than the median indicating a slight right skewness, which is borne out by the bow in the normal probability plot (axes are reversed from Figure 5.3). The straight line also shown in the normal probability plot is indicative of the plot we would expect if the data were perfectly normal. The box and whisker plot also shows a right skewness consisting mostly of high assessment ratio outliers marked with asterisks.

The range of 1.73 is 7.4 times the standard deviation, greater than the value of 6 times expected for a normal distribution. The interquartile range of 0.251 is 1.07 times the standard deviation, which is less than the 1.33 times expectation for normally distributed data. These metrics indicate that the data are more bunched about the central tendency than normal data and also more spread out at the tails than normal data.

---

4. See John Neter, William Wasserman, and Michael H. Kutner, *Applied Linear Statistical Models*, 3rd ed. (Homewood, Ill.: Irwin, 1990), 53. See also Daniel L. Rubinfeld, "Reference Guide on Multiple Regression" in *Reference Manual on Scientific Evidence*, 2nd ed. (Washington, D.C.: Federal Judicial Center, 2000). This topic is also discussed in more detail in Chapters 6 and 9 of this book.

Figure 5.4 Analysis of Distribution of Property Tax Assessment Ratios

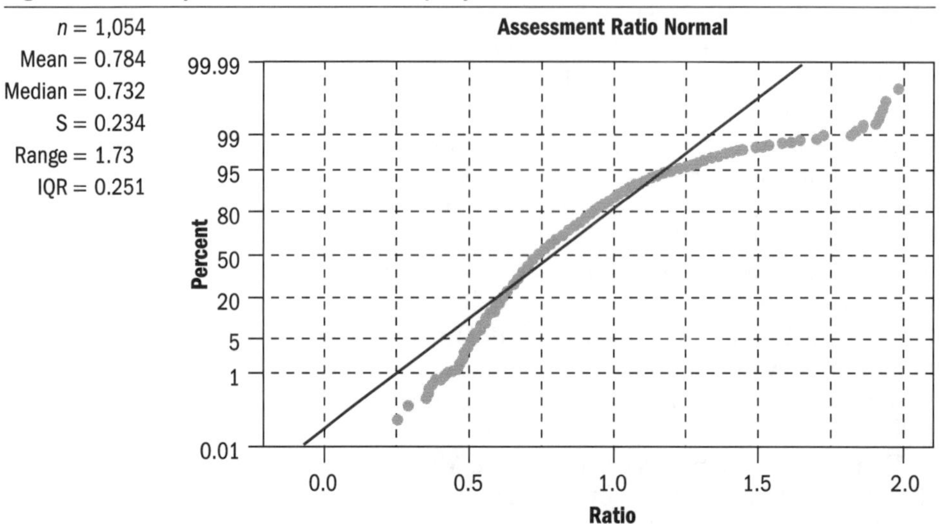

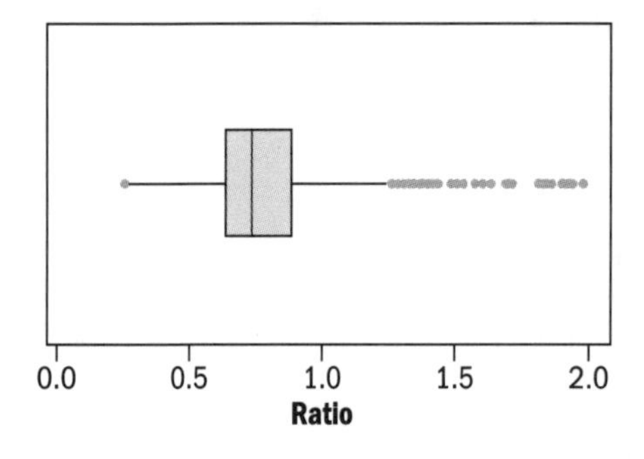

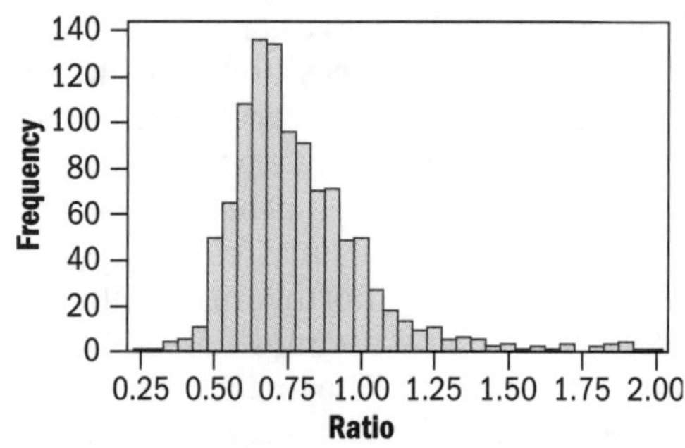

Normality Tests in SPSS and Minitab

In SPSS, a Kolmogorov-Smirnov test statistic is included with normal probability plots along with a significance level for the normality test. Note that small *p*-values* for normality tests indicate non-normal data. In SPSS, click the **Analyze** pull-down tab, click **Descriptive Statistics**, and then click **Explore**. Select the variable to analyze and then click on **Plots** in the **Explore Window** and check **Normality plots with tests.**

In Minitab, which was used to create the charts in Figure 5.4, click **Graph, Probability Plot,** and then select **Single** to assess normality for a single variable. Output includes a normal probability plot and a choice of three statistical tests for normality—**Anderson-Darling, Ryan-Joiner,** and **Kolmogorov-Smirnov.**

* Small *p*-values indicate non-normal data because the null hypothesis for this test is that the data are normal, which is rejected with a small *p*-value. If you are unfamiliar with the term "*p*-value" we will discuss it in Chapter 6. You might want to take a second look at the discussion here after you have read Chapter 6.

The histogram reveals that the data are densely packed around the central tendency. It also shows the wide spread at the tails, especially the right tail. In addition, although not shown here, there are statistical tests for normality included with SPSS and Minitab that indicate that these data are not normally distributed.

The simple question being addressed here is "Are the data normally distributed?" The answer is clearly "No." The more relevant question, which is less easy to assess, is "Are the data sufficiently close to normal to allow application of inferential tools that assume normality?" This answer isn't so clear. Given the large sample size, approximate symmetry of the data, and robustness of many inferential tests to violation of the normality assumption, it is probably acceptable in some circumstances to make inferences about the mean of these data under the assumption of normality. This conclusion is reasonable based on other criteria, despite the obvious departure of the data from normality. (In Chapter 7 we will examine why a large sample size and closeness to symmetry allow us to reach this conclusion.)

On the other hand, it would be improper to use the standard normal distribution as a proxy for the distribution of the *actual* tax assessment ratios. For example, more of the actual assessment ratios are closely packed around the mean than they would be if they were normally distributed. In this case, using the properties of the normal distribution to interpret these data would underestimate the probability of an assessment ratio being near the mean. The important lesson of this example is that decisions concerning how to employ the normal distribution to interpret data must be thoughtfully considered.

Probability and the Normal Distribution

The normal distribution is constructed so that the area under the normal curve from $-\infty$ to $+\infty$ (the area between the horizontal axis and the curve plotting the values of Y) is equal to 1. When the area under the curve is defined in this manner, the subset of the area under the curve between any two points along the horizontal axis represents the probability that a normally distributed outcome will fall within this range.

Take a look at **Figure 5.5**. Lines A and B are drawn vertically at the mean of the standard normal distribution, which is 0, and at a *Z* value of 1, which is one standard deviation above the mean. The shaded portion represents the area under the standard normal curve between these two points. The ratio of the shaded area to the entire area under the curve from -∞ to +∞ is the probability that a random variable that is normally distributed will attain a *Z* value between 0 and 1. Because we know the equation that defines the standard normal distribution, we can calculate the area of the shaded portion under the curve. To reiterate, this area is also the probability associated with this interval because the total area under the curve is equal to 1.

Figure 5.5 Area Under the Normal Curve

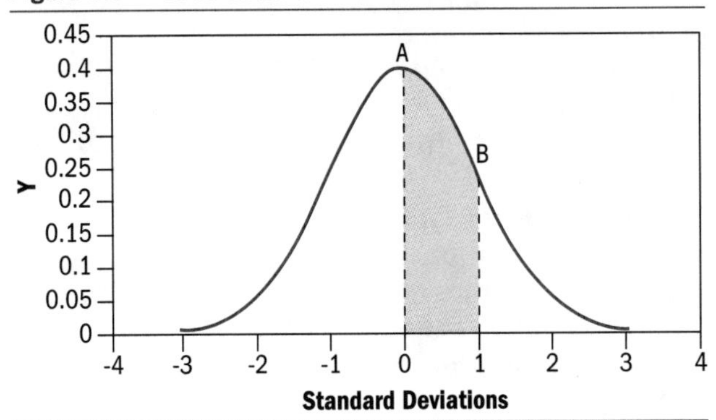

The shaded portion of Figure 5.5 has an area of approximately 0.3413. Therefore, the probability of a normally distributed random variable attaining a *Z* value between 0 and 1 is 0.3413. Also, because the normal curve is symmetrical, the area between -1 and 0 is also 0.3413. Therefore, the probability of attaining a *Z* value between -1 and 1 is $2 \times 0.3413 = 0.6826$. As a result, when assessing normality we can say that approximately 68% of the data should be within an interval of the mean ±1 standard deviation. This is true for all normally distributed data because a simple transformation converts any normal distribution to the standard normal distribution shown in Figure 5.5. As a reminder, the transformation equation is

$$Z = \frac{X - \mu}{\sigma}$$

Example Problems

Assume that residential annual heating and air-conditioning energy costs in a given market are normally distributed with a mean of $1 per square foot and a standard deviation of $0.10 per square foot. As long as this assumption is true, we can make statements about the probability of annual heating and air-conditioning energy costs being within certain intervals, as follows:

1. The probability that a home's annual heating and air-conditioning energy cost is less than $1.15 per square foot is _______.
2. The probability that a home's annual heating and air-conditioning energy cost is greater than $1.10 per square foot is _______.
3. The probability that a home's annual heating and air-conditioning energy cost is between $0.95 and $1.05 per square foot is ______.

In order to solve for the probabilities missing from these statements using the standard normal table, the stated energy cost amounts must be converted into *Z* values (i.e., standard deviations of the standard normal distribution), as follows:

$$Z_{\$1.15} = \frac{\$1.15 - \$1.00}{\$0.10} = 1.5$$

$$Z_{\$1.10} = \frac{\$1.10 - \$1.00}{\$0.10} = 1.0$$

$$Z_{\$1.05} = \frac{\$1.05 - \$1.00}{\$0.10} = 0.5$$

$$Z_{\$0.95} = \frac{\$95 - \$1.00}{\$0.10} = -0.5$$

Once this has been done the statements can be rephrased in standard normal terms.

- Statement 1 becomes the probability that *Z* is less than 1.5, which is 0.9332.
- Statement 2 becomes the probability that *Z* is greater than 1.0, which is 0.1587.
- Statement 3 becomes the probability that *Z* is between -0.5 and +0.5, which is 0.3829.

Z Values in Excel

Excel can provide these probabilities as well by using either the NORMSDIST function or the NORMDIST function. The NORMSDIST function returns the cumulative probability for *Z* values. Enter **=NORMSDIST(1.5)** in an Excel spreadsheet cell and the value will be 0.9332 (rounded to four decimal places). Enter **=NORMDIST(1.15, 1, 0.1,1)** in an Excel spreadsheet and the value will also be 0.9332. The NORMDIST function converts the input to a *Z* value and then calculates the cumulative probability. The function arguments here are 1.15 (the annual HVAC energy expense per square foot), 1 (the mean), 0.1 (the standard deviation), and 1 (signifying cumulative probability). Excel can also be used to calculate the probabilities associated with Statements 2 and 3 above. Give it a try and see if you can replicate the answers we derived using the standard normal table.

These solutions are derived from the standard normal table found in Appendix A, a portion of which is illustrated in **Figure 5.6**. For example, the probability that Z lies in the interval from 0 to 1.5 is 0.43319 (rounded in this analysis to 0.4332). We know that the probability that Z is less than 0 is 0.50 because the distribution is symmetric. (The area under the portion of the curve to the left of the mean is one-half of the total area under the curve.) Therefore, the probability that Z is less than 1.5 is the sum of the probability that Z is less than 0 and the probability that Z is between 0 and 1.5, or $0.50 + 0.4332 = 0.9332$.

Using the same logic, the probability that Z is greater than 1.0 is the same as $1 - P(Z < 1.0)$.

$$\mathbf{P(Z < 1.0) = 0.50 + 0.3413 = 0.8413}$$

$$\mathbf{P(Z > 1.0) = 1 - 0.8413 = 0.1587}$$

$$\mathbf{P(-0.5 < Z < 0.5) = 2 \times P(0 < Z < 0.5) = 2 \times 0.19146 = 0.3829}$$

Figure 5.6 **Standard Normal Probabilities**

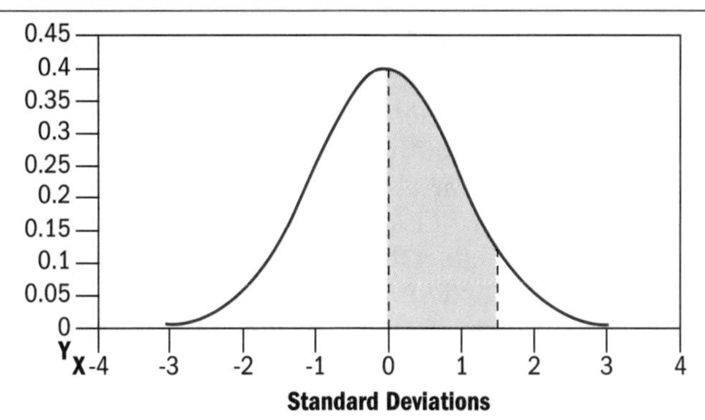

Area Under the Normal Curve from 0 to X

X	0.00	0.01	0.02	0.03	0.04	0.05	0.06	0.07	0.08	0.09
0.0	0.00000	0.00399	0.00798	0.01197	0.01595	0.01994	0.02392	0.02790	0.03188	0.03586
0.1	0.03983	0.04380	0.04776	0.05172	0.05567	0.05962	0.06356	0.06749	0.07142	0.07535
0.2	0.07926	0.08317	0.08706	0.09095	0.09483	0.09871	0.10257	0.10642	0.11026	0.11409
0.3	0.11791	0.12172	0.12552	0.12930	0.13307	0.13683	0.14058	0.14431	0.14803	0.15173
0.4	0.15542	0.15910	0.16276	0.16640	0.17003	0.17364	0.17724	0.18082	0.18439	0.18793
0.5	0.19146	0.19497	0.19847	0.20194	0.20540	0.20884	0.21226	0.21566	0.21904	0.22240
0.6	0.22575	0.22907	0.23237	0.23565	0.23891	0.24215	0.24537	0.24857	0.25175	0.25490
0.7	0.25804	0.26115	0.26424	0.26730	0.27035	0.27337	0.27637	0.27935	0.28230	0.28524
0.8	0.28814	0.29103	0.29389	0.29673	0.29955	0.30234	0.30511	0.30785	0.31057	0.31327
0.9	0.31594	0.31859	0.32121	0.32381	0.32639	0.32894	0.33147	0.33398	0.33646	0.33891
1.0	0.34134	0.34375	0.34614	0.34849	0.35083	0.35314	0.35543	0.35769	0.35993	0.36214
1.1	0.36433	0.36650	0.36864	0.37076	0.37286	0.37493	0.37698	0.37900	0.38100	0.38298
1.2	0.38493	0.38686	0.38877	0.39065	0.39251	0.39435	0.39617	0.39796	0.39973	0.40147
1.3	0.40320	0.40490	0.40658	0.40824	0.40988	0.41149	0.41308	0.41466	0.41621	0.41774
1.4	0.41924	0.42073	0.42220	0.42364	0.42507	0.42647	0.42785	0.42922	0.43056	0.43189
1.5 →	0.43319	0.43448	0.43574	0.43699	0.43822	0.43943	0.44062	0.44179	0.44295	0.44408

Practice Problems 5.4

Find the following probabilities:

1. $P(Z < 2)$ **2.** $P(Z < -2)$ **3.** $P(Z > 1.24)$ **4.** $P(1 < Z < 2)$

5. If the variable X is normally distributed with mean = 6 and standard deviation = 2, what is

a. $P(X < 4)$? b. $P(X > 8)$? c. $P(X > 3)$? d. $P(5 < X < 8)$?

Real-World Case Study: Part 3

The analysis of the real-world townhouse data found in Appendix G continues below, illustrating some of the statistical tools found in Chapter 5. The analysis focuses on the distribution of the price variable and comparisons to normality.

Recall that the first step in assessing normality is to chart the data distribution to see if it is generally bell-shaped and symmetrical. The **histogram below** plots price categories on the horizontal axis against percentage of observations in each category on the vertical axis. The chart was created in Minitab, which provides the option of also fitting a normal curve to the distribution based on the data's mean and standard deviation. Considering this chart alone, it is obvious that sale price is not normally distributed.

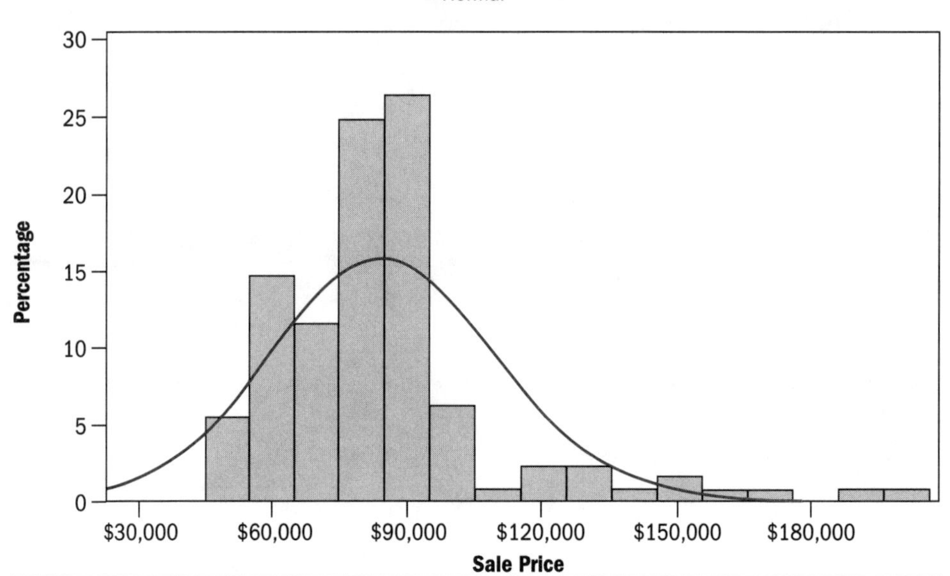

The conclusion of non-normality is confirmed by the Kolmogorov-Smirnov (KS) test for normality, which rejects an underlying hypothesis of normality, and a normal probability plot (**following page**) bent in a direction consistent with right skewness.

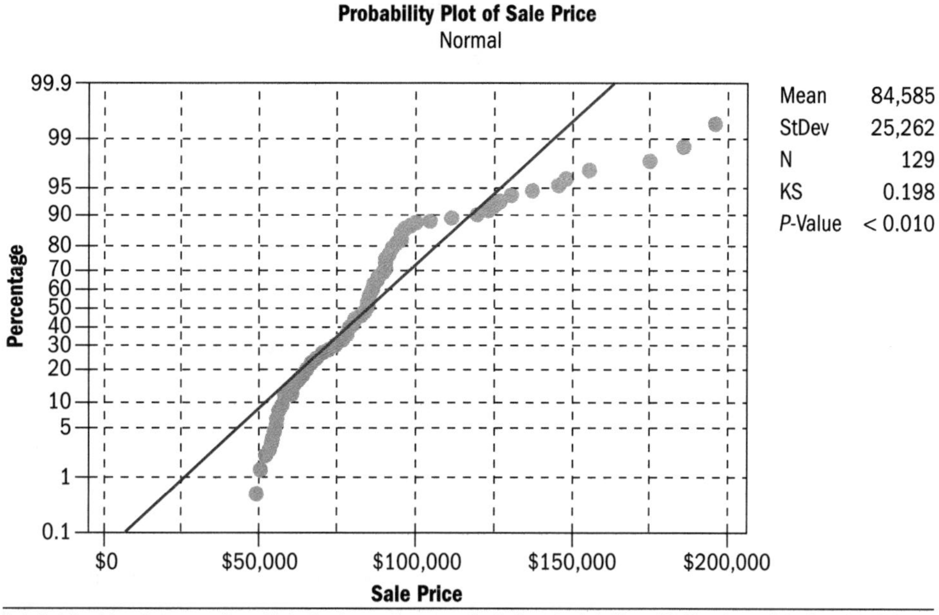

Probability Plot of Sale Price
Normal

Mean	84,585
StDev	25,262
N	129
KS	0.198
P-Value	< 0.010

Percentage

Sale Price

Note that the data values are on the vertical axis, whereas they are on the horizontal axis in Figure 5.3. Therefore, a right-skewed depiction here appears similar to the left-skewed depiction in Figure 5.3.

A second means of assessing normality is to look at summary descriptive statistics for the data, comparing the mean and median, the interquartile range and standard deviation, and the range and the standard deviation. These comparisons are shown in the **following table.**

Summary Descriptive Statistics

Mean and Median	
Mean Price	\$84,585
Median Price	\$83,500
Indication	Normal expectation is that mean = median, whereas mean > median implies right skewness
Interquartile Range and Standard Deviation	
S	\$25,262
IQR	\$22,250 (0.881 × S)
Indication	Not close to normal expectation of IQR = 1.33 × S
Range and Standard Deviation	
S	\$25,262
Range	\$147,000 (5.82 × S)
Indication	Close to normal expectation of Range = 6 × S

These tests indicate that the price data are right-skewed and more clustered around the central tendency than they would be if they were normally distributed. The range is about right for normal data, but a look at the histogram shows that the range is shifted to the right of where it would be if the data were normal.

The third means of assessing normality suggested in this chapter is to look at the distribution of the data, as shown in the **following table**.

Distribution Description	
Mean ± S	
Data points between the mean ± 1 standard deviation	99 (76.7%)
Expectation when $n = 129$	88 (approx 68%)
Mean ± 1.28 × S	
Data points between the mean ± 1.28 standard deviations	112 (86.8%)
Expectation when $n = 129$	103 (approx 80%)
Mean ± 2 × S	
Data points between the mean ± 2 standard deviations	122 (94.6%)
Expectation when $n = 129$	123 (approx 95%)

These three comparisons confirm that the data are more concentrated around the central tendency than one would expect if the distribution were normal. Also, the range of the data extremes appears to be close to the normal expectation. However, a closer examination will show that all seven of the observations that are outside of the Mean $\pm 2 \times S$ interval are on the upper end of the price range, again providing evidence of right skewness.

Because the townhouse price data are not normally distributed, the price data mean and standard deviation cannot be converted to a Z distribution as a means of assessing price probabilities. A frequency distribution with percentiles is better suited to this task, as shown in the **following table** (derived from SPSS).

Sale Price Percentiles	
Percentile	**Price**
10	$57,000
20	$64,990
30	$73,500
40	$79,000
50	$83,500
60	$85,000
70	$89,400
80	$91,900
90	$119,000

Interpretation of the percentile table above would be along the lines of

- The probability of price being less than $83,500 is 0.50.
- The probability of price being greater than $119,000 is 0.10.
- The probability of price being between $73,500 and $89,400 is 0.40.

Of course these probabilities are insensitive to any variation in property characteristics. For example, we know from the analysis we did in Chapter 4 that price is related to living area. Therefore, conditional price probabilities such as $P(\$ | SF)$ would differ from those in the percentile table. As a result, the percentile table above is better suited to describing the market using language like, "Most townhomes in this market area have historically sold for prices ranging from $64,990 to $91,900" or "Although they have occasionally occurred, prices in excess of $119,000 have been infrequent (roughly 1 in 10)." More precise statements about historical market prices can be obtained by including more percentiles. For example, reducing the percentile interval from 10% to 5% would provide a better estimate of the upper level of price rarity (1 in 20 sales were priced above $140,925).

Again, you are encouraged to work with the townhouse data in Appendix G. Try to replicate the results found here and see if you can obtain additional insight into how the data are distributed.

6 Research Design, Hypothesis Testing, and Sampling

Research Design and Hypothesis Testing

What Is the Question?

At its most elementary level, the application of inferential statistics boils down to answering questions. For example, we might ask, "Does the theory of diminishing marginal utility hold for this property type in this market area?" Or, "How does this particular rental market react to proximity to public transportation?" Or, "What, if any, is the influence of nearby street noise on home prices in the subject subdivision?" Ultimately, the question may be as fundamental as, "What is my opinion of market value for this property, and can the credibility of my opinion be bolstered by the application of inferential methods?"

Research questions like these may represent the entire scope of a valuation services assignment, or they may be small but important aspects of a more comprehensive study. In either case, the effective application of inferential methods requires a clear understanding of the relevant questions, explicit or implicit formulation of testable hypotheses, appropriate data, credible analysis, and valid interpretation of the analytical results.

In much the same way that the appraisal process serves as a systematic and organized way to design a work plan consistent with the scope of a specific assignment, statistical analysis research design provides a road map for moving from research question to insight.

From Research Question to Testable Hypotheses

Hypothesis testing relies on a principle often referred to as "Popperian falsification." Early twentieth-century philosopher Karl Popper held that inferential statistics cannot prove anything with absolute certainty. However, inferential methods can cast doubt on the veracity of an assertion of "truth." When sufficient doubt can be raised, an assertion of truth can be "falsified," at least to some degree. The degree of certainty associated with labeling a statement as false is related to "statistical significance" or merely "significance" in the language of statistics.

The process of forming and testing a hypothesis (i.e., a theory) is as follows:

1. Determine an appropriate expected outcome based on theory and experience. This is generally referred to in inferential statistics as a "research hypothesis."
2. Formulate a pair of testable hypotheses related to the research hypothesis: a "null hypothesis" and an "alternative (research) hypothesis." The testable hypotheses must be mutually exclusive and collectively exhaustive (conditions you will recall from the discussion of probability in Chapter 5). The hypothesis testing goal is to falsify or reject the statement of truth implied by the null hypothesis, leaving the research hypothesis as the only reasonable alternative.
3. Formulate a conclusion that falsifies (or fails to falsify) the null hypothesis.

Hypothesis Testing in the Real World

While the three-step process of forming and testing a hypothesis is easy to outline, it is generally much more difficult to apply. Let's consider a simple example and think about the complications that may arise.

Consider the effect of street noise on housing prices. This simple residential valuation issue illustrates the complications encountered in formulating and testing real world hypotheses. An appropriate research hypothesis could be a general statement like, "Exposure to street noise *affects* home price." Or, if the analyst makes a more specific supposition concerning the direction of the effect, the research hypothesis could be, "Exposure to street noise *reduces* home price." Depending on the scope of the assignment, the appropriate research question could be much more specific than this. More refinement might be required, resulting in research hypotheses such as "Exposure to street noise in excess of '*X* decibels' above the ambient noise level reduces home price" or "Exposure to street noise reduces home price, but the size of the reduction decreases with distance from an abutting street and becomes negligible at '*X* feet' from the abutting street." Based on these simple examples, it should be apparent that the testable hypotheses must be customized to the underlying research question.

For simplicity's sake, assume that the appropriate research hypothesis is "Exposure to street noise *reduces* home price in the subject property's market area." This statement becomes the alternative hypothesis–the hypothesis you believe the data will support. Remember that the null hypothesis and the alternative hypothesis are mutually exclusive and collectively exhaustive. The null hypothesis (the statement of truth you are attempting to falsify) would therefore be "Exposure to street noise either increases home price or has no effect on home price in the subject property's market area." These two statements are mutually exclusive (only one of them can be true), and they are collectively exhaustive (home price must go up, down, or stay the same).

In summary, the relevant hypotheses for this example are

- Research hypothesis: Exposure to street noise reduces home price in the subject property's market area.
- Testable hypotheses:
  - Null hypothesis (H_0): The street noise price effect is ≥ 0.
  - Alternative hypothesis (H_a): The street noise price effect is < 0.

Once these hypotheses have been formulated, a research plan must be devised that allows the analyst to credibly test the veracity of the null hypothesis. If the null hypothesis can be falsified with sufficient certainty, then the analyst can conclude that the alternative hypothesis is likely to be true.

It is important to recognize, however, that inferential statistical methods are not intended as means of supporting illogical, unreasonable, or atheoretical suppositions. The research and alternative hypothesis statements should be well reasoned and logical, keeping in mind that inferential methods are designed to support valid research hypotheses.

Validity and Reliability

Tests of the veracity of the null hypothesis are of no use unless the tests are credible (i.e., worthy of belief). Two concepts–validity and reliability[1]–are paramount to credible hypothesis testing. These concepts are deeply rooted in research design and scientific inquiry.

The concepts of reliability and validity can be confusing at first, but they are actually quite simple. For example, consider **Figure 6.1**, which illustrates the idea that reliability is analogous to clustering shots on a target. Shots that are scattered all over the target, as in the left panel, are unreliable. Shots that are tightly clustered but off center, as in the right panel, are reliable but invalid. Only shots that are tightly clustered and centered on the target are reliable (consistent) and valid (accurate).

Figure 6.1 Less than Ideal Target Shooting

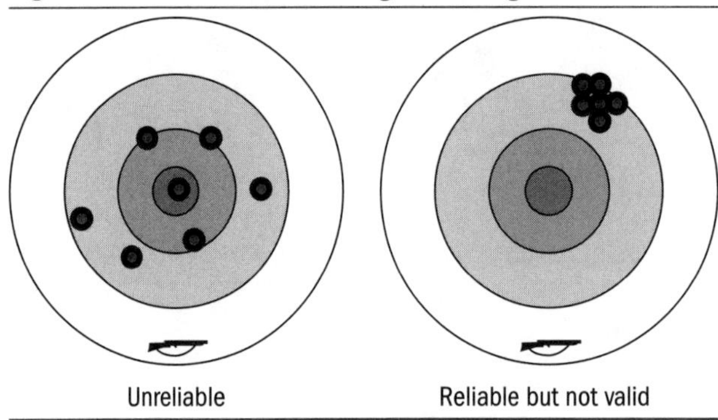

Because threats to reliability and validity erode credibility, credible research and valuation-related opinions are more likely to occur when analysts understand and assess the extent to which

1. A recommended source for a discussion of validity and reliability in research design and implementation is Mary L. Smith and Gene V. Glass, *Research and Evaluation in Education and the Social Sciences* (Boston: Allyn and Bacon, 1987).

the methods employed were both reliable and valid. Paying attention to a few simple criteria can go a long way toward ensuring credible results. For example, logical research designs, controlling measurement error, standardizing interview protocols, using representative data, and applying appropriate analytical tools are basic and essential elements of using statistical methods to support credible valuation opinions.

Validity

Strictly speaking, validity is the extent to which a statistical measure reflects the real meaning of what is being measured. Consider a scale that consistently indicates weights that are 95% of true weight. Obviously the result is not valid when the intent is to measure true weight. Although this sort of measurement error is correctable if the error is consistent and known, measurement error is usually neither consistent nor known in many situations.

Lack of research validity stems from many sources, and assessing the validity of research involves numerous considerations such as

- Logical validity
- Construct validity
- Internal validity
- External validity
- Statistical conclusion validity
- Bias

Logical Validity

A research design consists of several parts, such as a problem statement, a research hypothesis, selection and definition of variables, implementation of design and procedures, findings, and conclusions. Logical validity is satisfied when each part of the overall design flows logically from the prior step. If the overall design isn't logical, then the results aren't likely to be valid. Appraisers should already be familiar with the elements and logical flow of research design because the valuation process is a similar algorithm.

Construct Validity

Construct validity deals with how well actual attributes, characteristics, and features are being measured. For example, although tall people generally weigh more than short people, use of a weight scale is not a valid construct for measuring height. As this example illustrates, construct validity is a simple concept, but it can be quite nuanced in practice.

Concerns about construct validity are particularly applicable to the use of interviews and questionnaires. Precise definitions of variables and the elimination of ambiguity are important in ensuring that questions are not misinterpreted by respondents or researchers. Meanings assigned by respondents should be consistent with the meanings intended by the researcher. Furthermore, meaning should be consistent from respondent to respondent.

Construct validity is especially problematic when respondents must interpret technical or scientific language, as is the case in many interviews related to real property transactions (e.g., sales confirmation). Do not assume that persons being interviewed fully understand the meanings of technical terms such as capitalization rate, internal rate of return, net operating income, effective gross income, obsolescence, and the like.

Internal Validity

Internal validity requires that all alternative explanations for causality have been ruled out. Ruling out threats to internal validity is a laborious task because it requires explicit identification of each alternative explanation for causation along with the rationale for rejecting it. If all reasonable alternative causes cannot be ruled out, the research may be inconclusive and invalid.

External Validity

External validity exists when findings and conclusions can be generalized from a representative sample to a larger or different population. Random selection from a target population is the best means of obtaining a representative sample, subject to the vagaries of sampling error, which is ubiquitous. Therefore, when a random sample has been obtained, the analyst should assess the extent to which the characteristics of the sample match the characteristics of the target population.

Statistical Conclusion Validity

Statistical conclusions will not be valid if the statistical tests being applied are inappropriate for the data being analyzed. The researcher should be aware of the assumptions underlying each statistical test and how robust the test is if those assumptions are violated.

Bias

Bias occurs when there is a systematic error in research findings. Bias can come from several sources, and it can be classified into two categories: nonsampling error and sampling error.[2] *Nonsampling error* includes nonresponse bias, sample selection bias, and systematic measurement error. Sample selection bias may be encountered in real property studies, which often rely on observational samples (e.g., the occurrence of a comparable sale cannot be assumed to have been a random event). *Sampling error* stems from the fact that a random sample can differ from the underlying population simply by chance.

Reliability

Reliability is the extent to which "the same data would have been collected each time in repeated observations [measurements] of the same phenomenon."[3] A reliable model would produce results that can be thought of as consistent, dependable, and predictable.

As an example, assume that six appraisers are asked to measure the same 1,400-sq.-ft. house and calculate its improved living area. A set of estimates consisting of 1,360 square feet, 1,420 square feet, 1,400 square feet, 1,450 square feet, 1,350 square feet, and 1,340 square feet would not be reliable, even though they tend to bracket the true floor area. However, in comparison, a set of estimates consisting of 1,440 square feet, 1,435 square feet, 1,445 square feet, 1,445 square feet, 1,440 square feet, and 1,435 square feet would be more

2. David M. Levine, Timothy C. Krehbiel, and Mark L. Berenson, *Business Statistics: A First Course*, 3rd ed. (Upper Saddle River, N.J.: Prentice Hall, 2003), 23-25.
3. Earl Babbie, *The Practice of Social Research*, 6th ed. (Belmont, Calif.: Wadsworth, 1992).

reliable, despite not bracketing the true floor area. Although the second set of living area estimates is more reliable (i.e., predictable and consistent), the floor area calculations exhibit a systematic, upward bias, making this set of estimates invalid. An ideal set of estimates would be highly consistent (reliable) and accurate (valid), such as 1,398 square feet, 1,402 square feet, 1,400 square feet, 1,395 square feet, 1,405 square feet, and 1,400 square feet.

Reliability can be difficult to attain, especially when data come from sources beyond the analyst's control. For example, subjective assessments of condition, construction quality, and curb appeal provided by third parties may be unreliable, especially if more than one person is rendering opinions. What appears to be "excellent" to one person may be viewed as being "above average" or merely "average" to another.

Because reliability can be difficult to assess and control, it is good practice to think about possible threats to reliability that may be encountered. When data comes from an outside source, ask if a standardized measurement or categorization protocol was employed. Find out if more than one person was involved in making quality or condition assessments. Think about how errors in scoring or measurement may occur, and make random checks for measurement error. Look for the use of ambiguous questions, ambiguous instructions, or idiosyncratic (technical) language that might be difficult for respondents to comprehend.

Sampling

As defined earlier in this book, a *sample* is a subset of a larger population selected for study. When the research goal is to better understand the larger population, the sample should be as similar to the larger population as possible. Statisticians use the term "representative" to indicate the similarity of a sample to the larger population. When a sample is not representative, it is difficult to assert that the characteristics of the sample are indicative of the characteristics of the larger target population.

While sampling is a simple concept, it can be a challenging process in application. The first challenge is obtaining a *sample frame*, which is a list of items in, or members of, the population you want to study. Sometimes full or partial lists exist. Often they do not exist at all, or the compilers of the lists are unwilling to allow access to them. For example, if you were interested in knowing what percentage of lake homes in your state are serviced by central sewer systems and how many have on-site septic systems, you could develop a representative sample of lakeshore properties to obtain an estimate of the population proportions. However, obtaining a comprehensive list of lakeshore properties (the sample frame) in order to draw the sample could be difficult. Compiling an owner's list yourself from county records is one option, but it would be time-consuming.

Samples can be broadly divided into two categories–probability samples and nonprobability samples. Probability samples are characterized by knowledge of the probability that an item in the population will be selected. As you would expect, the probability that an item will be selected is unknown in a nonprobability sample. Statistical inferences formed through the analysis of probability samples are preferred because inferences drawn from nonprobability samples may be unreliable and inaccurate.

Probability Samples

Numerous probability sampling methods exist, and the most common include

- Simple random samples
- Stratified random samples
- Systematic random samples
- Cluster samples

Simple Random Samples

In a simple random sample, every item in a population has the same probability of selection. A simple random sample may be selected either *with* replacement or *without* replacement. When sampling with replacement, the probability of selection for each member of the population is $1/N$ each time a selection is made, where N represents total population size. When sampling without replacement the probability of selection increases as items are selected. The probability of selection for the first item selected is $1/N$, reducing to $1/(N-1)$ for the second item selected, $1/(N-2)$ for the third item selected, and so forth as the unsampled population size is reduced through the sample selection process.

Think of sampling with replacement as picking a card from a full deck, replacing the card, shuffling the deck, and picking another card. In contrast, think of sampling without replacement as being dealt a hand of poker, with each new card in your hand being dealt from a smaller and smaller deck.

Because each item in a population has an equal probability of selection on a given draw, simple random samples are considered to be highly representative. Nevertheless, it is still possible to randomly select a nonrepresentative sample merely by chance. This possibility is referred to as *sampling error*. Although sampling error cannot be totally eliminated, it can be minimized through the selection of larger samples.

Stratified Random Sample

Creating a stratified random sample begins by dividing the population into subgroups (known as *strata*) based on one or more essential characteristics. Once this has been done, you can select random samples from each stratum. Stratified samples ensure that the sample proportion for the stratifying characteristic is identical to the population proportion, reducing sampling error and improving the accuracy of inferences.

As a simple example of the value of stratified random sampling, assume that you want to use sampling to make a statement about the mean apartment rent in a market area. Assume also that the apartment population contains many floor plans with different bedroom and bath counts. If a simple random sample were used, you would have no assurance (due to sampling error) that the floor plan mix of the sample would be identical to the floor plan mix of the population. Although the mix would, on average, be the same with repeated random samples, the mix is apt to differ from the population in any single sample. Use of a stratified sample allows you to control the proportion of the sample being drawn from each apartment unit type, thereby controlling for unit-mix sampling error. When this is done, sample mean rent is a more accurate estimate of population mean rent.

When the parameter of an important population proportion is known, a stratified random sample mirroring the population proportion usually provides the most accurate inferences. You might be wondering, "If stratified random sampling improves the accuracy of inferences, why isn't it done more often?" The primary reason is insufficient understanding of the population proportion for one or more important characteristic. For instance, the simple stratification in the preceding paragraph could not be done if the population unit mix proportions were unknown.

Systematic Random Sample

A *systematic sample* is just what its name implies– a system employed to select the sample from the frame. Systematic sampling typically involves sorted data such as accounting records filed by date or medical records filed alphabetically. For example, if you want to sample 1,000 files out of a population of 30,000 files, you could decide to select every 30th file. You could then randomly select a file from the first 30 files as a starting point and then select every 30th file after the starting point. If you randomly chose to start with file 14, your sample would consist of files 14, 44, 74, 104, and so forth.

While systematic sampling may seem convenient, it can pose problems when there is a systematic pattern associated with how the data were sorted. If this is the case, the sample could be biased. Say, for example, you are auditing your company and randomly choose to look at accounting records from the 4th and 23rd day of each month. You would not be happy to learn, after the fact, that a part-time employee who helped out on the 14th and 15th of each month had been embezzling money. Because you randomly chose the wrong days to audit, the theft would have gone undiscovered. Had a random sample been drawn from each month of the year, there would have been an 80% probability of picking the 14th or 15th day of at least one month.[4] The pattern in the

4. Assuming a 30-day month, the probability of randomly picking the 14th or 15th each month is 2 ÷ 30, or 1/15th. Over 12 months this sums to 12/15ths, or 80%.

data, along with the systematic sample's unfortunate starting point, biased the sample by inadvertently excluding all of the dates when criminal activity occurred.

Use of a systematic sample requires an assessment of the likelihood of the existence of a pattern in the data in the sample frame that could bias the sample. When in doubt, use a different sampling method.

Cluster Sampling

Cluster sampling is often used for geographic data such as real estate where clusters are naturally occurring. City blocks, subdivisions, census tracts, and zip codes are examples of naturally occurring geographic clusters. Random selection of clusters and of items within each selected cluster constitutes a random sample.

Consider the apartment sample referred to in the earlier discussion of stratified random samples. If there were no available sample frame, you could draw a sample by randomly selecting geographic clusters (e.g., census tracts) within the study area, identifying all of the apartments within each selected cluster, and randomly selecting a sample from the identified apartments in each cluster. The resulting sample would be representative of the population if the selected clusters were representative of the market and the properties chosen from each cluster were representative of their cluster.

The problem with this method is one that appraisers are familiar with from other contexts–namely, compounding error. Achieving the state of "representativeness" becomes a multilayered construct in the use of cluster sampling. If the coarser selection layer–the clusters–is not representative, then the sample will not be representative regardless of how well the selected properties represent their clusters. If the coarser, cluster layer is representative, the more focused granular selection layer–properties within each cluster–may still not be fully representative if some or all of the selected properties do not represent their cluster. Due to these issues, sample size in terms of number of clusters and items selected from each cluster should be greater than the sample size required for a simple random sample or stratified sample.

When a sample frame is unavailable, cluster sampling may be the only alternative. Care should be taken however to ensure that the clusters are as representative of the population as possible. With geographic data this often entails selecting clusters that incorporate all of a market area's important geographic variables. Depending on the situation, important geographic variables might include

- School districts
- Municipalities
- Counties
- Age of neighborhoods
- Relative household incomes
- Length of commutes

Self-Selection and the Appraiser's Quandary

Recall from the earlier discussion of validity that sample selection bias may be encountered in real property studies, which often rely on observational samples because a self-selection process separates properties that are offered for sale from those that are not offered for sale. Property owners are not randomly chosen to sell their homes each month. Therefore, a sample of homes "for sale" or "sold" may not be representative of the population of all similar properties in a market.

Self-selection may or may not be a problem, depending on how and if sold properties differ from unsold or not-for-sale properties. Generally speaking, in broader and more active markets self-selection is less likely to be a relevant issue. For example, the housing market is more active than the shopping center market and is less likely to exhibit systematic differences between properties offered for sale and properties not for sale. Nevertheless, some residential neighborhoods could be affected by a localized externality such as an environmental hazard, plant closing, or change in access.[5] In such cases data from an affected loca-

5. Localized externalities differ from marketwide externalities affecting all properties. For example, the 2008/2009 residential foreclosure wave determines the market in many locales, and a representative sample would legitimately be expected to include foreclosed properties and foreclosure price effects, if any.

tion might not be representative of properties in unaffected locations.

The retail sector provides a good example of how self selection can affect real property transaction data. Suppose a prominent and common anchor tenant is ceasing operation or reorganizing through bankruptcy, and several of the market area's shopping centers occupied by this anchor tenant are offered for sale. If and when these properties sell, they probably would not be representative of the remaining shopping centers in the market that were not affiliated with this tenant. Statistical analysis of market transaction data biased by inclusion of these sales might misrepresent the segment of the retail population unaffected by the store closings. The same logic applies to comparable rents associated with retail centers having dark anchor stores.

Because real property offered for sale or rent is a self-selected sample rather than a random sample, appraisers should take care to ensure that the transaction data being analyzed is truly representative of the subject property's competitive market. Experienced appraisers should be able to determine the influence, if any, of self-selection in a market that may preclude some data from inclusion in a given analysis or study. Furthermore, competent appraisers know that unfamiliarity with a market and an inability to assess the existence of self-selection bias within it require the assistance of someone who understands the market in order to credibly assess transaction data.

Nonprobability Samples

Nonprobability samples are less useful for inference than the sorts of probability samples we have been talking about so far in this chapter because the conclusions we can reach through statistical analysis of a nonprobability sample are sample-specific. Information obtained from the sample data may not be applicable to the larger population because there is no guarantee that the sample data are representative of the population.

For example, Internet surveys where users of a site are asked their opinion on matters as varied as election outcomes, results of sports contests, or whether or not an economic recession is looming are nonprobability samples. The results of such a survey only tell us how the proportion of a Web site's users who responded felt about the issue. We do not know if survey respondents are representative of all of the site users. Nor do we know if the opinions of the respondents mirror the opinions of the general population. The survey results could be applicable to the general population, but no statistical measure has been provided of the relationship of such a nonprobability sample to the general population.

This is why expert, professional appraisal judgment is necessary when applying statistical analysis of comparable sale or rental data to a subject property or subject market. Because the generation of comparable data is largely a self-selection process, valuation expertise is required to assess whether or not comparable data items are representative of the subject of a study. When unrepresentative data items are identified, the items can either be removed from the analysis or flagged for later treatment (i.e., attempting to statistically control and adjust for the aspects of the transactions that cause them to be unrepresentative of the subject market).

A decision to exclude data or to apply some form of statistical control depends on the amount of available data and the reason for conducting the study. When a reduced data set excluding unrepresentative data is large enough for a valid study, the unrepresentative data may be excluded. For example, in a residential context it is usually preferable to exclude luxury home sales and entry-level home sales from a study of *mid-priced* home values. While it is possible to statistically control for the differences between luxury homes, entry-level homes, and mid-priced homes, figuring out which controls to employ may not be a simple task. However, if we needed to know the effects of an externality such as street noise or power line proximity on an *entire* residential market, then we would most likely want to understand the effect of the externality across all price categories–entry-level, mid-priced, and luxury. The study of the entire residential market might also include apartments, condominiums, and townhomes in the sample, depending on the scope of work applicable to the assignment.

Sampling Error

Sampling error occurs when the sample differs from the population. In hypothesis testing, sampling error may result in rejection of a null hypothesis that is actually true or it may result in failure to reject a null hypothesis that is actually false. Either of these results will lead to an inappropriate conclusion. In the first case, the research hypothesis is flawed, but the data indicate that it is not. In the second case, the research hypothesis is correct, but the analysis doesn't support it.

The outcomes of hypothesis testing can be reduced to four possibilities:

- H_0 is true — fail to reject H_0 — correct result
- H_0 is false — reject H_0 — correct result
- H_0 is true — reject H_0 — erroneous result
- H_0 is false — fail to reject H_0 — erroneous result

Rejecting a true null hypothesis is referred to as *Type I Error*. With Type I error the study supports the research hypothesis even though it is based on a false supposition. The probability of rejecting a true null hypothesis is symbolized by the lowercase Greek letter alpha (α), which is called the *significance level*. The probability of not making a Type I error ($1 - \alpha$) is referred to as the *confidence level*. The probability of Type I error can be controlled by selecting the significant level α prior to performing a statistical test of the null hypothesis. The researcher decides on an acceptable probability of rejecting a true null hypothesis and rejects the null hypothesis if the statistical results are at or better than the predetermined α threshold. For example, if α is set at 5% and the statistical result is an α of 3%, the result is considered to be significant and the null hypothesis is rejected.

If this seems confusing, look at it from a confidence level perspective. Setting α at 5% is the same as saying, "If the data allow me to be 95% confident that my research supposition is correct, then I am going to reject the null hypothesis and accept the research hypothesis." For example, if the analysis results in α of 3%, the corresponding confidence level is 97%. In this case we have exceeded the 95% confidence level threshold, supporting the validity of our research hypothesis.

One way to guard against the erroneous rejection of a null hypothesis that is actually true is to take care in the construction of the research hypothesis. Note that the null hypothesis is true only if the research hypothesis is false. Better reasoning, logic, and understanding of underlying phenomena will help guard against flawed research designs that attempt to support false suppositions.

The erroneous failure to reject a null hypothesis that is actually false is known as *Type II Error*. In this case the study fails to support the research hypothesis even though it is based on a true supposition. The probability of Type II error is symbolized by the lowercase Greek letter beta (β). Unfortunately β cannot be known with certainty unless you know the true population parameter you are attempting to infer. (If you know β, why would you be attempting to infer it?)

Consider the example of the effect of traffic noise on home price. If the effect of noise is substantial, then the probability of failing to support the research hypothesis is small. If the effect does exist but is not substantial, then we will have more difficulty demonstrating the effect statistically, meaning that β will be relatively large. When the effect of traffic noise is small, the statistical analysis must be more "powerful," increasing the probability of demonstrating the effect. The *power* of a statistical test is symbolized by $1 - \beta$.

Statistical power can be increased in three ways:

1. Relax α. This choice may not be very satisfactory, if the initial logic behind the initial decision on α has not changed.[6]
2. Increase the size of the sample. Small effects are much easier to uncover with more data.
3. Eliminate confounding effects. In the street noise example, the street noise effect may be

6. Because the choice of a high level of significance reduces β, the effect on β should have been considered in the initial selection of α.

masked if lots abutting a thoroughfare are generally larger than lots in the interior of the same subdivision. Controlling for lot size in the analysis should improve a model's ability to detect the effect of street noise.

Relating Choice of α to the Standard Normal Distribution

Choice of α is a way of stating how far–in statistical distance–the sample mean must be from what the population mean would be if the null hypothesis were true in order to reject the null hypothesis. Consider a simple pair of statistical test hypotheses:

H_0: $\mu = 0$

H_a: $\mu \neq 0$

If we select $\alpha = 0.05$ and the population to which μ applies is normally distributed, then we are saying, "If the sample mean is 1.96 standard deviations or more from 0, I will reject the null hypothesis that the population mean is 0."

Why 1.96 standard deviations? If we look up $Z = 1.96$ in the standard normal table, we will find a probability of $Z \leq 1.96 = 0.975$. Additionally when we look up -1.96 on the standard normal table, we will find a probability of $Z \leq -1.96 = 0.025$. Therefore,

$$\mathbf{P(-1.96 \leq Z \leq 1.96) = 0.975 - 0.025 = 0.95.}$$

The confidence level of 95% is associated with $\alpha = 0.05$ (5%). To be 95% confident that the null hypothesis of $\mu = 0$ is false, $\overline{x}$ must be at least 1.96 standard deviations from 0. This concept is illustrated pictorially in **Figure 6.2**.

Figure 6.2 The Standard Normal Distribution

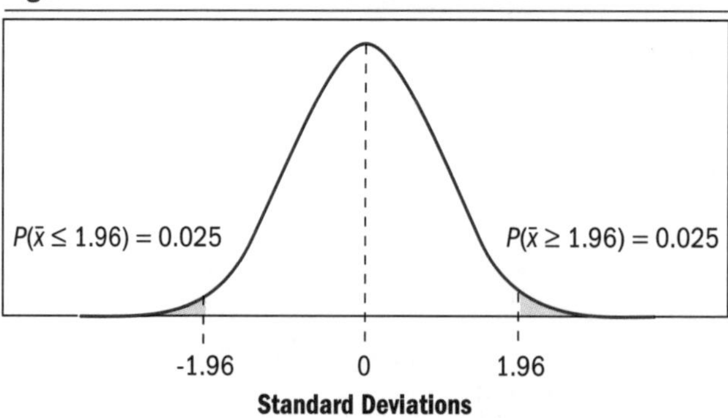

Figure 6.2 shows the standard normal curve along with the locations of 0 standard deviations (the middle of the curve) and ±1.96 standard deviations. Recall that the area under the curve is equal to 1 (100%) and the area under a portion of the curve is equal to the probability of a

sample mean ($\overline{x}$) being in that location when the true population mean is 0. By looking up -1.96 in the standard normal table we find that the area under the curve to the left of -1.96 is 0.025 (or a 2.5% probability of $\overline{x}$ being in this location if $\mu = 0$). By looking up 1.96 in the standard normal table we find that the area under the curve to the left of 1.96 is 0.975 (97.5%), leaving 0.025 to the right of 1.96 (2.5% probability of $\overline{x}$ being in this location if $\mu = 0$). Therefore, with $\alpha = 0.05$, we can reject the null hypothesis that $\mu = 0$ if $\overline{x} \leq -1.96$ standard deviations from 0 or if $\overline{x} \geq 1.96$ standard deviations from 0.

Hypothesis testing is a skill that is developed through practice. So, let's work through an example problem. Suppose we interviewed a representative of a fast food franchise and were told that the chain's average restaurant floor area is 2,400 square feet. Other sources indicate that the average floor area for this particular fast food concept has grown over time as menus have expanded and adjusted to new consumption patterns. This fast food concept is fairly new to your state and we suspect that the average floor area here exceeds 2,400 square feet. We decide to use a random sample of floor areas to test our hypothesis and decide also that if we can be 90% confident we will conclude that the average floor area in this state exceeds 2,400 square feet.

First we state the research, null, and alternative hypotheses and the significance level required to reject the null hypothesis with 90% confidence.

Research Hypothesis: Average floor area exceeds 2,400 square feet.

H_0: $\mu \leq 2{,}400$ square feet

H_a: $\mu > 2{,}400$ square feet

$\alpha = 0.10$

Notice that the null hypothesis in this example contains the "≤" symbol rather than the "=" symbol because the research hypothesis is stated as "exceeds." Remember that the null and alternate hypotheses must be mutually exclusive and collectively exhaustive, therefore H_0 must cover all of the possibilities that differ from H_a.

Next, we calculate the sample mean and assume for now that we know the population standard deviation σ. (We'll learn how to handle this problem using the sample standard deviation later in the book.)

$\overline{x} = 2{,}560$ square feet

$\sigma = 114$ square feet

$$Z = \frac{\overline{x} - \mu}{\sigma} = \frac{2{,}560 - 2{,}400}{114} = 1.40$$

Is the sample mean of 2,560 far enough from 2,400 in statistical terms to reject the hypothesis that the mean floor area is 2,400 square feet or less? We can address this question in one of two ways:

1. Select the Z value associated with $\alpha = 0.10$ and compare 1.40 to this significance level threshold.
2. Assess the probability of $\overline{x}$ being 1.40 standard deviations or more from the hypothesized mean, and compare this result to the required α level of 10%.

Let's do it both ways.

The Z value associated with $\alpha = 0.10$ is the value–call it "*B*"–where $P(Z \leq B) = 0.90$. The standard normal table indicates that this occurs with a value of approximately 1.28 standard deviations. The value of 1.28 is called the *critical value* of Z because an $\overline{x}$ result of this amount or more is required to reject the null hypothesis.[7] Because 1.40 is greater than the critical value of 1.28 you can reject the null hypothesis.

Alternately, the standard normal table shows that the probability of Z being less than or equal to 1.40 is 0.919. Therefore, the significance level α indicated by the sample is 0.081, which is less than 0.10, so we can reject the null hypothesis and state with at least 90% confidence (or, more precisely, 91.9% confidence) that the mean floor area in this state is greater than 2,400 square feet. In statistics the Z value probability of 0.081 is referred to as a *p*-value, which is the probability of the $\overline{x}$ result being 1.40 standard deviations from 2,400, assuming the null hypothesis is true.

7. Try calculating the critical value associated with 0.90 using the "=NORMINV" macro in Excel to derive a more precise critical value of 1.2816.

In this example we rejected the null hypothesis based on what is called a *one-tailed test.* The null hypothesis contains a ≤ statement, so we need only be concerned with the right tail of the Z distribution to test the validity of the null hypothesis. Similarly, if the null hypothesis contained a ≥ statement, then we would only concern ourselves with the left tail of the Z distribution (also a one-tailed test). When the null hypothesis contains an = statement, it can be rejected at either tail of the Z distribution, which is referred to as a *two-tailed test.*

As a practical matter this exercise, though quite simple in statistical terms, could be useful in assessing whether or not an old floor plan is significantly smaller than new store requirements in support of an assessment of functional obsolescence. Or it could support a highest and best use analysis of a pad site, adjusting requisite floor area ratio to the current trend in store size.

Sample Size

Once you decide to gather sample data for a statistical study, you are immediately confronted with the issue of how much data you need. The resolution of this issue can be simple or complex, depending on the situation. If you intend to study sample means or sample proportions, the calculation of sample size may be a straightforward result of selecting the accuracy you expect to achieve and plugging that information into a simple equation. If the study involves data collection by survey, the sample size will have to be adjusted for nonresponders and inappropriate responders. Of course you may not know how many of these you will encounter until you have completed the survey.[8]

We will discuss regression modeling later. However, be aware that if you plan to employ a regression model the sample will have to be large enough to accommodate all of the variables you may need to include in the model. Unfortunately, you may

8. Perhaps the best reference for survey sampling design and maximizing response rate is by Don Dillman, who has written a series of books on the topic and is a highly regarded expert. His latest book is *Mail and Internet Surveys: The Tailored Design Method* (Hoboken, N.J.: John Wiley & Sons, 2007). If you are not anticipating conducting a Web-based survey, then one of his older books would be sufficient.

not know how many variables are needed in the model in advance, which is a confounding issue.

Sample Size for Estimating Means

Suppose you want to estimate the mean rent for one-bedroom apartments from a sample representative of all one-bedroom apartments in your city. Required sample size can be calculated once you make three decisions:

1. Level of confidence you require
2. Degree of accuracy you expect to achieve
3. An estimate of the standard deviation of one-bedroom rents in the city

The level of confidence you require is $1 - \alpha$. Therefore, this decision determines α, which is required to estimate sample size. The degree of accuracy you expect to achieve is stated in terms of units of measure. For instance, if you are estimating mean rent, the degree of accuracy is stated in dollars. Degree of accuracy is called *sampling error*, which is symbolized as *e*. The standard deviation (σ) of the variable being estimated will be unknown and must be estimated. Methods of estimating σ include referencing prior studies, conducting a small pilot study, or investigating the range of the variable of interest (the range will often be approximately 6 times σ for a normal distribution).

The equation for estimating sample size required to estimate a population mean is

$$n = \frac{Z^2\sigma^2}{e^2}$$

Picking up the one-bedroom apartment rent example again, let's assume you decide on a 95% confidence level, expect to be accurate within ±$10.00, and estimate the range of monthly rent for one-bedroom apartments in the market area at $120 ($650 to $770). Based on these factors, you select $Z = 1.96$ based on $\alpha = 0.05$ and the standard normal distribution, $e = 10$ and $\sigma = 20$ ($120 ÷ 6). The required sample size is

$$n = \frac{Z^2\sigma^2}{e^2} = \frac{1.96^2 \cdot 20^2}{10^2} = 15.36$$

Sample size calculations are generally rounded up, so you would want to draw a random sample of at least 16 one-bedroom apartment rents.

Suppose you require more precision than an estimate of mean rent ±$10. For example, you may need more statistical power to compare mean rents for two types of one-bedroom apartment (say, those with and without a private balcony). Assume you need to decrease sampling error from $10 to $5 in order to have enough statistical power to detect the effect of private balconies. What does this requirement do to sample size?

$$n = \frac{Z^2\sigma^2}{e^2} = \frac{1.96^2 \cdot 20^2}{5^2} = 61.47$$

Sample size essentially quadruples to 62. This emphasizes an important point:

> Increases in statistical power are "costly" when cost is stated in terms of sample size.

Cutting sampling error in half quadruples sample size, and reducing sampling error to one-quarter of the amount illustrated in this example ($2.50) would increase sample size 16-fold. The relationship between sample size and sampling error is exponential due to the e^2 term in the denominator of the sample size equation.

Sample Size for Estimating Proportions

As Americans, we are accustomed to reading about proportion estimates at election time. The following was reported by Reuters on the eve of the January 8, 2008, New Hampshire presidential primary election:

> A Reuters/C-SPAN/Zogby poll showed Obama with a 10-point edge on Clinton in the state, 39 percent to 29 percent, as he gained a wave of momentum from his win in Iowa.

The margin of error (*e*) for the statement above was reported elsewhere to have been ±4.4%. Assuming a significance level of 0.05, the poll taker was 95% confident that Obama's proportion of the vote was between 34.6% and 43.4% and Clinton's proportion was between 24.6% and 33.4%. Based on this information we can deduce that there were approximately 496 respondents, as we will see shortly.

The equation for the sample size required to estimate a population proportion is

$$n = \frac{Z^2\, p(1 - p)}{e^2}$$

where Z is the standard normal value associated with the confidence level, e is the margin of error, and p is an estimate of the population proportion. For most proportion estimates p is set at 0.50 because this proportion maximizes the value of $p(1 - p)$, ensuring that the sample is sufficiently large regardless of the true population proportion. Returning to the New Hampshire presidential primary poll, we can apply the equation for sample size to deduce the number of respondents:

$$n = \frac{Z^2\, p(1 - p)}{e^2} = \frac{1.96^2 \cdot 0.50(1 - 0.50)}{0.44^2} = 496$$

Now let's look at a more practical problem for an appraiser. Suppose you want to estimate the proportion of recent in-migrants to a city opting for rental housing rather than home ownership during their first year of residency. Assuming you could obtain a list of recent in-migrants from which to draw a sample (e.g., from electric company records), you could determine the number of respondents you would need by deciding on a confidence level and the margin of error. If you set $\alpha = 0.05$ and $e = 2\%$, the number of randomly chosen responses you would need is

$$n = \frac{1.96^2 \cdot 0.50(1 - 0.50)}{0.02^2} = 2{,}401$$

Comparing the presidential primary poll to the housing tenure choice sample above, we can see that reductions in the margin of error (4.4% to 2%) dramatically increase sample size and related costs. Therefore, we must carefully consider the question of what is a sufficient margin of error and the associated amount time and money to devote to data gathering.

Practice Problems 6.1

1. Assume that the mean debt coverage ratio (DCR) for apartment mortgages financed through ABC Bank was 1.15 from 2000 to 2005. You have sample DCR data for 2006 and 2007 and want to test a supposition that the mean DCR has changed. What are the appropriate null and alternative hypotheses?
2. How would you write the null and alternative hypotheses in the preceding question if you wanted to test a supposition that the mean DCR had declined in 2006 and 2007?
3. How would you write the null and alternative hypotheses in the preceding question if you wanted to test a supposition that the mean DCR had increased in 2006 and 2007?
4. Looking again at Problems 1, 2, and 3 above, which will be one-tailed tests and which will be two-tailed tests?
5. Setting $\alpha = 0.10$, what critical values are indicated for Z in Problems 1, 2, and 3 above?
6. Take the following information as given:

   H_0: $\mu = 6.5\%$ annual percentage rate (APR)

   H_a: $\mu \neq 6.5\%$ APR

   $\bar{x} = 6.23\ \%$ APR

   $\sigma = 0.12\%$

   Can the null hypothesis be rejected at $\alpha = 0.10$, $\alpha = 0.05$, and $\alpha = 0.01$? What type of error do you risk as you reduce the size of α?
7. Take the following information as given:

   H_0: $\mu \geq 8.0\%$ vacancy

   H_a: $\mu < 8.0\%$ vacancy

   $\bar{x} = 7.20\%$ vacancy

   $\sigma = 0.65\%$

   Can you reject the null hypothesis that vacancy is greater than or equal to 8.0% at a 10% significance level?
8. You are doing a market study to assess the housing preferences of university students. The university sponsoring the study will give you access to a list of all of their students containing address and telephone number information, class rank (freshman, sophomore, junior, senior, graduate student), and whether they attend full-time or part-time. What is the most efficient way to sample the university's students?
9. As a continuation of Problem 8 above, assume you are also interested in estimating the amount of rent students are currently paying with an accuracy of ±$10 per month. What minimum sample size will you require to be 95% confident of your rent estimate if you rely on prior studies that indicated standard deviations of monthly student rent of $50? What would the minimum sample size become if you expected a 40% response to your student survey questionnaire?
10. A home builder is interested in investigating the proportion of homes sold with three-car garages in order to finalize design features for a new tract home subdivision. If the builder wants to be 90% confident in the proportion estimate with a margin of error of no more than ±3%, what sample size will you require if you are consulting with the developer regarding this issue?

Inferences About Population Means and Proportions

Sampling Distribution of the Mean

Thorough comprehension of statistical inference is impossible without an understanding of the sampling distribution of the mean. Consequently, this important and essential topic sets the stage for the remainder of this chapter and the chapters that follow.

The *sampling distribution of the mean* refers to the distribution of a sample's arithmetic average, $\bar{x}$. Empirically modeling the distribution of the sample mean entails the following

1. Taking numerous random samples from a population
2. Calculating the mean of each sample
3. Examining the distribution of the means derived from the numerous samples

Fortunately, mathematical statisticians have developed a complete understanding of the sampling distribution of the mean. Therefore, we can use what is already known about this statistical measure to make inferences about the populations we are interested in studying.

Constructing a Distribution of Sample Means

A simple but somewhat laborious example will help us understand the development of the sampling distribution of the mean more fully. Assume each of the five houses in a small neighborhood has a different number of bedrooms:

- House 1 has one bedroom.
- House 2 has two bedrooms.

- House 3 has three bedrooms.
- House 4 has four bedrooms.
- House 5 has five bedrooms.

The population bedroom mean (μ) is therefore (1 + 2 + 3 + 4 + 5) ÷ 5 = 3, and the population bedroom standard deviation (σ) is 1.414214.

Now let's calculate the mean of all possible samples of size 2, taken with replacement, from this population of five houses. **Table 7.1** lists the 25 different samples possible ($5^2 = 25$) and the sample mean ($\overline{x}$) of each possible sample. We can look at a frequency distribution of the 25 sample means (**Table 7.2**) and calculate the mean and standard deviation of the sample means.

The probability of selecting a one-bedroom house from this small population is 1 in 5 or 0.20, which is also the probability of selecting any of the other houses. However, the probability of selecting a *sample* with a bedroom mean of 1 is 1 in 25 or 0.04. A sample mean equal to 1 bedroom occurs only when the two-house sample consists of selecting the one-bedroom house on the first random selection and selecting it again on the second random draw. (Recall from Chapter 5 that *P*(1BR & 1BR) = *P*(1BR) × *P*(1BR) = 0.20 × 0.20 = 0.04.) In contrast, the probability of selecting a sample with a bedroom mean equal to 3 is 5 in 25 or 0.20 (see Table 7.2). This suggests that sample means

Table 7.1 25 Possible Samples of Size $n = 2$ from a Population of Five Houses

Sample	Bedrooms	Sample Mean ($\overline{x}$)	Sample	Bedrooms	Sample Mean ($\overline{x}$)
1	1,1	1.0	14	3,4	3.5
2	1,2	1.5	15	3,5	4.0
3	1,3	2.0	16	4,1	2.5
4	1,4	2.5	17	4,2	3.0
5	1,5	3.0	18	4,3	3.5
6	2,1	1.5	19	4,4	4.0
7	2,2	2.0	20	4,5	4.5
8	2,3	2.5	21	5,1	3.0
9	2,4	3.0	22	5,2	3.5
10	2,5	3.5	23	5,3	4.0
11	3,1	2.0	24	5,4	4.5
12	3,2	2.5	25	5,5	5.0
13	3,3	3.0			

Table 7.2 Frequency Distribution of Sample Means

Sample Mean ($\overline{x}$)	Frequency	Cumulative Frequency
1.0	1	1
1.5	2	3
2.0	3	6
2.5	4	10
3.0	5	15
3.5	4	19
4.0	3	22
4.5	2	24
5.0	1	25

Mean of Sample Means = 75 ÷ 25 = 3

Standard Deviation of Sample Means = 1.0

nearer to the population mean are more likely to occur than sample means farther from the population mean, indicating that sample mean dispersion ($\sigma = 1$) is less than the dispersion of the population from which the samples were drawn ($\sigma = 1.414$).

Notice also that the mean of the sample means is equal to the population mean. This is a logical outcome because, if the sample is representative, we would *expect* the sample mean to be an unbiased estimator of the population mean.

Figure 7.1 pictorially illustrates the two different distributions of bedroom counts we have been discussing. Notice that the population bedroom count histogram shows a symmetric, uniform distribution of bedroom counts ranging from 1 to 5 with each having an equal 0.20 probability of occurring. In contrast, the sample mean histogram also shows a symmetric distribution of bedroom counts ranging from 1 to 5, but it is concentrated

Figure 7.1 Population and Sample Mean Histograms

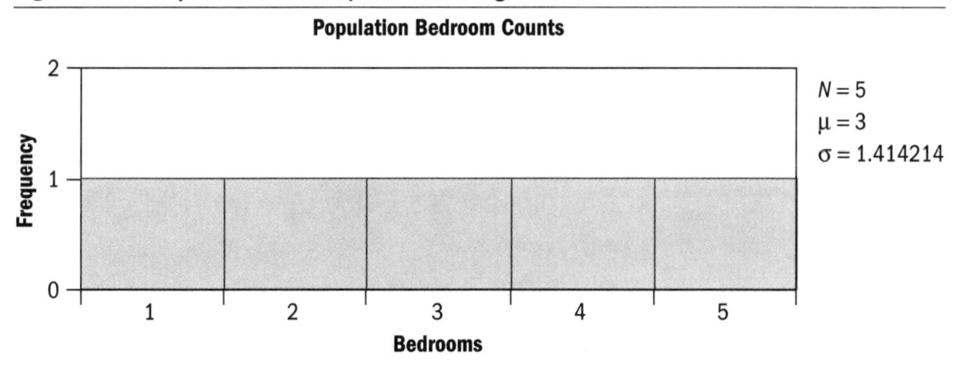

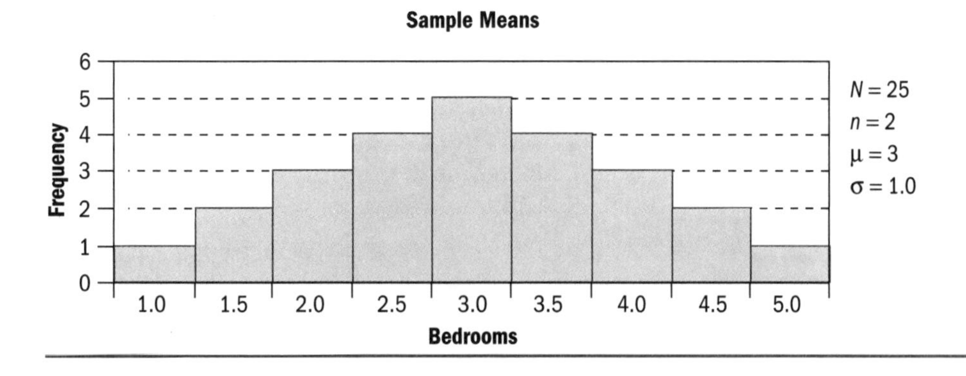

more densely around the population mean of three bedrooms consistent with its smaller standard deviation ($\sigma = 1$) in comparison to the population standard deviation ($\sigma = 1.414214$).

Quantifying the Sampling Distribution of the Mean

As mentioned earlier, mathematical statisticians have developed a complete understanding of the sampling distribution of the mean. The benefit of this knowledge is not having to go through an exercise similar to the example we just completed every time we encounter a different population and sample. Two equations capture the relationship between a population and the sampling distribution of means drawn from that population.

First, the expected value of the distribution of sample means ($\mu_{\bar{x}}$) is equal to the population mean (μ_x):

$$\mu_{\bar{x}} = \mu_x$$

Second, the standard deviation of the distribution of sample means, called the *standard error of the mean*, is less than the population standard deviation except for the trivial case of a sample size of 1:

$$\textbf{Standard Error of the Mean} = \sigma_{\bar{x}} = \frac{\sigma_x}{\sqrt{n}}$$

Note that the square root of sample size is the denominator in the calculation of the standard error of the mean, so the standard error decreases as sample size increases. As a consequence the sample mean becomes a more precise (i.e., less dispersed) estimator of the population mean as sample size increases. Logically, then, a good way to reduce estimation error is to draw larger samples, which is a concept worth remembering.

Let's revisit the bedroom count example from Table 7.2 to demonstrate computation of the sampling distribution of the mean:

$$\mu_{\bar{x}} = \mu_x = 3$$

which is the mean of the sample means we calculated in the example, and

$$\sigma_{\bar{x}} = \frac{\sigma_x}{\sqrt{n}} = \frac{1.414214}{\sqrt{2}} = 1.0$$

which is the standard error of the mean from Table 7.2.

Sampling from Normally Distributed Populations
When a sample is randomly drawn from a normally distributed population, the sample mean will be normally distributed with $\mu_{\bar{x}} = \mu_x$ and $\sigma_{\bar{x}} = \sigma_x/(\sqrt{n})$, which is often written in mathematical shorthand as

$$\bar{x} \sim N\left(\mu_x, \frac{\sigma_x}{\sqrt{n}}\right)$$

This shorthand notation says, "*x*bar is distributed normally with mean equal to 'mu sub *x*' and standard deviation equal to 'sigma sub *x*' divided by the square root of *n*." Because $\bar{x} \sim N$, any value of $\bar{x}$ can be converted into a Z statistic when the underlying population parameters (μ_x and σ_x) are known using the following equation:

$$Z = \frac{\bar{x} - \mu_x}{\frac{\sigma_x}{\sqrt{n}}}$$

As an example, assume that residential customer water use (W) in a large southwestern U.S. city is distributed normally with a mean of 12,000 gallons per month and a standard deviation of 2,500 gallons per month, i.e., $W \sim N(12{,}000, 2{,}500)$. Therefore,

$$\bar{W} \sim N\left(12{,}000, \frac{2{,}500}{\sqrt{n}}\right)$$

First, let's look at the probability that a water use sample mean based on a sample with size $n = 16$ will be 11,200 gallons or less:

$$Z_{11,200} = \frac{11{,}200 - 12{,}000}{\frac{2{,}500}{\sqrt{16}}} = \frac{-800}{625} = -1.28$$

Therefore,

$$P(\overline{W} \leq 11{,}200) = P(Z \leq -1.28) = 0.1003$$

Next, let's calculate the probability that a randomly chosen individual residence's water usage will be 11,200 gallons or less:

$$Z_{11,200} = \frac{11{,}200 - 12{,}000}{2{,}500} = \frac{-800}{2{,}500} = -0.32$$

Therefore,

$$P(W \leq 11{,}200) = P(Z \leq -0.32) = 0.3745$$

How do these two inferences differ? In the first instance we are making an inference about the sample mean, and the appropriate measure of statistical distance is the standard error of the mean, which is 625 gallons per month. In the second instance we are making an inference about a single residence from the entire population of residences, and the appropriate measure of statistical distance is the population standard deviation, which is 2,500 gallons per month. With a sample size of 16, the standard error of the mean is one-fourth the size of the population standard deviation. Therefore, a nominal distance of 800 gallons from the mean is much greater in statistical terms when evaluating probabilities associated with $\overline{W}$ than when evaluating probabilities associated with W. The two normal distributions of residential water use are compared in **Figure 7.2.**

Figure 7.2 Comparison of Sampling Distribution of the Mean to the Related Population Distribution

(n = 16, μ = 12,000 Gallons per Month, σ = 2,500 Gallons per Month)

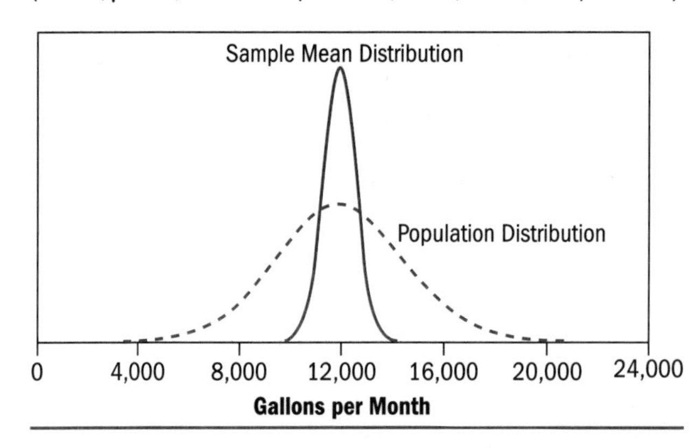

Practice Problems 7.1

1. Four associates of a small appraisal firm take the following number of hours of qualifying education each year:

15 30 7 45

a. Sampling with replacement, select all possible samples of size 2 and construct a table showing the sampling distribution of the mean.
b. Compute the mean of the sample means and the population mean. Are these two values equal? Why or why not?
c. Using Excel, compute the population standard deviation and the standard error of the mean. Are they equal? Should they be equal?

2. Assume that market rent for a standard two-bedroom apartment is distributed normally with the following mean and standard deviation:

$\mu = \$1,000$ per month $\sigma = \$200$ per month

a. What is the probability that an individual standard two-bedroom apartment will rent for $1,150 or more per month?
b. If a random sample of size $n = 25$ is drawn from the standard two-bedroom apartment population, what is the probability of the sample mean being $1,150 per month or more?
c. If a random sample of size $n = 9$ is drawn from the standard two-bedroom apartment population, what is the probability of the sample mean being $1,150 per month or more?
d. Why do the probabilities differ when n changes from 25 to 9?

Sampling from Non-Normal Populations

Central Limit Theorem

The Central Limit Theorem is the theoretical basis for the ability to apply inferential statistical methods to nearly any population because the theorem enables us to make inferences about a population without knowing how the population is distributed. In particular, it allows us to relax the assumption we made in the previous section of the chapter concerning population normality.

Central Limit Theorem
Regardless of the distribution of the population, the sampling distribution of the mean is approximately normal when sample size is 30 or more.

In practice, the Central Limit Theorem allows us to make inferences about population means relying on the normal distribution when a) the population is normal or b) when $n \geq 30$. As a practical matter, the sampling distribution of the mean will be approximately normal when $n \geq 15$ *and* the population is symmetrically distributed. However, appraisers usually know very little about the shape of population distributions of price, property attributes, financing arrangements, and the like. Therefore, the $n \geq 30$ criterion generally applies to real property valuation work.

The Central Limit Theorem can be illustrated using the real-world townhouse data from zip code 89015 found in Appendix G. **Figure 7.3** is a histogram of time on market, measured in days, for the 129 townhouse sales, which we will treat as a population rather than a sample solely for the purposes of this illustration. The distribution of time on market is fairly typical of this variable in many real estate markets. It is far from being normally distributed, being asymmetrical and highly right skewed. Mean time on market is 51.48 days and the standard deviation is 57.39 days, which you can verify by analysis of the data found in Appendix G.[1]

Figure 7.4 illustrates the distribution of sample means for 200 random samples of size 30 drawn from the time on market data. As the illustration's normal distribution overlay demonstrates, the

Figure 7.3 Time on Market for Zip Code 89015 Townhouses

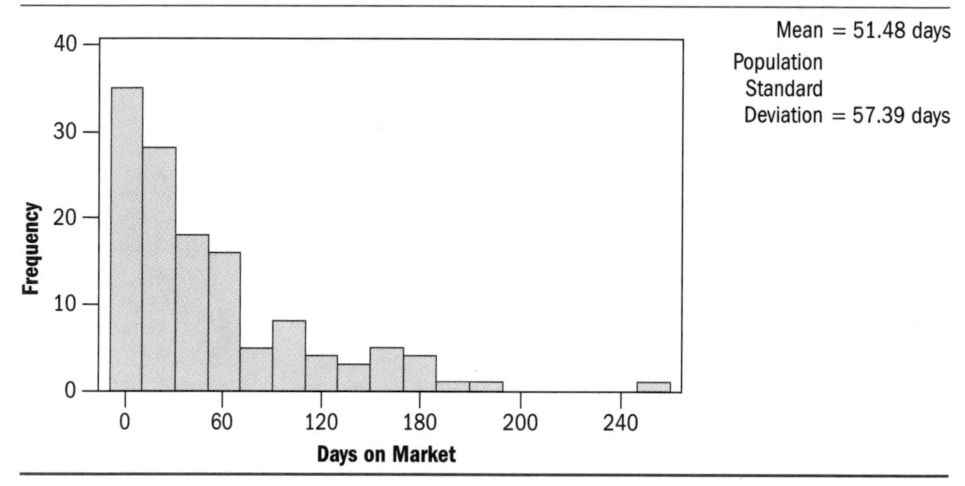

1. We use the STDEVP Excel macro (population standard deviation) here because we are treating the sample as a population for the purposes of this illustration. The sample standard deviation of time on market is 57.61 days, which can be calculated using the STDEV macro in Excel.

Figure 7.4 Mean Time on Market Estimates from 200 Random Samples of Size 30

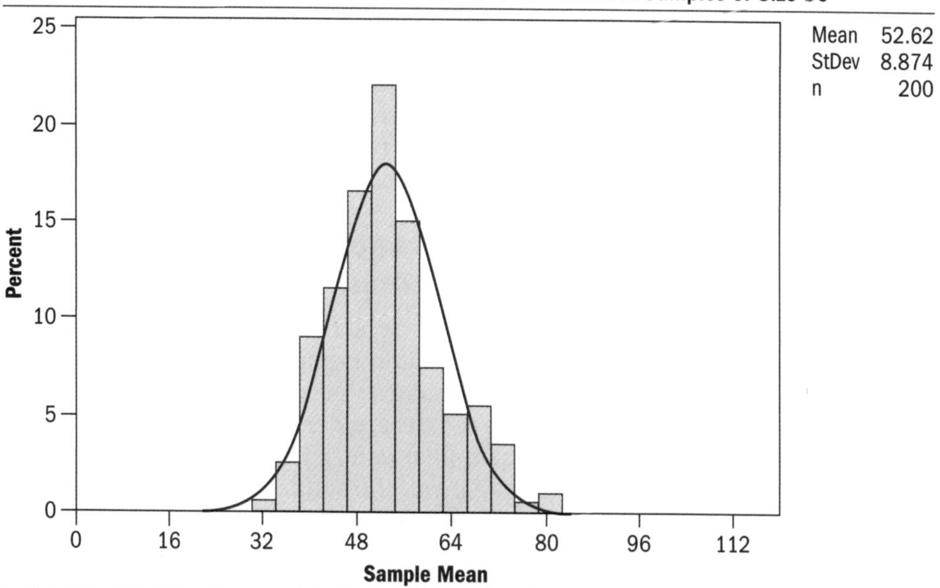

distribution of the sample means is approximately normal. As expected, the mean of the sample means is close to the underlying data's mean time on market (52.62 days for the former and 51.48 days for the latter). Notice also that the standard error of mean time on market is much smaller than the standard deviation of time on market (8.874 days for the former and 57.39 days for the latter), which is consistent with expectations for the sampling distribution of the mean.

As this exercise demonstrates, samples of size $n = 30$ drawn from a highly skewed, asymmetric population of time on market data yield a fairly symmetric and approximately normal sampling distribution of the mean. This experiment relied on 200 sample means of size $n = 30$. If more sample means had been randomly drawn, the distribution of the sample means would have looked more like a normal distribution, the mean of the sample means would have moved closer to the expected value of 51.48, and the standard error of the mean would have approached its expected value of 10.48 (i.e., $57.39 \div \sqrt{30}$).

A Note about Sample Size

Two sample-size criteria have been discussed thus far. The first criterion, introduced in Chapter 6, dealt with controlling sampling error (or margin of error) when making inferences about population means or proportions. The second criterion dealt with being able to rely on the Central Limit Theorem when the population probability distribution is not known to be normal. Generally, the criterion resulting in a larger required sample size will prevail.[2]

Inferences Concerning the Population Mean

Generally speaking, statistical inferences concern an unknown population mean and an unknown population standard deviation. Inferences about population means are derived from three statistics: the sample mean ($\overline{x}$), the sample standard deviation (S), and the sample size (n). When using these statistics, we can develop inferences about the population mean either by analysis of a confidence interval or by conducting a hypothesis test. We will explore both of these methods.

Working with the Student's *t* Distribution

Since the population standard deviation (σ) is rarely known, the sample standard deviation (S) is almost always used to estimate σ. Whenever S is used as an estimate of σ, inferences must be based on the Student's t distribution instead of the standard normal Z distribution.

Remember that Z is calculated for the sampling distribution of the mean as follows:

$$Z = \frac{\overline{x} - \mu_x}{\frac{\sigma_x}{\sqrt{n}}}$$

By comparison, Student's t is calculated for making inferences about the mean when σ is unknown and S is used as an estimate of σ in a similar fashion, where

$$t_{n-1} = \frac{\overline{x} - \mu_x}{\frac{S_x}{\sqrt{n}}}$$

2. Nonparametric statistical methods (Chapter 8) provide alternative inferential methods for small samples drawn from non-normal populations.

The Student's t distribution is symmetric and similar in shape to the standard normal distribution, and as sample size increases the Student's t distribution and the standard normal distribution converge. (The distributions are practically the same when $n > 120$.) The subscript $n - 1$ in the Student's t equation refers to *degrees of freedom*, which is 1 less than the sample size when using the t distribution for a single sample test. Unlike Z, which is insensitive to sample size, a different Student's t distribution exists for each value of $n - 1$.

Figure 7.5 contains a portion of the Student's t distribution table found in Appendix B. Each column in the table, ranging from $\alpha = 0.10$ to $\alpha = 0.001$, corresponds to the area under the right tail of the t distribution having a probability equal to α. The t distribution is symmetric, so the same values are associated with the left tail of the curve. This t table is designed to apply to one-tailed tests. For two-tailed tests, read "α" as "$\alpha \div 2$." For example, at a significance level level of 0.05, select t

Figure 7.5 Portion of the Student's *t* Distribution Table

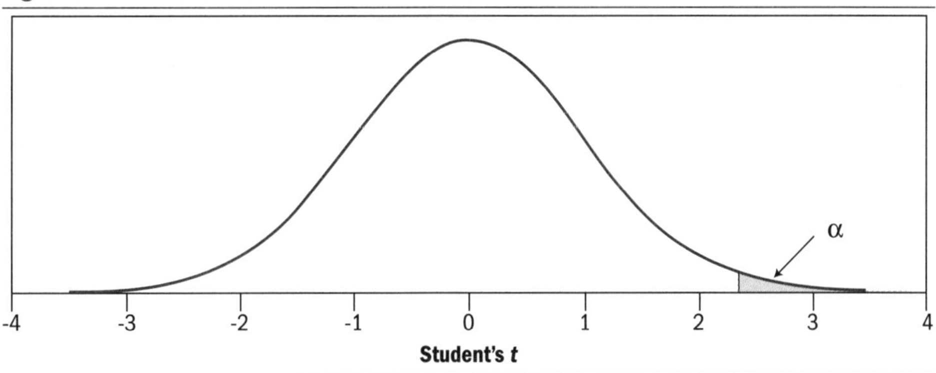

	Student's *t* Values Associated with α Probabilities					
$n - 1$	$\alpha = 0.10$	0.05	0.025	0.01	0.005	0.001
1.	3.078	6.314	12.706	31.821	63.657	318.313
2.	1.886	2.920	4.303	6.965	9.925	22.327
3.	1.638	2.353	3.182	4.541	5.841	10.215
4.	1.533	2.132	2.776	3.747	4.604	7.173
5.	1.476	2.015	2.571	3.365	4.032	5.893
6.	1.440	1.943	2.447	3.143	3.707	5.208
7.	1.415	1.895	2.365	2.998	3.499	4.782
8.	1.397	1.860	2.306	2.896	3.355	4.499
9.	1.383	1.833	2.262	2.821	3.250	4.296

values from the column under the heading "0.05" for a one-tailed test, but for a two-tailed test at the same significance level, select t values from the column under the heading "0.025," or $0.05 \div 2$.

The column labeled $\alpha = 0.025$ and the row labeled $n - 1 = 8$ are highlighted in Figure 7.5. The column and row intersect at a Student's t value of 2.306. Therefore, when $n = 9$, there is a 97.5% probability that $t \leq 2.306$ (i.e., a 2.5% probability that $t \geq 2.306$). Also, due to the t distribution's symmetry, there is a 95% probability that $-2.306 \leq t \leq 2.306$. As Figure 7.5 demonstrates, the logic involved in interpreting the Student's t distribution table is very similar to the logic of the standard normal table.

Student's *t* in Excel

Student's t values and associated probabilities can also be calculated in Excel, which many prefer to use instead of a t distribution table. For example, enter "=TDIST(2.306,8,1)" into an Excel cell. The displayed probability of 0.025 is the area under the right tail of the t distribution with 8 degrees of freedom and $t = 2.306$. More generally, the TDIST macro in Excel provides the probability associated with a given t value, degrees of freedom ($n - 1$ in the current context), and information indicative of either a one-tailed or two-tailed probability. Conversely, entering "=TINV(0.05,8)" into Excel yields a cell display of 2.306.

Note that the TINV macro assumes the probability entry in the parentheses is associated with a two-tailed significance level. In our "table versus Excel" comparison, we are interested in deriving a t statistic associated with a one-tailed α value of 0.025. Therefore, in order to use TINV macro the one-tailed α probability of 0.025 was doubled and entered into TINV as 0.05.

Confidence Interval Estimate for the Mean (σ not known)

When a sample is representative of a population, the sample mean is an unbiased estimator of the true population mean. That is, the expected value of the sample mean is the population mean, and with numerous trials the mean of the sample means will equal the population mean. In practice, however, we can seldom afford the luxury of drawing multiple random samples from the population being studied.The mean value of a single sample does not provide an exact estimate of the underlying population mean. So the best we can do is derive a *confidence interval*, which is an interval around the sample mean accompanied with a statement of how confident we are that the true population mean lies within the interval. The width of the confidence interval depends on the sample standard deviation

(S) and sample size (n). All else equal, the larger the sample the narrower the confidence interval.

The following expression provides a confidence interval for the mean based on sample statistics $\overline{x}$ and S:

$$\overline{x} - t_{\alpha/2,\, n-1}\left(\frac{S}{\sqrt{n}}\right) \le \mu \le \overline{x} + t_{\alpha/2,\, n-1}\left(\frac{S}{\sqrt{n}}\right)$$

The value of $t_{\alpha/2,\, n-1}$ is determined by sample size and our choice of confidence level $(1 - \alpha)$. Another way of expressing the same idea is to say that we are $(1 - \alpha)$ percent confident that μ lies within the interval

$$\overline{x} \pm t_{\alpha/2,\, n-1}\left(\frac{S}{\sqrt{n}}\right)$$

Example Problem 1

Jennifer surveyed 32 randomly chosen managers of garden apartment complexes with more than 100 units located in a major U.S. city. In addition to other information relevant to her study, she obtained expense ratios as a percentage of effective gross income (excluding replacement reserves) for each of the apartment complexes she contacted. The sample mean expense ratio ($\overline{x}$) was 37.8% and the sample standard deviation (S) was 2.9%. Jennifer would like to form an opinion, with 95% confidence, concerning an interval within which the true population expense ratio is likely to fall.

The solution to this problem involves Jennifer's choice of α and the subsequent selection of the correct t statistic value. Since $1 - \alpha = 0.95$, $\alpha = 0.05$ and $\alpha/2 = 0.025$. Sample size is 32, so $n - 1 = 31$. Looking in the Student's t distribution table, the 0.025 column and the $n - 1 = 31$ row intersect at a value[3] of 2.040. Therefore, she can be 95% confident that the true mean expense ratio lies within the interval

$$\overline{x} \pm t_{0.025,\, 31}\left(\frac{S}{\sqrt{n}}\right) = 37.8\% \pm 2.040\left(\frac{2.9\%}{\sqrt{32}}\right) = 37.8\% \pm 1.05\%$$

Another way to express this result would be to say that Jennifer is 95% confident that 36.75% ≤

3. Using Excel, TINV(0.05,31) = 2.039515.

the true population mean expense ratio ≤ 38.85%. Or she might choose to round the extremes to one decimal place, consistent with the sample mean, and say, "Based on my survey research, I am 95% confident that the mean expense ratio for garden apartments with 100 or more units in this market lies between 36.8% and 38.9%."

Could Jennifer obtain a more precise estimate of the population expense ratio? She could, by increasing the size of her sample. If her sample size was 49 rather than 32 and her sample mean and sample standard deviation remained the same, her confidence interval would narrow to

$$\bar{x} \pm t_{0.025,\,48}\left(\frac{S}{\sqrt{n}}\right) = 37.8\% \pm 2.011\left(\frac{2.9\%}{\sqrt{49}}\right)$$

$$= 37.8\% \pm 0.83\%$$

Jennifer would have to decide whether the cost associated with contacting and interviewing 17 additional apartment managers justifies the benefit of a 21% reduction in the confidence interval width from 2.10% to 1.66%. Notice that the increase in precision results from two factors. One factor is the decrease in the standard error of the mean caused by dividing *S* by the square root of 49 rather than the square root of 32. The other is a slight reduction in the *t* statistic from 2.040 to 2.011 due to an increase in the value of $n - 1$ from 31 to 48.

Example Problem 2

John drew a random sample of size 9 from a normally distributed population. The sample mean was 20 and the sample standard deviation was 3. What is the 90% confidence interval for the population mean?

$$\frac{\alpha}{2} = 0.05$$

$$S = 3$$

$$\bar{x} = 20$$

$$n = 9$$

$$n - 1 = 8$$

$t_{0.05, 8} = 1.860$ Alternatively, using the Excel macro and α, TINV(0.10,8) = 1.8595.

$$\bar{x} \pm t_{0.05, 8}\left(\frac{S}{\sqrt{n}}\right) = 20 \pm 1.86\left(\frac{3}{\sqrt{9}}\right) = 20 \pm 1.86$$

John can use the *t* distribution here, even though sample size is less than 30, because the population is normally distributed. Had the population distribution been unknown, John would have needed a larger sample, or he could have used nonparametric statistics (see Chapter 8).

Using the *t* Statistic to Test Hypotheses about the Mean

Suppose we want to test a hypothesis that a vancy rate in a particular market is greater than 5%, or that the market capitalization rate for a certain type of property is less than 7.5%. Or we may want to test a hypothesis and have no preconception regarding how a market parameter compares to some benchmark value.

Suppose, for example, that loan closing costs on moderately priced homes in a given market average 1.5% of the loan amount, and we would like to know if closing costs on higher-priced homes are lower than 1.5% (due to economies of scale) or greater than 1.5% (because of less efficient jumbo mortgage markets). We can test the relationship of the mean value of a variable to a benchmark value using inferential methods that rely on the *t* statistic. This methodology is referred to in statistics as a *one-sample t test.* As in the prior discussion of confidence intervals, the *t* statistic is used in a one-sample test of the mean when the population standard deviation is unknown. The *t* test statistic is calculated as follows:

$$t_{n-1} = \frac{\bar{x} - \mu_x}{\frac{S_x}{\sqrt{n}}}$$

Example Problem

Assume you have random sample data from a large city that indicates a mean vacancy rate of 9% for ancillary (non-anchor) tenants in 130 shopping centers included in your sample. You suspect, however, that retail centers experience significantly lower vacancy rates if they have the best anchor tenants in terms of the quantity and quality of consumers they attract.

You have stratified your data, separating the vacancy rates associated with your definition of "retail centers with the best anchor tenants" from the rest of the data, and you want to test your research hypothesis that the mean vacancy rate for ancillary tenant space in these centers is less than the market average of 9%. The sample mean vacancy rate for ancillary tenants in the 36 shopping centers in your "best" category is 7.5% with a standard deviation of 3.2%. Armed with this information, you conduct the following test of your research hypothesis with 95% confidence:

H_0: Vacancy rate ≥ 9%

H_a: Vacancy rate < 9% *(left-tailed test)*

$$\alpha = 0.05$$

$$S = 3.2\%$$

$$\bar{x} = 7.5\%$$

$$n = 36$$

$$n - 1 = 35$$

$$t_{0.05,\,35} = 1.69$$ (This is a one-tailed test, so we do not use $\alpha/2$.)

Decision Rule: Reject the null hypothesis that the vacancy rate is ≥ 9% and accept the alternative hypothesis that the vacancy rate is < 9% if the *t* statistic calculated from the data is less than $-t_{0.05,\,35} = -1.690$.

Test Statistic:

$$t_{0.05,\,35} = \frac{\bar{x} - \mu}{\frac{S_x}{\sqrt{n}}} = \frac{7.5\% - 9\%}{\frac{3.2\%}{\sqrt{36}}} = \frac{-1.5\%}{0.533\%} = -2.81$$

Conclusion: The test statistic of -2.81 is less than the critical value of -1.69. Reject the null hypothesis and conclude that the vacancy rate for the "retail centers with the best anchor tenants" is less than the indicated marketwide vacancy rate of 9%.

This exercise justifies selection of a market vacancy rate of 7.5% for properties that satisfy the best anchor tenants criterion.

For the sake of completeness, note that this same analysis can be done by developing a 95% confidence interval for the mean vacancy rate at the "best" shopping centers. The 95% confidence interval is

$$\bar{x} \pm t_{0.025,\,35}\left(\frac{S}{\sqrt{n}}\right) = 7.5\% \pm 2.030\left(\frac{3}{\sqrt{36}}\right) = 7.5\% \pm 1.08\%$$

which is 6.42% to 8.58%.

The 95% confidence interval does *not* include 9%, so the null hypothesis is rejected. Note that the *t* statistic used to derive the confidence interval is different from the 1.69 critical value used in the one-tailed hypothesis test because development of a confidence interval is always a two-tailed endeavor since a confidence interval calculation always includes the ± operator. The confidence interval analysis provides more information than the hypothesis test analysis because it confirms the research hypothesis and also provides a vacancy rate confidence interval applicable to "retail centers with the best anchor tenants."

Practice Problems 7.2

1. Use a *t* distribution table or Excel to derive the critical *t* value associated with a two-tailed test with 59 degrees of freedom and a 10% significance level (90% confidence level). What is the critical *t* value if the test is one-tailed?
2. What is the right-tailed probability associated with a *t* statistic of 2.763 and a sample size of 29? What is the two-tailed probability associated with this *t* statistic and sample size?
3. The lending side of operations at a large bank is interested in updating the amount to anticipate as an appraisal fee on a conventional, conforming home loan. The chief review appraiser randomly selected a random sample of 81 residential appraisals conducted on behalf of the bank for conventional, conforming loans within the previous month. The mean appraisal fee was $375 with a standard deviation of $28. What is the 95% confidence interval for the mean conventional, conforming residential loan appraisal fee?
4. The county assessor uses a rating system scaled from 0 to 10 to evaluate the curb appeal of residences in the assessment district, with 0 indicating "no curb appeal," 5 indicating "average curb appeal," and 10 indicating "superb curb appeal." A random sample of 98 residential curb appeal ratings was drawn from the assessor's files to test the research hypothesis that the curb appeal rating system is biased. Bias is defined as a mean curb appeal rating that is different from the purported average rating of 5. The sample mean curb appeal rating was 5.5 with a standard deviation of 0.5. The statistical null hypothesis is $\mu = 5$, and the alternative hypothesis is $\mu \neq 5$. Can the assessor reject the null hypothesis at a 90% confidence level and conclude that the curb appeal rating is biased?

5. Price per square foot has been increasing at a modest rate in the town of Usual, Minnesota. According to information provided by the local real estate board, the average single-family detached house resale price per square foot in Usual is now $155.00. John Smiley, real estate agent, claims that resale homes served by the locally acclaimed Better-than-Usual Middle School sell at prices above the Usual average of $155 per square foot. Sue Analytic, being a conscientious real estate appraiser, decides to check John Smiley's claim by randomly selecting a sample of 33 recent resale homes served by Better-than-Usual Middle School. Her sample mean is $157.00 per square foot and the sample standard deviation is $8.50 per square foot. Assuming she sets her significance level at 5%, can Sue accept John's claim? Why or why not?

Inferences about the Population Proportion

Confidence Interval Estimate for the Proportion

Inferences are often made concerning binary categorical data where some proportion of the population (p) fits a category and the remaining proportion does not. Examples include homes with or without a garage, homes with or without a swimming pool, and lots fronting a golf course or not fronting a golf course. The standard normal (Z) distribution can be used to estimate confidence intervals for the population proportion (p) when the expected number of sample observations in each of the two categories is at least 5. That is, $np \geq 5$ and $n(1 - p) \geq 5$.

Under these circumstances the confidence interval estimate for the proportion is

$$\bar{p} \pm Z\sqrt{\frac{\bar{p}(1-\bar{p})}{n}}$$

where $\bar{p}$ signifies the sample proportion.

Example Problem

A bank's chief review appraiser has been working diligently to develop a program to better inform and train staff and fee appraisers who provide mortgage collateral appraisals on the bank's behalf. In order to test the effectiveness of this program, the review appraiser wants to obtain an estimate of the proportion of "problem appraisals" the bank has encountered over the six months prior to the program's initiation. This will provide a baseline estimate of the pre-program proportion of problem appraisals, which will be used to assess the program's effectiveness.

The chief review appraiser randomly selected 150 appraisals conducted over the past 6 months and discovered that 15 of the selected appraisals fit into the bank's "problem appraisal" classification. Using this information, the chief appraiser developed a 95% confidence interval estimate for the true proportion of problem appraisals prior to initiation of the new information and training program as follows:

$$\bar{p} \pm Z\sqrt{\frac{\bar{p}(1-\bar{p})}{n}} = 0.10 \pm 1.96\sqrt{\frac{0.10(1-0.10)}{150}}$$

$$= 0.10 \pm 0.048$$

The baseline confidence interval for the true percentage of problem appraisals is therefore 5.2% to 14.8%.[4]

Using the *Z* Statistic to Test Hypotheses about the Proportion

The corollary to the one-sample *t* test of the mean is the *one-sample Z test of the proportion.* This test can be used to test a hypothesis about the proportion based on a single sample, under the previously stated requirements that $np \geq 5$ and $n(1-p) \geq 5$. The test statistic for the proportion is a *Z* statistic, computed as follows:

$$Z \cong \frac{\bar{p} - p}{\sqrt{\frac{p(1-p)}{n}}}$$

where the distribution is approximated by the standard normal distribution and *p* is the hypothesized population proportion.[5]

Example Problem

The mortgage and foreclosure crisis of 2008 and 2009 caused homeowner concern in many neigh-

4. The standard normal distribution (*Z*) is appropriate for use here because we estimate $np = 15$ and $n(1-p) = 135$ using the sample proportions as estimates of the population proportions.
5. See David M. Levine, Timothy C. Krehbiel, and Mark L. Berenson, *Business Statistics: A First Course*, 3rd ed. (Upper Saddle River, N.J.: Prentice Hall, 2003), 304.

borhoods across the United States. The subprime mortgage delinquency rate in one southwestern city approached 15% but was much higher in newly developed neighborhoods and much lower in older, established neighborhoods. To test the hypothesis that subprime delinquency rates were significantly higher in newly developed neighborhoods, a loan underwriter randomly selected a neighborhood from a list of the area's newly developed neighborhoods. He then randomly selected 80 residences in the newly developed neighborhood and discovered that 42 had subprime mortgages and 11 of the 42 subprime mortgages were delinquent. His hypotheses, chosen significance level, and results are as follows:

H_0: $p \leq 15\%$

H_a: $p > 15\%$ *(right-tailed test)*

$$\alpha = 0.05$$

$$n = 42$$

$$\bar{p} = 26.2\%$$

$$p = 15\%$$

Decision Rule: Reject the null hypothesis that the neighborhood's subprime delinquency rate is $\leq 15\%$ and accept the alternative hypothesis that the newly developed neighborhood subprime delinquency rate is $> 15\%$ if the Z statistic calculated from the data is ≥ 1.64.

Test Statistic:

$$Z \cong \frac{\bar{p} - p}{\sqrt{\frac{p(1-p)}{n}}} = \frac{26.2\% - 15\%}{\sqrt{\frac{15\%(1-15\%)}{42}}} = \frac{11.2\%}{5.51\%} = 2.03$$

Conclusion: Reject the null hypothesis and conclude that the newly developed neighborhood subprime delinquency rate is greater than the marketwide rate of 15%.

Practice Problems 7.3

1. Cisterns are becoming a popular feature of city homes in the some areas of the arid Southwest because they allow homeowners to trap rainwater and use it for landscape irrigation instead of relying completely on potable water sources. A random sample of 200 single-family residences conducted in one medium-sized southwestern city found that 40 had cisterns and used the water for yard irrigation purposes. What is the 90% confidence interval for the proportion of single-family residences in this city that use cisterns for yard irrigation?
2. Jack claims that "90% of the homes in Phoenix have swimming pools." Rita believes that is a common misconception, based on her experience as an appraiser. To make her point Rita randomly selects 50 Phoenix homes and discovers that 32 of them have pools. Conduct a hypothesis test to examine Rita's belief with 95% confidence.

Two-Sample Tests of Means

So far in this chapter we have investigated single-sample methods that can be used to make inferences about an interval that is likely to contain the value of a population parameter or about the validity of a hypothesis concerning the relationship between a population parameter and a stated value, such as ≠, >, or <. These inferences were based on one sample.

We can now turn our attention to comparing samples drawn from two presumably different populations. These new procedures will allow us to investigate questions such as

- Can we say with confidence that the means of two populations are different? For example, we can compare office rents in Class A buildings to Class B office building rents and make a statement concerning whether they are significantly different.
- Can we say with confidence that a treatment had the desired effect? For example, we could measure some attribute of a population prior to a treatment and measure the same attribute after the treatment and make a statement concerning whether the treatment was effective. Similarly, we could take before and after measurements of individual people or items and determine if there was a treatment effect.
- Can we say with confidence that a population has changed over time? For example, we could look at operating expense per square foot at one point in time and look at it at an earlier point in

time and make a statement about whether or not operating expense per square foot had changed.

In fact, these procedures allow us to compare common attributes of nearly any two populations by learning a few simple techniques.

Comparing Two Independent Samples

The two-sample test applies when random samples have been drawn from two different populations for the purpose of making a comparison on the basis of a common variable. For instance, we could sample capitalization rates derived from regional malls and neighborhood shopping centers and address the question of whether or not they are the same.

The Z statistic can be used for this test only if the standard deviations of the two populations are known, otherwise the t statistic applies. Of course, population standard deviations are rarely, if ever, known in practice, so we will limit our analyses to the t distribution. The t statistic calculation differs depending on whether or not the variances (standard deviations squared) of the two independent populations are equal.

Equality of Two Variances *F* Statistic

Before we can look at the equations for pooled variance and separate variance t statistics, we must determine whether or not the variances of the two samples are equal. We do this using an F test. The F distribution results from the ratio of two variances, and in this application we analyze two independent sample variances:

$$F_{n_1 - 1, n_2 - 1} = \frac{S_1^2}{S_2^2}$$

where $n_1 - 1$ and $n_2 - 1$ represent numerator and denominator degrees of freedom, respectively.

The ratio of two variances will be 1 or close to 1 if the two variances are equal or approximately equal, allowing use of the pooled variance t statistic. If S_1^2 is significantly smaller than S_2^2, the F ratio will be small, and if S_1^2 is significantly larger than S_2^2, the F ratio will be large. Therefore, a two-tailed test of a null hypothesis that the variances are equal will be rejected with either large or small values of F.

The F distribution is not symmetric, so upper and lower critical values are calculated separately. **Figure 7.6**, taken from the F distribution table in Appendix C, illustrates the asymmetric shape of the F distribution probability density function (PDF) with $\alpha = 0.05$. For example, if the confidence level were 95% ($\alpha = 0.05$), the numerator sample size were 10 (9 degrees of freedom), and the denominator sample size were 9 (8 degrees of freedom), the upper-tail critical value for the F statistic variance ratio would be 3.388, as shown in **Figure 7.7**.

Lower-tail critical values can be calculated from the upper-tail table as well. This is done by switching numerator and denominator degrees of freedom, and then taking the reciprocal of the value read from the table. Continuing the example, (9, 8) numerator and denominator degrees of freedom become (8, 9) numerator and denominator degrees of freedom. The corresponding F value from the table is 3.230, and the reciprocal is $1 \div 3.230 = 0.310$.

Figure 7.6 Curve Illustrating the Shape of the *F* Distribution

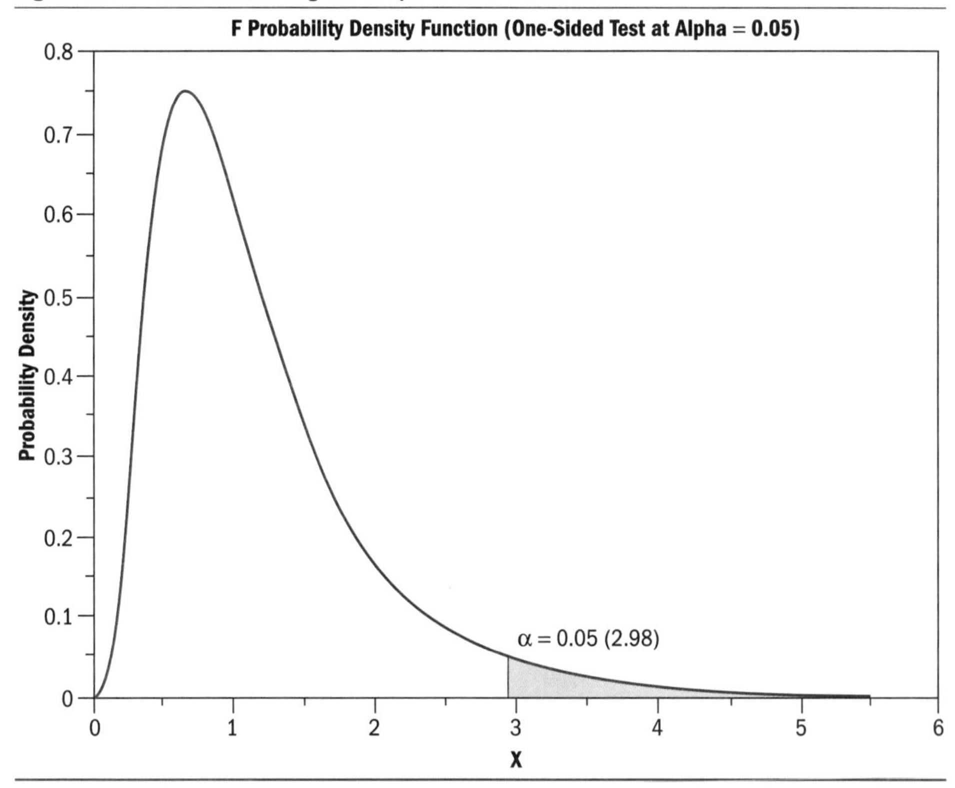

Figure 7.7 Portion of the *F* Distribution Table

Upper critical values of the *F* distribution for df_1 numerator degrees of freedom and df_2 denominator degrees of freedom

$F_{0.05}(df_1, df_2)$

$df_2 \backslash df_1$	1	2	3	4	5	6	7	8	9	10
1	161.448	199.500	215.707	224.583	230.162	233.986	236.768	238.882	240.543	241.882
2	18.513	19.000	19.164	19.247	19.296	19.330	19.353	19.371	19.385	19.396
3	10.128	9.552	9.277	9.117	9.013	8.941	8.887	8.845	8.812	8.786
4	7.709	6.944	6.591	6.388	6.256	6.163	6.094	6.041	5.999	5.964
5	6.608	5.786	5.409	5.192	5.050	4.950	4.876	4.818	4.772	4.735
6	5.987	5.143	4.757	4.534	4.387	4.284	4.207	4.147	4.099	4.060
7	5.591	4.737	4.347	4.120	3.972	3.866	3.787	3.726	3.677	3.637
8	5.318	4.459	4.066	3.838	3.687	3.581	3.500	3.438	3.388	3.347
9	5.117	4.256	3.863	3.633	3.482	3.374	3.293	3.230	3.179	3.137
10	4.965	4.103	3.708	3.478	3.326	3.217	3.135	3.072	3.020	2.978
11	4.844	3.982	3.587	3.357	3.204	3.095	3.012	2.948	2.896	2.854
12	4.747	3.885	3.490	3.259	3.106	2.996	2.913	2.849	2.796	2.753
13	4.667	3.806	3.411	3.179	3.025	2.915	2.832	2.767	2.714	2.671

Therefore, when S_1^2 is derived from a sample of size $n = 10$ and S_2^2 is derived from a sample of size $n = 9$, you would reject the null hypothesis that the variances are equal if the ratio of the two variances (i.e., the F statistic) were ≤ 0.310 or ≥ 3.388. Rejection of the null hypothesis that the variances are equal results in using the separate variance *t* statistic. Otherwise, the pooled variance *t* statistic may be used.

Pooled Variance and Separate Variance *t* Statistics
Equations for the pooled variance (variances equal) and separate variance (unequal variance) *t*-test statistics follow:

Pooled Variance *t* Statistic

$$t_{n_1 + n_2 - 2} = \frac{(\bar{x}_1 - \bar{x}_2) - (\mu_1 - \mu_2)}{\sqrt{S_p^2\left[\frac{1}{n_1} + \frac{1}{n_2}\right]}}$$

and

$$S_p^2 = \frac{(n_1 - 1)S_1^2 + (n_2 - 1)S_2^2}{(n_1 - 1) + (n_2 - 1)}$$

where S_p^2 = pooled sample variance, $\bar{x}_1$ and $\bar{x}_2$ are the two sample means, S_1^2 and S_2^2 are the two sample variances, n_1 and n_2 are the two sample sizes, and $n_1 + n_2 - 2$ represents the degrees of freedom associated with the *t*-test statistic.

Separate Variance *t* Statistic

$$t_{df} = \frac{(\bar{x}_1 - \bar{x}_2) - (\mu_1 - \mu_2)}{\sqrt{\frac{S_1^2}{n_1} + \frac{S_1^2}{n_2}}}$$

and

$$df = \frac{\left(\frac{S_1^2}{n_1} + \frac{S_2^2}{n_2}\right)^2}{\frac{\left(\frac{S_1^2}{n_1}\right)^2}{(n_1 - 1)} + \frac{\left(\frac{S_2^2}{n_2}\right)^2}{(n_2 - 1)}}$$

where *df* = degrees of freedom associated with the *t*-test statistic, and all of the other variables are defined the same as in the pooled variance *t* statistic equation. Do not be intimidated by the complexity of these equations. Excel, Minitab, and SPSS perform these calculations effortlessly.

Example Problem 1

The sample data in **Table 7.3** are derived from two populations of residential building lots. Population 1 consists of lots in the interior of a subdivision not affected by noise from a busy street. Population 2 consists of perimeter lots affected by noise from an abutting thoroughfare. For the purposes of this example, assume that the two populations are normally distributed and that the lots are similar in all price-determining aspects except exposure to street noise. Given this assumption, is there evidence at the 95% confidence level that lots affected by noise from an abutting busy street sell for a lower price per square foot?

We will use a two independent samples *t*-test to solve this problem, performing an initial test for equality of variance prior to deciding which independent sample *t*-test statistic to calculate:

Table 7.3 Sale Prices per Square Foot of Two Samples

Population 1 ($/SF)	Population 2 ($/SF)
6.50	6.26
6.83	6.20
6.39	6.42
6.75	6.09
6.46	6.27
6.30	6.05
6.84	6.39
6.58	5.95
6.59	
6.26	
6.75	

$$S_1^2 = 0.0424$$

$$S_2^2 = 0.0271$$

$$F_{10,7} = \frac{S_1^2}{S_2^2} = \frac{0.04242}{0.02714} = 1.563$$

According to Figure 7.7 the critical value of *F* is 3.637 (the lower-tail critical value is irrelevant because the calculated ratio > 1), and we cannot reject the null hypothesis that the population variances are equal at the 95% confidence level. Therefore, we can use the pooled variance *t*-test.

The test for the equality of two variances is available in Excel by selecting **Tools, Data Analysis,** and then **F-Test for Two Sample Variances** from the menu in the **Data Analysis** window. The *F* test window will open and prompt you to select input and output ranges. The output will look like **Figure 7.8**, which provides information on the numerator sample in the left column and the denominator sample in the right column. Excel's output provides a good deal of information, including a *p*-value for the *F* test. The *p*-value for this test is 0.284, indicating that we can only be 71.6% confident that the two population variances are different.[6]

Figure 7.8 Excel Output Related to the *F* Test for Equality of Variance

F Test Two-Sample for Variances

	Population 1 ($/SF)	*Population 2 ($/SF)*
Mean	6.568	6.204
Variance	0.042	0.027
Observations	11	8
df	10	7
F	1.563	
P(F<=f) one-tail	0.284	
F Critical one-tail	3.637	

We now have all of the information we need to conduct the pooled variance *t* test (thanks to the sample mean information provided in the Excel output). The first step is to calculate the pooled variance, which is essentially a degree-of-freedom weighted average of the two variances:

$$S_p^2 = \frac{(n_1 - 1)S_1^2 + (n_2 - 1)S_2^2}{(n_1 - 1) + (n_2 - 1)}$$

$$= \frac{(11 - 1) \times 0.04242 + (8 - 1) \times 0.02714}{(11 - 1) + (8 - 1)}$$

6. Another perspective on *p*-value is: if the two population variances were actually equal, we would expect sample variances of two samples of this size to differ by this amount or more 28.4% of the time.

$$= \frac{0.61418}{17}$$

$$= 0.0361$$

Then we incorporate the pooled variance result into the t-test statistic equation, as follows:

$$t_{n_1 + n_2 - 2} = \frac{(\bar{x}_1 - \bar{x}_2) - (\mu_1 - \mu_2)}{\sqrt{S_p^2\left(\frac{1}{n_1} + \frac{1}{n_2}\right)}}$$

$$= \frac{(6.5682 - 6.2038) - 0}{\sqrt{0.361\left(\frac{1}{11} + \frac{1}{8}\right)}}$$

$$= \frac{0.3644}{0.0883}$$

$$= 4.13$$

Note that we do not know the values of μ_1 or μ_2. However, the numerator is designed to compare the sample mean difference to the hypothesized population mean difference, so we can substitute 0 for $(\mu_1 - \mu_2)$ because we are working under the assumption that the two population means are equal for the purposes of this calculation.

Now that we have completed all of the calculations, let's formalize the test.

Research Hypothesis: Mean price per square foot for lots affected by noise from an abutting busy street (Population 2) is less than mean price per square foot for lots not affected by abutting street noise (Population 1).

$H_0: \mu_1 - \mu_2 \leq 0$

$H_a: \mu_1 - \mu_2 > 0$

Critical Value of t statistic: $t_{0.05, 17} = 1.740$

Decision Rule: Reject the null hypothesis and accept the alternative hypothesis that $\mu_1 - \mu_2 > 0$ if the pooled variance t statistic is ≥ 1.740.

Conclusion: Since $4.13 > 1.74$, the analysis supports the research hypothesis that mean price per square foot for lots affected by noise from an abutting busy street is less than mean price per square foot for lots not affected by abutting street noise.

The *t*-Test in Excel, Minitab, and SPSS

The pooled variance *t*-test can be conducted in Excel and in most statistical software programs. Conduct this test in Excel by selecting **Tools, Data Analysis,** then selecting **t-Test: Two Sample Assuming Equal Variances** from the menu in the Data Analysis window. The *t*-test window will open and prompt you to select input and output ranges. The output will look like the table below. Excel does not know whether or not you are conducting a one-tailed or two-tailed test, so it provides critical values and *p*-values for both tests. In Example Problem 1, we are interested in the one-tailed critical value of 1.7396, which is the same as the 1.740 value from the *t* table (the table value is rounded). The *p*-value of 0.0004 indicates that sample mean differences this extreme would be expected only 0.04% of the time if the population means were truly equal.

Excel Output Related to a Pooled Variance Two-Sample *t*-Test

t-Test: Two-Sample Assuming Equal Variances

	Population 1 ($/SF)	*Population 2 ($/SF)*
Mean	6.5682	6.2038
Variance	0.0424	0.0271
Observations	11	8
Pooled Variance	0.0361	
Hypothesized Mean Diff	0	
df	17	
t Stat	4.1264	
P(T<=t) one-tail	0.0004	
t Critical one-tail	1.7396	
P(T<=t) two-tail	0.0007	
t Critical two-tail	2.1098	

This analysis can be accomplished in Minitab by selecting **Stat**, then **Basic Statistics**, and then **2-Sample t.** Minitab will handle sample data in either a one-column or two-column format. Minitab requires the user to select whether or not to assume equal variance. The Minitab test for variance equality is found in a separate **2 Variances** menu within **Basic Statistics.** In SPSS select **Analysis**, then **Compare Means**, then **Independent-Samples t Test.** In SPSS the data must be entered into a single column and the classifying variable (identifying population 1 or 2) entered into another column. SPSS tests for equality of variance within its *t*-test routine and provides two *t* statistics, one for equal variance and one for unequal variance.

Example Problem 2

Table 7.4 lists random samples of vacancy rates at a certain point in time for large Class A and Class C office buildings in service for at least two years in a major U.S. city. Assuming for the purposes of this problem that vacancy rates are normally distributed, is there evidence, at the 95% confidence level, that Class A office space vacancy differs from Class C office space vacancy?

Minitab provides two test statistics for the equality of variances as shown in **Figure 7.9**: one is the *F* test we have been using up to this point in the chapter, and the other is Levene's test, which is not as restrictive concerning the normality of the population distributions. (Levene's Test is not available in Excel.)

The null hypothesis that the two population variances are equal is rejected using either the F test statistic of 5.99 (p-value = 0.012) or the Levene's test statistic of 15.43 (p-value = 0.001). Therefore, we use the separate variance t-test to investigate whether or not the vacancy rates are equal. We have not specified which vacancy rate we expect to be greater than the other, so we will conduct a two-tailed test.

Table 7.4 Vacancy Rates of Two Samples

Class A Vacancy (%) $n = 12$	Class C Vacancy (%) $n = 10$
12.5	7.8
9.4	8.5
5.7	7.5
14.3	6.5
7.4	9.0
14.9	5.8
12.3	9.8
8.1	6.5
11.5	8.9
7.8	9.2
6.0	
13.8	

Figure 7.9 Test for the Equality of Variances in Minitab

```
F-Test (normal distribution)
Test statistic = 5.99, p-value = 0.012

Levine's Test (any continuous distribution)
Test statistic = 15.43, p-value = 0.001
```

The Minitab output in **Figure 7.10** does not specify that this is a separate (unequal) variance t-test, but this is the default in Minitab. The output will notify the user when pooled variance has been used in the test. The Minitab output summarizes the sample data on vacancy, the difference in sample means (2.35833%), and provides both a

Figure 7.10 Two-Sample *t*-Test in Minitab

```
Two-sample T for Class A Vacancy (%) vs Class C Vacancy (%)

                   N   Mean  StDev  SE Mean
Class A Vacancy   12  10.31   3.30     0.95
Class C Vacancy   10   7.95   1.35     0.43


Difference = mu (Class A Vacancy (%)) - mu (Class C Vacancy (%))
Estimate for difference:   2.35833
95% CI for difference:   (0.13503, 4.58163)
T-Test of difference = 0 (vs not =) :
T-Value: 2.26 P-Value = 0.039 DF = 15
```

confidence interval for the difference and a two-tailed *t*-test along with an abbreviated summary of the test hypotheses. This information allows us to do a formal write-up of our test.

Research Hypothesis: The mean vacancy rate for Class A office space differs from the mean vacancy rate for Class C office space.

$H_0: \mu_1 - \mu_2 = 0$

$H_a: \mu_1 - \mu_2 \neq 0$

Decision Rule: Reject the null hypothesis and accept the alternative hypothesis that $\mu_1 - \mu_2 \neq 0$ if the separate variance *t*-test *p*-value[7] is ≤ 0.05.

Conclusion: The *p*-value of 0.039 is less than 0.05, so the analysis supports the research hypothesis that mean vacancy rates for Class A and Class C office space are not equal.[8]

Comparing Paired Samples

Paired samples occur when repeated measures are obtained from the same items or people, or the items or people in one group have been matched to the items or people in another group based on a common characteristic. An example of this sort of test is testing each person in a sample's skill or knowledge level prior to exposure to training (i.e., treatment) and then testing each person's level post-training to estimate the change, if any, in skill or knowledge attributable to training. Another example would be to compare sample item prices at two different points in time to assess whether prices had changed over the time period.

These sorts of tests rely on the *t-test for mean difference*, also referred to as a "paired *t* test" (Minitab), "paired samples *t* test" (SPSS), and "*t* test: Paired Two Sample for Means" (Excel).

---

7. A *p*-value statement of critical value is employed in this decision rule because we did not calculate degrees of freedom for the separate variance *t*-test (it was reported post-test as *df* = 15 in the Minitab results). Therefore, we could not calculate the critical value of *t* in advance of the test. Decision rules based on *p*-value interpretation are becoming more common as statistical software applications replace hand calculations of statistical test statistics.
8. Minitab will accommodate right- and left-tailed tests by selection of the desired option from a pull-down menu located in the program's two-sample *t*-test window.

t-Test for Mean Difference

$$t_{n-1} = \frac{\bar{d} - \mu_d}{\frac{S_d}{\sqrt{n}}}$$

where

$$\bar{d} = \frac{\Sigma d_i}{n}$$

$$S_d = \sqrt{\frac{\Sigma(d_i - \bar{d})^2}{n - 1}}$$

and

$\bar{d}$ = sample mean difference

μ_d = hypothesized population mean difference (usually, but not necessarily 0)

S_d = standard deviation of sample differences

d_i = sample item difference where $i = 1$ to n

n = sample size (number of difference measures in the sample)

$n - 1$ = degrees of freedom for the *t*-test statistic

As with all *t*-tests the underlying assumption is that population of differences are normally distributed, but the test can be applied to non-normal population differences with sufficiently large samples. With this in mind, consider the following real estate related example.

Example Problem

Table 7.5 lists median resale detached home prices by zip code for east and central San Diego County, California, for the periods January 2007 and January 2008. The difference column is the January 2008 median price for each zip code minus the January 2007 median price for each zip code. Although it appears obvious from a cursory look at the data that there had been a remarkable price decline in east and central San Diego County, the *t*-test quantifies the significance of the price decrease.

The formal statement of the test is as follows.

Research Hypothesis: Median resale price of detached homes in central and east San Diego

Table 7.5 Median Resale Prices by Zip Code for January 2007 and January 2008

Zip Code	January, 2007 Median Price	January, 2008 Median Price	Difference
92120	579,000	555,000	-24,000
92105	380,000	275,000	-105,000
92117	456,250	486,250	30,000
92115	472,000	380,000	-92,000
92118	1,300,000	975,000	-325,000
92101	563,000	563,000	0
92114	413,500	290,000	-123,500
92102	455,000	309,000	-146,000
92103	960,000	660,000	-300,000
92116	515,000	492,500	-22,500
92037	1,570,000	1,432,500	-137,500
92111	500,000	430,000	-70,000
92113	340,000	325,000	-15,000
92126	487,500	410,000	-77,500
92109	770,000	757,500	-12,500
92110	670,000	647,500	-22,500
92104	540,000	485,000	-55,000
92107	840,000	829,500	-10,500
92139	470,000	289,750	-180,250
92106	781,000	852,500	71,500
92119	570,000	388,500	-181,500
92131	730,000	674,500	-55,500
92123	450,000	408,750	-41,250
92121	691,500	655,000	-36,500
92124	587,500	590,000	2,500
92122	650,000	640,000	-10,000
91901	540,000	430,000	-110,000
91905	360,000	329,000	-31,000
91906	353,250	311,500	-41,750
92019	571,500	430,000	-141,500
92020	457,500	515,000	57,500
92021	448,500	370,000	-78,500
91935	590,000	624,000	34,000
91941	550,000	440,000	-110,000
91942	465,000	400,000	-65,000
92040	510,000	340,000	-170,000
91945	441,500	340,500	-101,000
91978	602,500	446,000	-156,500
92071	445,000	360,000	-85,000
91977	417,000	310,000	-107,000

$\bar{d} = -\$76{,}131$

$S_d = \$84{,}196$

$n = 40$

$$t_{n-1} = \frac{\bar{d} - \mu_d}{\frac{S_d}{\sqrt{n}}} = \frac{-\$76{,}131 - 0}{\frac{\$84{,}196}{\sqrt{40}}} = \frac{-\$76{,}131}{\$13{,}313} = -5.72$$

Data source: http://www.dqnews.com/ZIPSDUT.shtm

County changed between January 2007 and January 2008.

$H_0: \mu_d = 0$

$H_a: \mu_d \neq 0$

Critical value of t statistic: $t_{0.05,\,39} = 1.685$

Decision Rule: Reject the null hypothesis and accept the alternative hypothesis that $\mu_d \neq 0$ if the t-test statistic is ≥ 1.685 or ≤ -1.685.

Conclusion: The result of the t-test = -5.72 (Table 7.5), which is < -1.685, so the analysis supports the research hypothesis that median resale detached home price in central and east San Diego County changed between January 2007 and January 2008.

Paired Samples *t*-Test in SPSS, Minitab, and Excel

This analysis can also be accomplished in SPSS, Minitab, and Excel. The SPSS output is presented below. The SPSS procedure for this test is to select **Analyze, Compare Means,** and then **Paired Samples t Test.** The results are the same as the arithmetic calculations we just completed. SPSS also provides a confidence interval for the average median-price difference, which is -$49,204 to -$103,058.

Paired Samples Test

		Paired Differences					t	df	Sig. (2-tailed)
					95% Confidence Interval of the Difference				
		Mean	Std. Deviation	Std. Error Mean	Lower	Upper			
Pair 1	January 2008 Median Price - January 2007 Median Price	-76131.3	84195.89466	13312.54	-103058	-49204.1	-5.719	39	.000

To perform this test in Minitab, select **Stat**, then **Basic Statistics**, and then **Paired *t*.** In Excel select **Tools, Data Analysis**, and then **t-Test: Paired Two Sample for Means** from the list in the **Data Analysis** window.

Practice Problems 7.4

1. Many people have a difficult time interpreting metes and bounds legal descriptions, especially if the property boundary includes curves. A real estate school is in the final stages of designing a classroom module dealing with how to read and interpret metes and bounds legal descriptions and has conducted a test of the proposed instruction module's effectiveness. The test involves random selection of 15 participants from a list of licensed appraisal trainees, administering a pre-treatment examination on metes and bounds interpretation, exposure to classroom education including the proposed metes and bounds module, and a post-treatment examination on metes and bounds interpretation. Pre- and post-treatment examination results were as follows:

Student	Pre-Treatment Score	Post-Treatment Score
1	74	82
2	50	60
3	42	40
4	68	84
5	88	90
6	92	90
7	66	78
8	100	100
9	35	54
10	80	84
11	78	78
12	58	66
13	40	52
14	64	72
15	38	44

Assuming that score difference is normally distributed, can the course developer be 95% confident that the classroom module on metes and bounds interpretation was effective?

2. A development company is considering inclusion of three-car garages rather than two-car garages as a standard feature in its next residential housing development. This feature will be included only if it will allow the developer to be 90% confident that she can increase the market price of her base 1,800-sq.-ft. house by $3.50 per square foot or more. The developer conducted a study to determine whether or not to include the additional garage. She collected random samples of new home sale prices from several directly competitive developments (some with two-car garages and some with three-car garages), and adjusted the sample prices to reflect a common floor area and typical upgrade features. Her sample statistics were as follows:

Homes with two-car garages:	Sample Size	40
	Sample Mean Price	$137.00/sq. ft.
	Sample Standard Deviation	$6.25/sq. ft.
Homes with three-car garages:	Sample Size	36
	Sample Mean Price	$142.00/sq. ft.
	Sample Standard Deviation	$6.50/sq. ft.

Conduct a statistical test designed to assess whether the three-car garage home market price exceeds the two-car garage home market price by at least $3.50 per square foot.

3. The following random samples of prices relate to standard 10,000-sq.-ft. lakefront lots and lots fronting a golf course recently sold at a large development. Using a significance level of 5%, assess whether or not the two population mean prices differ. Assume that lakefront and golf-front lot prices are normally distributed.

Market Price	
Lakefront Lots	**Golf Lots**
$150,000	$178,000
$147,500	$127,500
$160,000	$186,350
$180,000	$132,800
$137,500	$205,500
$157,900	$126,600
$159,900	$154,200
$162,500	$187,000
$175,000	$138,600
$180,000	$210,900
$153,400	$165,000
$139,000	$119,900
$167,300	$167,500
$165,000	$156,000
$138,600	$145,000
$192,450	$198,600
	$135,000
	$128,200
	$167,300

Test of Means from More Than Two Samples

Sometimes it is useful to compare means from more than two samples using a method–Analysis of Variance (ANOVA)–that is similar to the *t*-test of the difference between two means. As pictured in **Figure 7.11**, ANOVA "partitions" total variance in a sample (*SST*) into variation among the three or more groups (*SSF*) and variation within the three or more groups (*SSE*). This method tests the hypothesis that $\mu_1 = \mu_2 = \mu_3 = \ldots = \mu_i$ where i groups are being compared.

The between-group variation is called the "treatment effect" or "factor effect," which is the effect ANOVA is analyzing. This is called "factor" variance in Minitab and "between groups" variance in Excel and SPSS. The remaining within-group variation includes both systematic variation within each group and any random variance. The remaining variance is simply called "error" variance in Minitab and "within groups" variance in Excel and SPSS.

The three or more means analysis illustrated in Figure 7.11 is referred to variously as "one-way analysis of variance" or "single factor analysis of variance." These names indicate that the hypothesized difference in means is due to a single influence, or factor. **Figure 7.12** illustrates the two forms of variance dealt with in a single-factor ANOVA. The three bell-shaped curves each represent a different group, defined by one of three levels of a single factor. In a real property appraisal context the factor could be three different locations, three different types of retail center, three different classes of office space, and the like.

Figure 7.11 Partitioning of Total Variance in ANOVA

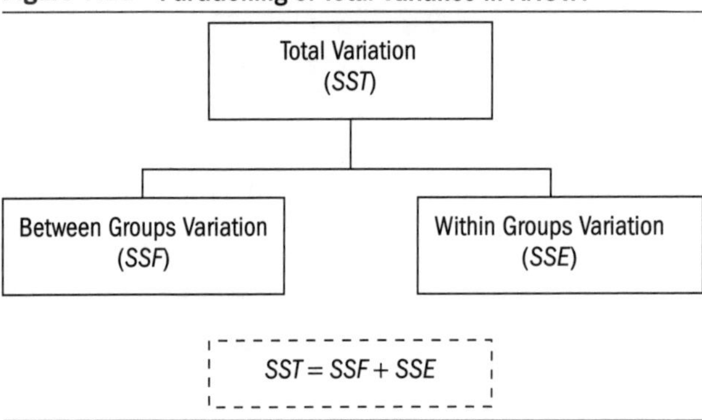

Figure 7.12 Variance of Three Means

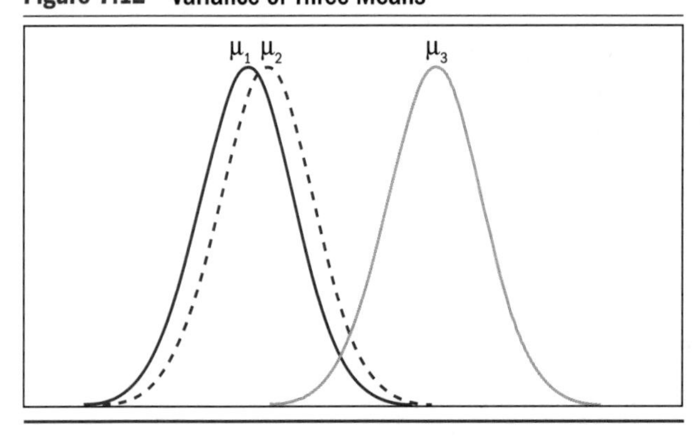

For simplicity's sake, let's assume Figure 7.12 illustrates price per square foot for similar properties in three different locations: Location 1, Location 2, and Location 3. As drawn here, variation within the groups (*SSE*) is essentially equal and appears normal (i.e., the curves are similar in size and shape). SSE captures uncertainty in estimating μ_1, μ_2, and μ_3. The among-groups variance (*SSF*) is due to differences in μ_1, μ_2, and μ_3 caused by the factor (location). ANOVA separates total variance

(SST) into these two forms of variance, allowing us to assess whether or not SSF is great enough to reject the null hypothesis that $\mu_1 = \mu_2 = \mu_3$.

Example Problem

The data in **Table 7.6** represent three random samples of vacancy rates for Class A, Class B, and Class C office space in a major U.S. city. For the purposes of this example, assume that office space vacancy is normally distributed. Single-factor analysis of variance allows us to analyze the research hypothesis that mean vacancy rate does vary by the single factor of class–i.e., Class A, Class B, and Class C office space. It is important to recognize that this research statement is true if only one of the mean vacancy rates differs from the others. Therefore, ANOVA tests the null hypothesis that all three means are equal against the alternative hypothesis that not all are equal. Other tests are required to investigate where the paired differences occur.

The output in **Figure 7.13** was generated using Minitab. It is executed by selecting **Stat**, then **ANOVA**, and then either **One-Way** or **One-Way (Unstacked)** depending on how the data is organized. The

Table 7.6 Vacancy Rates in Class A, B, and C Office Space

Class A Vacancy (%)	Class B Vacancy (%)	Class C Vacancy (%)
12.5	10.4	7.8
9.4	9.6	8.5
9.8	11.1	7.5
14.3	11.5	6.5
8.9	9.3	9.0
13.2	13.6	5.8
12.3	11.9	9.8
8.1	9.5	6.5
11.5	10.4	8.9
12.9	12.0	9.2
10.2	8.4	5.7
13.1	12.8	7.8

Figure 7.13 One-way ANOVA in Minitab

One-way ANOVA: Class A Vacancy (%), Class B Vacancy (%), Class C Vacancy (%)

```
Source  DF      SS     MS      F      P
Factor   2   91.80  45.90  16.58  0.000
Error   33   91.34   2.77
Total   35  183.15

Level             N    Mean  StDev
Class A Vacancy  12  11.350  1.998
Class B Vacancy  12  10.875  1.551
Class C Vacancy  12   7.750  1.380

Pooled StDev = 1.664
```

ANOVA in Excel and SPSS

In Excel, run the ANOVA procedure by selecting **Tools, Data Analysis,** and then the **ANOVA: Single Factor** procedure from the menu in the **Data Analysis** window. In SPSS, select **Analyze, Compare Means,** and then the **One-Way ANOVA** procedure.

data in Table 7.6 are "unstacked" because the vacancy data are in three columns rather than one column. If the data were "stacked" there would be one vacancy column with another parallel column identifying the office class for each adjoining vacancy entry. SPSS only accepts stacked data and Excel only accepts unstacked data. Minitab accepts either data entry format.

ANOVA relies on an *F* statistic derived from a ratio of mean variance. The *F* ratio for this example problem is

$$F_{i-1, n-i} = \frac{MSF}{MSE} = \frac{\frac{91.80}{2}}{\frac{91.34}{33}} = \frac{45.90}{2.77} = 16.58$$

where *MSF* is the estimated mean variance due to the factor (Class A, B, or C) and *MSE* is the estimated mean variance within each factor. Note that *i* represents the number of levels of the factor (3 in this example) and *n* is the total number of observations in the sample (36 in this example).

$$MSF = \frac{SSF}{i-1} = \frac{91.80}{2} = 45.90$$

$$MSE = \frac{SSE}{n-i} = \frac{91.43}{33} = \ 2.77$$

The Minitab ANOVA output shows an *F* statistic of 16.58, which results in a *p*-value of 0.000 with 2 numerator degrees of freedom and 33 denominator degrees of freedom. If this analysis were being conducted at a 5% significance level, then we could write it up formally as follows:

Research Hypothesis: Class A, Class B, and Class C office vacancy rates are not equal.

$H_0: \mu_A = \mu_B = \mu_C$

H_a: All three means are not equal

Critical Value[9] of F statistic: $F_{2,33} = 3.285$

Decision Rule: Reject the null hypothesis and accept the alternative hypothesis that all means are not equal if the F test statistic is ≥ 3.285.

Conclusion: Since $16.58 > 3.285$, the analysis supports the research hypothesis that Class A, Class B, and Class C office vacancy rates are not equal.

Where Does the Difference Occur?

In our example the null hypothesis was rejected in favor of the hypothesis that not all of the means are equal. The next step may be to analyze which means differ. Although it is tempting to conduct three *t*-tests of means (A vs. B, A vs. C, and B vs. C), multiple tests such as these suffer from what statisticians refer to as an "inflated error rate problem," which compromises the validity of the three separate *t*-tests. The best way to overcome the inflated error rate problem is to apply a test designed for multiple comparisons, such as a Bonferroni procedure, which is based on a *t* statistic that adjusts the observed significance level for the fact that multiple comparisons are being made. SPSS provides this option (the Bonferroni output is presented in **Figure 7.14**).[10]

Vacancy rate difference between Class A and Class C is significant (p-value = 0.000, Rows 2 and 5 in the table results), and vacancy rate difference between Class B and

Figure 7.14 Bonferroni Output in SPSS

Multiple Comparisons

Dependent Variable: Vacancy

Bonferroni

(I) 1=A, 2=B, 3=C	(J) 1=A, 2=B, 3=C	Mean Difference (I-J)	Std. Error	Sig.	95% Confidence Interval	
					Lower Bound	Upper Bound
1.00	2.00	.47500	.67921	1.000	-1.2381	2.1881
	3.00	3.60000*	.67921	.000	1.8869	5.3131
2.00	1.00	-.47500	.67921	1.000	-2.1881	1.2381
	3.00	3.12500*	.67921	.000	1.4119	4.8381
3.00	1.00	-3.60000*	.67921	.000	-5.3131	-1.8869
	2.00	-3.12500*	.67921	.000	-4.8381	-1.4119

*. The mean difference is significant at the .05 level.

9. The lower-tail F statistic is irrelevant here because we are only interested in knowing if the MSF to MSE ratio is sufficiently large.

10. Choice of statistical software can be complicated by differences in the routines and procedures each provides. If you plan to do a lot of ANOVA and will need to make multiple comparisons, SPSS does this conveniently. If not, either Minitab or SPSS is a reasonable choice. Excel will run most of the tests you need, but it is very light on diagnostic procedures. Excel is primarily a spreadsheet program with good charting capabilities. It is not a well-developed inferential statistics program. Nevertheless, Excel can be an adequate choice in some situations.

Class C is also significant (p-value = 0.000, Rows 4 and 6 in the table results). Class A and Class B vacancy rates are not significantly different (Rows 1 and 3 in the table results).

Minitab provides a nice chart (**Figure 7.15**) showing all of the data points and their means. This chart reinforces the SPSS Bonferroni test results and illustrates the superior vacancy rate performance of Class C office space.

Assumptions Underlying ANOVA

Three assumptions are implicit in an ANOVA procedure:

1. Normality. Each sample is assumed to have been drawn from a normally distributed population. This assumption is usually not too constraining if the population is approximately normal or the samples are sufficiently large.
2. Homogenous variance. Because pooled variance (and standard deviation) is used in the procedure, variances are assumed to be equal for each population. This constraint is less binding when sample sizes from each population are equivalent. The sample sizes of 12, 12, and 12 in the example are equal. Class B and C vacancy rates have slightly less variance (2.41 and 1.90) than Class A vacancy (3.99), but this small difference does not appear to be significant. Minitab includes variance tests for more than two variances in its **ANOVA** menu, one of which assumes normality and one of which does not. The null hypothesis of equal variance for the data in the office vacancy example cannot be rejected with either test, as shown by the Minitab output in **Figure 7.16**. SPSS includes a homogeneity of variance test in the **Options** menu attached to the

Figure 7.15 Individual Value Plot of Class A, B, and C Office Space Vacancy in Minitab

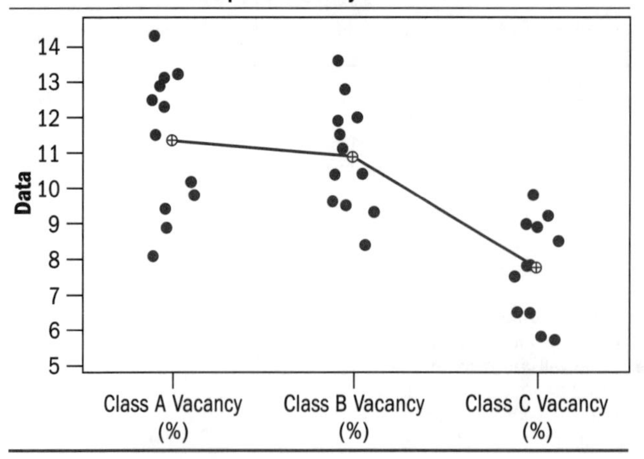

One-Way ANOVA window.

3. The samples are representative (randomly drawn) and independent. Choice of a sample element from one population must not in any way influence selection of sample elements from another population (or populations).

Figure 7.16 Variance Tests in Minitab

```
Bartlett's Test (normal distribution)
Test statistic = 1.56, p-value = 0.458

Levene's Test (any continuous distribution)
Test statistic = 1.38, p-value = 0.265
```

Practice Problem 7.5

The table below is a random sample of capitalization rates for industrial warehouse space from four different U.S. cities. Assuming capitalization rates are normally distributed, test the research hypothesis at 95% confidence that the capitalization rates derived from the four cities differ. If you have software that can run additional tests, test for homogeneity of variance and test each pair to determine which pairs are different.

Office Warehouse Cap Rates			
City A	**City B**	**City C**	**City D**
8.5	9.3	9.6	7.9
7.6	7.9	10.3	7.6
9.3	8.4	10.7	8.2
8.1	7.8	8.9	7.4
7.4	9.5	9.7	8.4
8.8	8.4	11.1	8.1
8.2	8.1	9.2	7.3
9.1	9.4	10.6	8.6
9.5	8.5	12.1	7.8
8.3	8.8	10.1	7.9

Two-Sample Test of Proportions

Sometimes it is necessary to compare categorical data across subsets of respondents. Examples might include floor plan preferences of young families and of seniors, or mortgage term choice for initial purchase financing or for refinancing. In these cases a two-sample Z test of proportions can provide inferences concerning the preferences of two populations. The Z statistic for this test is

$$Z = \frac{(p_{S1} - p_{S2}) - (p_1 - p_2)}{\sqrt{p^*(1 - p^*)\left(\frac{1}{n_1} + \frac{1}{n_2}\right)}}$$

where p^* is the pooled estimate of the population proportion

$$p^{*} = \frac{x_1 + x_2}{n_1 + n_2}$$

and

p_{S1} = proportion in Sample 1

p_{S2} = proportion in Sample 2

n_1 = size of Sample 1

n_2 = size of Sample 2

$p_1 - p_2$ = hypothesized population proportion difference

x_1 = within category count in Sample 1

x_2 = within category count in Sample 2

Example Problem

It has been postulated that first-time home buyers are more likely to opt for a multistory floor plan than home buyers in the 50+ age cohort. Logically, older homeowners are more concerned about mobility within their homes whereas first-time home buyers are more concerned about maximizing house size given the constraints of household income and the affordability of the home.

A random sample of households who recently purchased a home yielded 60 respondents in the first-time home buyer and age 50 or less category (Group 1), 42 of whom purchased a multistory home. Additionally, the sample included 40 home buyer respondents in the 50+ age category (Group 2), 15 of whom had purchased a multistory home. Test the hypothesis that age 50 or less first-time home buyers prefer multistory homes when compared to the 50+ home buyer age group at a 95% confidence level.

$H_0: p_1 - p_2 \leq 0$

$H_a: p_1 - p_2 > 0$

$$n_1 = 60$$

$$n_2 = 40$$

$$x_1 = 42$$

$$x_2 = 15$$

$$p_{S1} = 0.70$$

$$p_{S2} = 0.375$$

$$p^* = \frac{x_1 + x_2}{n_1 + n_2} = \frac{42 + 15}{60 + 40} = 0.57$$

$$Z = \frac{(p_{S1} - p_{S2}) - (p_1 - p_2)}{\sqrt{p^*(1 - p^*)\left(\frac{1}{n_1} + \frac{1}{n_2}\right)}}$$

$$= \frac{(0.70 - 0.375) - 0}{\sqrt{0.57(1 - 0.57)\left(\frac{1}{60} + \frac{1}{40}\right)}}$$

$$= \frac{0.325}{\sqrt{0.0102125}}$$

$$= 3.216$$

Critical Value of Z: $Z = 1.64$

Decision Rule: Reject the null hypothesis and accept the alternative hypothesis that $p_1 - p_2 > 0$ if the Z test statistic is ≥ 1.64.

Conclusion: Since $3.216 > 1.64$, the data indicate that first-time home buyers are more likely to purchase a multistory home than home buyers in the 50+ age category.

Practice Problem 7.6

Two random samples of mortgage consumers are shown below. Test the hypothesis that the proportion of fixed-rate mortgages is different for refinancing than for initially purchasing a home at the 95% confidence level.

Purchase Mortgage	Refinance Mortgage
Fixed	Fixed
ARM	Fixed
Fixed	Fixed
ARM	ARM
ARM	ARM
Fixed	ARM
ARM	Fixed
Fixed	Fixed
ARM	ARM
ARM	ARM
Fixed	Fixed
ARM	Fixed
ARM	Fixed
Fixed	ARM
ARM	ARM
ARM	Fixed
Fixed	Fixed
ARM	ARM
ARM	Fixed
Fixed	Fixed
ARM	Fixed
Fixed	ARM
Fixed	Fixed
ARM	ARM
ARM	Fixed
Fixed	ARM
Fixed	Fixed
ARM	Fixed
ARM	ARM
ARM	Fixed
Fixed	ARM

Real-World Case Study: Part 4

Price per Square Foot

In the analysis of the townhouse data, an interesting and uncertain issue is whether or not townhouse price per square foot remains constant as floor area changes. Theory is inconclusive here. On one hand we would expect price per square foot to decrease as floor area increases because fixed costs of development such as land, lot development (streets and utilities), and design and regulatory costs are spread over a larger floor area. On the other hand, it is conceivable that smaller units are associated with more austere projects whereas larger units are associated with more luxurious projects. In addition, other factors may mitigate any fixed-cost price reduction per square foot associated with larger unit size.

The **scatter plot below** compares sale price to unit size in square feet. Clearly price rises as unit size increases, as expected. The correlation is direct, meaning these two variables tend to move up and down together, and the Pearson correlation coefficient is significant at 0.72.

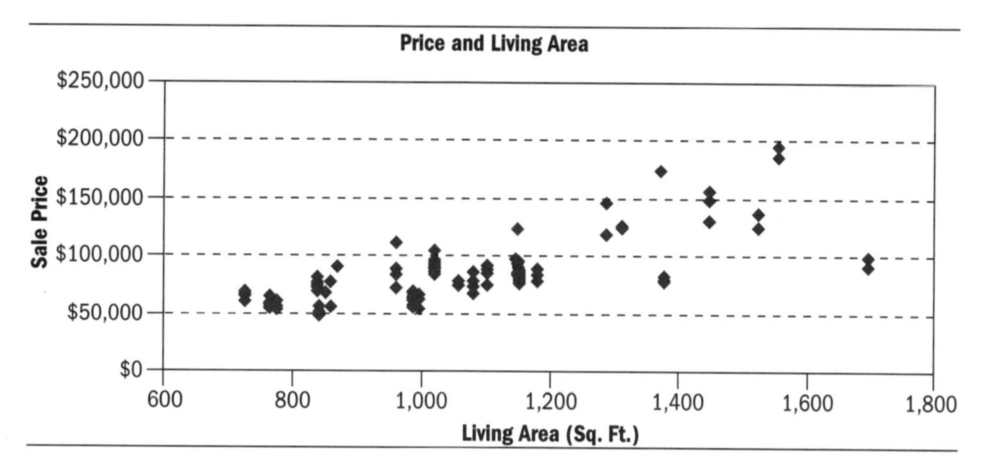

The relationship between price per square foot and living area is unclear, however, as shown in **the next scatter plot**. The correlation is positive, but the Pearson correlation coefficient is low at 0.113 and not statistically significant (p-value = 0.204). Spearman's rho is nearly zero, at 0.002. Note that there are three observations at 1,376 square feet and two observations at 1,696 square feet that

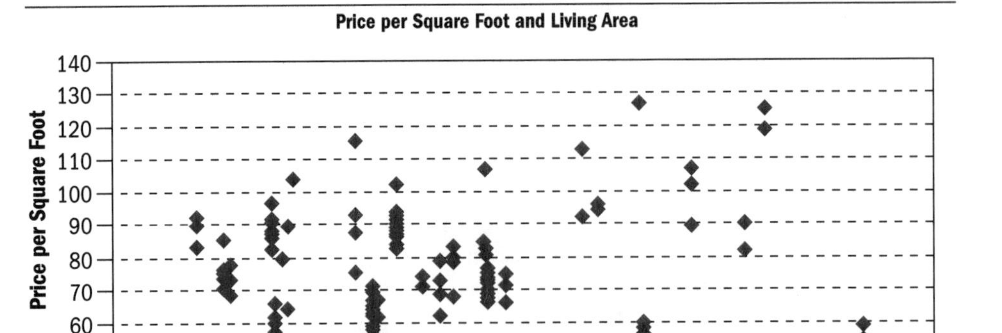

sold for less than $60 per square foot. By comparison, the remaining units in excess of 1,200 square feet in area appear to have sold for a higher average price per square foot than the average price per square foot for the less than 1,200-sq.-ft. sized units.

Another way to address the price per square foot vs. living area issue is to compare price per square foot for above-median-sized units to price per square foot for below-median-sized units. Median unit size is 1,019 square feet. There are 19 median-sized units, 46 below-median-sized units (724 square feet to 992 square feet), and 64 above-median-sized units (1,056 square feet to 1,696 square feet). The **following Excel outputs** compare price per square foot for the 46 smaller units to the 64 larger units.

F Test Two-Sample for Variances

	Smaller Units	*Larger Units*
Mean	75.34	78.49
Variance	198.72	238.60
Observations	46.00	64.00
df	45.00	63.00
F	0.83	
P(F<=f) one-tail	0.26	
F Critical one-tail	0.63	

Based on the *F*-test, we cannot reject the null hypothesis that the price per square foot variances are equal. Therefore, we use the pooled variance *t*-test to compare the two unit-size subsamples, as **shown on the following page**.

t-Test: Two-Sample Assuming Equal Variances

	Smaller Units	*Larger Units*
Mean	75.34	78.49
Variance	198.72	238.60
Observations	46.00	64.00
Pooled Variance	221.98	
Hypothesized Mean Difference	0.00	
df	108.00	
t Stat	-1.09	
P(T<=t) one-tail	0.14	
t Critical one-tail	1.66	
P(T<=t) two-tail	0.28	
t Critical two-tail	1.98	

The larger-unit mean price per square foot of \$78.49 is not significantly different from the smaller-unit mean price of \$75.34 per square foot (*p*-value = 0.28), including the five previously identified larger-unit sales at prices less than \$60 per square foot in the calculation of the larger-units subsample mean. The result is different when the five lowest unit-price sales of the larger-unit residences are excluded from the sample. When they are excluded the larger-townhouse sample mean increases to \$80.26 per square foot. As **shown below**, we can be at least 90% confident that the two mean prices per square foot are different (*p*-value = 0.09) when these five townhouses are excluded from the analysis.

t-Test: Two-Sample Assuming Equal Variances (with five larger units omitted)

	Smaller Units	*Larger Units*
Mean	75.34	80.26
Variance	198.72	217.75
Observations	46.00	59.00
Pooled Variance	209.43	
Hypothesized Mean Difference	0.00	
df	103.00	
t Stat	-1.73	
P(T<=t) one-tail	0.04	
t Critical one-tail	1.66	
P(T<=t) two-tail	0.09	
t Critical two-tail	1.98	

Effect of Occupancy

The data include information concerning three types of occupancy–owner, tenant, and vacant. Price per square foot can be partitioned into these three categories to investigate the effect of occupancy type. As the **SPSS output below** illustrates, mean price per square foot varies with type of occupancy.

Price SF

	n	Mean	Std. Deviation
Owner	54	81.5274	18.02784
Vacant	70	77.2880	11.00659
Tenant	5	74.9521	13.81145
Total	129	78.9721	14.51807

A one-way analysis of variance can be used to investigate whether or not the differences in these three mean prices per square foot are significant. The **following SPSS output** reports an *F*-statistic of 1.511 with a p-value of 0.225. This means that we can only be 77.5% confident (100% – 22.5%) that price per square foot varies with type of occupancy. This confidence level is lower than generally required to reject the null hypothesis that the three occupancy type mean prices per square foot are equal.

ANOVA

Price SF

	Sum of Squares	df	Mean Square	F	Sig.
Between Groups	631.934	2	315.967	1.511	.225
Within Groups	26347.198	126	209.105		
Total	26979.132	128			

Failure to reject the null hypothesis that the mean prices per square foot of the three occupancy types are equal could result from a number of causes. First, the means actually may not be different. Second, other factors may be confounding the results, such as differences in property characteristics across occupancy types. Third, the small tenant-occupancy subsample will have a relatively large standard error of the mean. Also, due to there being only five observations in this subsample, we cannot rely on the Central Limit Theorem to control for violation of the normality assumption. Therefore, the tenant-occupied subsample may be too small to produce reliable analysis.

Comments

This chapter's townhouse data discussion probably raises more questions than it answers. For example, are the five units with the lowest price per square foot in the larger-unit subsample different from the remaining units in that subsample? This issue can probably be addressed by investigating how the five larger-unit sales that were priced below $60 per square foot differ from the remaining units in that subsample. If these five units should be excluded from the subsample, what characteristics of the larger- and smaller-unit subsamples are responsible for the significant difference in price per square foot? Other questions might include

- Are important variables missing from the data?
- Is a lack of support of a research hypothesis that price per square foot varies with occupancy type credible?

Project age, type of occupant, land-to-building ratio, or garage parking may be correlated with unit size. In addition, other unknown factors may be at play, which may be discovered through inspection of the sold properties or interviews with knowledgeable parties during sales confirmation. The data set could be expanded to include information concerning common amenities available at each project, association fee differences, each sold unit's curb appeal, and condition. Other relevant factors might be inclusion of furnishings in the sale transaction, atypical financing, and unusual buyer or seller motivations. Finally, there simply may not be enough available market information to reliably test the effect of tenant occupancy on townhouse sale price in this zip code. In order to address this question we might need to geographically extend the study area to include more townhouse data.

8 Nonparametric Tests

As used here *nonparametric statistics* refers broadly to tests that do not rely on the assumption that sample data are drawn from a normal population. Nonparametric methods are useful when sample size is too small to assume that the sampling distribution of the mean is normal, and they are often applicable to ordinal data. In particular, knowledge of nonparametric methods can be important to real estate analysts due to the small data sets often encountered.

Nonparametric statistics is often viewed as a separate subdiscipline of statistics due to the nature of the underlying mathematics and the large variety of nonparametric tests that exist. This chapter introduces several nonparamateric methods that do not rely on normality and are functionally equivalent to many of the tests covered in the prior chapter. We will limit ourselves to a few basic tests, recognizing that many additional nonparametric tools are available when needed.[1]

Before we begin, though, it is important to note that nonparametric tests are not as powerful as their parametric counterparts–that is, they are not as good at detecting small differences between samples. For this reason, when there is a large amount of numerical data, it is advisable to opt for a parametric test (t statistic, Z statistic, or F statistic), lim-

1. A fairly complete and applied nonparametric reference is W.J. Conover, *Practical Nonparametric Statistics*, 3rd ed. (New York: John Wiley & Sons, 1999). Also, if you can find a copy, the small handbook by Jean D. Gibbons, *Nonparametric Statistics: An Introduction* (Newbury Park, Calif.: Sage Publications, 1993) is a very practical reference book.

iting the use of less-powerful, nonparametric tests to small samples where unintended violations of an assumption of normality are most problematic.

Two-Sample Test of Proportions: Chi-Square

In Chapter 7, we derived a Z statistic to test whether proportions differ for two samples. A similar test, which does not rely on the normally distributed Z statistic, is the χ^2 (chi-square) test for the difference between two proportions. This test relies on dividing each of two samples of sizes n_1 and n_2 into two subsets we will refer to as the number of "successes" (x_i) and "failures" ($n_i - x_i$). The successes and failures are displayed in a 2×2 contingency table, as shown in **Table 8.1**.

Defining $p_1 = x_1/n_1$ and $p_2 = x_2/n_2$, the χ^2 statistic allows us to test a null hypothesis of the equality of p_1 and p_2. For illustrative purposes, consider an example problem we looked at in Chapter 7. As you will recall, a random sample of local households who recently purchased a home yielded 60 respondents in the category of first-time home buyers of age 50 or less (Sample 1), 42 of whom had purchased a multistory home. Additionally, the sample included 40 respondents in the 50+ age category (Sample 2), 15 of whom had purchased a multistory home. Defining a "success" as having purchased a multistory home, we can divide each sample into "success" and "failure" subcategories as shown in **Table 8.2**.

Table 8.1 Layout of 2×2 Contingency Table

	Sample		
	Sample 1	**Sample 2**	**Total**
Successes	x_1	x_2	X
Failures	$n_1 - x_1$	$n_2 - x_2$	$n - X$
Total	n_1	n_2	n

Table 8.2 First-Time Home Buyer Single-Story and Multistory Home Purchase by Age Cohort

	First-Time Home Buyers		
	Age 50 or less	**Age 50+**	**Total**
Multistory home (success)	42	15	57
Single-story home (failure)	18	25	43
Total	60	40	100

The chi-square test for equality of proportions can be used to test whether or not the multistory home purchase sample proportions (p_1 and p_2) are equal. The statistical null hypothesis is

$$H_0: p_1 = p_2$$

and the alternative research hypothesis is

$$H_a: p_1 \neq p_2$$

The chi-square statistic with 1 degree of freedom is

$$\chi^2_{df=1} = \sum_1^4 \frac{(f_o - f_e)^2}{f_e}$$

where f_o is the observed frequency in each cell, f_e is the expected frequency for each cell (the frequency that would occur if the sample proportions were equal), and 1 through 4 represent cell numbers. It is important to note that this test relies on the requirement that the expected frequency in each cell is at least 5.

For Sample 1 the expected proportion of successes and failures are

$$f_e \text{ (success)} = n_1 \frac{X}{n}$$

and

$$f_e \text{ (failure)} = n_1 - n_1 \frac{X}{n}$$

Likewise, for Sample 2 the expected proportion of successes and failures are

$$f_e \text{ (success)} = n_2 \frac{X}{n}$$

and

$$f_e \text{ (failure)} = n_2 - n_2 \frac{X}{n}$$

For the multistory home purchase example in Table 8.2, $X/n = 57/100 = 0.57$. **Table 8.3** shows the observed and expected frequencies for each of the four cells in Table 8.2, and the resulting chi-square statistic.

Using Excel's "=CHIDIST" function,[2] we can calculate the p-value for

Table 8.3 Chi-Square Calculations for First-Time Home Buyer Analysis

Cell	f_o	f_e	$(f_o - f_e)$	$(f_o - f_e)^2$	$\frac{(f_o - f_e)^2}{f_e}$
1	42	34.2	7.8	60.84	1.779
2	18	25.8	-7.8	60.84	2.358
3	15	22.8	7.8	60.84	2.668
4	25	17.2	-7.8	60.84	3.537
					$\chi^2 = \Sigma = 10.342$

2. The p-value of 0.0013 is computed using the cell input =CHIDIST(10.342, 1).

our calculated chi-square value of 10.342 with 1 degree of freedom as 0.0013. By this test, we reject the null hypothesis that the proportions are equal in favor of the research hypothesis that the proportions of multistory purchases differ for the two categories. This result is consistent with the result from the prior chapter's similar Z statistic test, without relying on a normality assumption.

A comparison of the Z statistic for this problem from Chapter 7 reveals an important relationship between Z and χ^2. The Z statistic for these data in Chapter 7 was 3.216, while the χ^2 statistic of 10.342 for the same data in Table 8.3 is equal to the square of the Z statistic (accounting for rounding in the calculations) because the corresponding Z statistic is always the square root of the χ^2 value when χ^2 has been calculated with 1 degree of freedom. As we will discover later in the chapter, an advantage of the χ^2 test is that it will accommodate a comparison of more than two proportions, whereas the Z test from the prior chapter will not.

Example Problem

An occupancy rate is merely the *proportion* of a building's rentable space that is occupied. It can be expressed as a ratio of rented square feet to total available square feet, a ratio of occupied units to the total number of units, or in some cases a ratio of gross revenue to potential gross revenue. Suppose you would like to test the hypothesis that small apartment project occupancy varies depending on proximity to public transportation. You randomly sample several small apartment buildings within walking distance to a city's mass transit system and find a total of 250 units, with 240 being occupied. You also randomly sample several similar small apartment buildings that are too far from mass transit to be considered within walking distance. The second sample consists of 200 units, with 180 being occupied.

Sample proportions are $p_1 = 0.96$ and $p_2 = 0.90$. Now test the hypothesis that occupancy rates differ based on proximity to mass transit by investigating whether the sample proportion differences are sufficient to support this proposition with 95% confidence.

The statistical null hypothesis is

$H_0: p_1 = p_2$

and the alternative research hypothesis is

$H_a: p_1 \neq p_2$

The 2 × 2 contingency table for this example is presented in **Table 8.4.**

Cell frequencies and expected frequencies are shown in **Table 8.5,** where $X/n = 420/450 = 0.93\overline{3}$. No expected frequency is less than 5, so we can apply the χ^2 test. Using "=CHIDIST" in Excel, the *p*-value for a χ^2 of 6.428 with 1 degree of freedom is 0.011. Therefore, since 0.011 ≤ 0.05, we reject the null hypothesis. We can state that we are 95% confident that small apartment building occupancy rates are higher when the building is within walking distance of mass transit.[3]

Table 8.4 Apartment Occupancy by Proximity to Mass Transit

	Proximity to Mass Transit		
	Within Walking Distance	Not Within Walking Distance	Total
Occupied	240	180	420
Unoccupied	10	20	30
Total	250	200	450

Table 8.5 Chi-Square Calculations for Proximity to Mass Transit Analysis

Cell	f_o	f_e	$(f_o - f_e)$	$(f_o - f_e)^2$	$\frac{(f_o - f_e)^2}{f_e}$
1	240	233.33	6.67	44.44	0.190
2	10	16.67	-6.67	44.44	2.667
3	180	186.67	-6.67	44.44	0.238
4	20	13.33	6.67	44.44	3.333
					$\chi^2 = \Sigma = 6.428$

More-Than-Two-Sample Test Of Proportions: Chi-Square

The distinction between the two-sample test of proportions and a test of more than two samples is apparent from the following null and alternative hypotheses:

$H_0: p_1 = p_2 = \ldots = p_k$

H_a: Not all *k* proportions are equal

The χ^2 test statistic is

$$\chi^2_{df = k-1} = \sum_1^{2k} \frac{(f_o - f_e)^2}{f_e}$$

3. This analysis is intentionally simplistic for pedagogical reasons. In actuality, we would want to rule out any other possible causes for the significant difference in occupancy rates (e.g., school district effects).

Table 8.6	Layout of 2 × 3 Contingency Table			
		Sample		
	Sample 1	Sample 2	Sample 3	Total
Successes	X_1	X_2	X_3	X
Failures	$n_1 - X_1$	$n_2 - X_2$	$n_3 - X_3$	$n - X$
Total	n_1	n_2	n_3	n

The 2 × 3 contingency table ($k = 3$) in **Table 8.6** shows how to set up a more-than-two-sample test of proportions. The expected value for each cell, assuming the null hypothesis is true, is computed in the same manner as the 2 × 2 contingency table shown earlier. For any of the k samples the expected proportion of successes is

$$\mathbf{f_e\ (success) = n_k \frac{X}{n}}$$

and

$$\mathbf{f_e\ (failure) = n_k - n_k \frac{X}{n}}$$

Example Problem

One observable market trend for single-family residential property in a mid-sized southwestern U.S. city is the increase of floor area over time with little or no change in lot area. This trend should be reflected in changes over time in the proportion of multistory homes on a constant lot size. (Multistory homes offer more square footage in small-lot, tract home developments.) Random samples of homes built in the 1970s are compared to developments with similar lot sizes built in the 1980s and 1990s to test this theory at $\alpha = 0.05$.

The null hypothesis is

H_0: $p_{1970s} = p_{1980s} = p_{1990s}$

and the alternative research hypothesis is

H_a: Not all three proportions are equal

Table 8.7	Single-Story and Multistory by Decade			
		Sample		
	1970s Sample	1980s Sample	1990s Sample	Total
Single-story	80	60	20	160
Multistory	30	60	80	170
Total	110	120	100	330

Results of the samples from the three decades are shown in **Table 8.7**. Observed and expected frequencies and the resulting χ^2 statistic calculation are shown in **Table 8.8**, where $X/n = 160/330 =$

0.485. Notice that none of the expected frequencies are less than 5, so we can apply the χ^2 test.

Degrees of freedom for this test are $k - 1 = 3 - 1 = 2$. Therefore, we enter "=CHIDIST(58.478, 2)" in Excel to find the p-value for the χ^2 test. The p-value is 0.000 since the χ^2 test value of 58.478 far exceeds the 5.99 χ^2 critical value required for 95% confidence.[4] We reject the null hypothesis and conclude that the proportion of single-story tract homes in new developments has not been constant over the three decades being analyzed.

Table 8.8 Chi-Square Calculations for Single-Story Proportion by Decade Analysis

Cell	f_o	f_e	$(f_o - f_e)$	$(f_o - f_e)^2$	$\frac{(f_o - f_e)^2}{f_e}$
1	80	53.33	26.67	711.11	13.333
2	30	56.67	-26.67	711.11	12.549
3	60	58.18	1.82	3.31	0.057
4	60	61.82	-1.82	3.31	0.053
5	20	48.48	-28.48	811.39	16.735
6	80	51.52	28.48	811.39	15.750
					$\chi^2 = \Sigma = 58.478$

This result seems self-evident given that the single-story proportion was approximately 73% in the 1970s compared with 50% in the 1980s and 20% in the 1990s. Even so, the statistical analysis adds value because it confirms that the difference in proportions is so extreme that the probability of this being a chance occurrence is essentially nil (based on the p-value of 0.000). **Figure 8.1** reinforces this conclusion by use of a graphic image.

Figure 8.1 Single-Story and Multistory Development by Decade

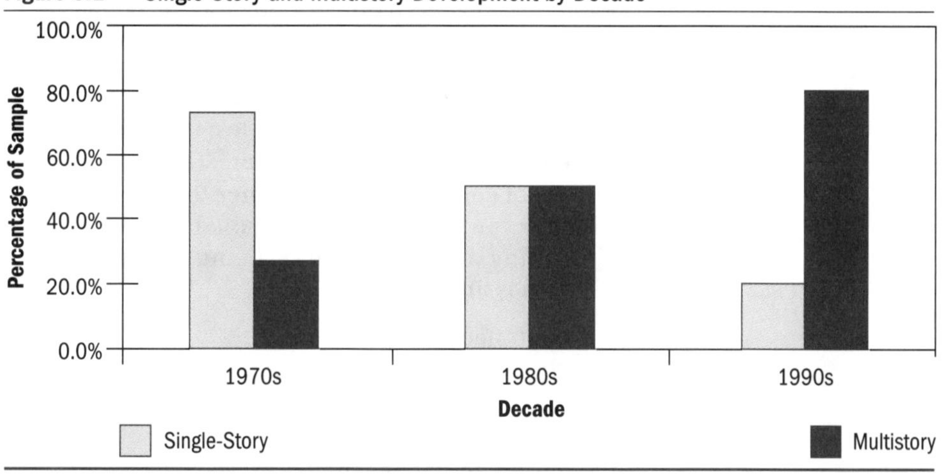

4. The critical value for this test, at 95% confidence, is 5.99 computed in Excel as =CHIINV(0.05,2).

Practice Problems 8.1

1. Use Excel to calculate *p*-values for the following χ^2 test statistics:

   $\chi^2 = 2.60$, degrees of freedom = 1

   $\chi^2 = 4.72$, degrees of freedom = 2

   $\chi^2 = 12.3$, degrees of freedom = 4

2. Calculate the χ^2 statistic and *p*-value for the following 2 × 2 contingency table.

	Sample		
	Sample 1	**Sample 2**	**Total**
Occupied	14	21	35
Unoccupied	8	14	22
Total	22	35	57

3. Three real estate schools—School A, School B, and School C—operate in a large city. Random samples of 100 state license exam records for each school show the following success counts (success = a passing score): School A = 86, School B = 90, School C = 92. Conduct a χ^2 test to determine if the differing success rates are attributable to school choice or are likely to be indicative of random variation in license exam pass rates.

Two-Sample Test of Central Tendency: Independent Samples

The Wilcoxon sum of ranks test and the Mann-Whitney U statistic represent nonparametric corollaries to the independent samples *t*-tests presented previously in Chapter 7. These two tests provide similar results. As a practical matter, the choice of one over the other is a matter of personal preference.

Wilcoxon Sum of Ranks Test

As is typical of nonparametric tests of central tendency, this is a test of medians (M) as representations of central tendency rather than means (μ). The test can be set up as one-tailed or two-tailed, depending on the hypothesis being examined. Three possibilities exist:

1. One-tailed (left-tailed)

   $H_0: M_1 \geq M_2$

   $H_a: M_1 < M_2$

2. One-tailed (right-tailed)

   $H_0: M_1 \leq M_2$

   $H_a: M_1 > M_2$

3. Two-tailed

   H_0: $M_1 = M_2$

   H_a: $M_1 \neq M_2$

We begin by recognizing that the sum of n consecutive positive integers starting at the number 1 is equal to $[n(n + 1)]/2$. For example, the sum of 1, 2, 3, 4, and 5 is 15, which is equal to

$$\frac{5(5 + 1)}{2} = \frac{30}{2} = 15$$

If two independent samples are pooled and ranked, the sum of the ranks for the combined sample is equal to $[n(n + 1)]/2$. In addition the total sum of the ranks of one of the samples (T_1) plus the total sum of the ranks of the other sample (T_2) adds up to the sum of the ranks of the combined sample.[5] That is,

$$T_1 + T_2 = \frac{n(n + 1)}{2}$$

Traditionally, as presented in critical value tables, the smaller sized sample is referred to as Sample 1 and the larger sample is referred to as Sample 2. (The critical value table for this test in Appendix F conforms to this tradition.)

Logically, when the central tendencies of the two samples are equal, T_1 and T_2 are expected to be equal. If the central tendency for Sample 1 is less than the central tendency for Sample 2, then T_1 is expected to be less than T_2. If the central tendency for Sample 1 is greater than the central tendency for Sample 2, then T_1 is expected to be greater than T_2. Statistical measures have been developed to assess when the relative value of T_1 is small enough or large enough to be significant.

Example Problem

Solar tint on exterior windows is promoted as a means to control air-conditioning cost by reducing the interior heating effect of solar radiation. The

5. The ranking rule for ties is to assign each tied observation the average of the ranks that would have been assigned if there were no ties.

independent, random samples of electricity costs in **Table 8.9** were obtained during July for houses with similar floor area, location, and quality rating. The samples consist of tinted-window (Sample 1) and untinted-window (Sample 2) categories.

These data can be used to test the research hypothesis that electricity cost is lower when windows have been treated with solar tint. The appropriate null and alternative hypotheses are

H_0: $M_1 \geq M_2$

and

H_a: $M_1 < M_2$

Table 8.10 shows the sum of ranks calculations: $T_1 = 35$, which is considerably smaller than the expected value of 68, assuming the population medians are equal as stated in the null hypothesis.[6]

Figure 8.2 shows part of the Wilcoxon sum of ranks critical values table from Appendix F. It highlights the lower critical values for T_1 applicable to this example problem based on $n_1 = 7$ and $n_2 = 9$. The highlighted row shows the lower, one-tailed critical values for T_1 of 35 at $\alpha = 0.005$, 37 at $\alpha = 0.01$, 40 at $\alpha = 0.025$, 43 at $\alpha = 0.05$, and 46 at $\alpha = 0.10$. Therefore, this test's T_1 value of 35 indicates that we can be

Table 8.9 Monthly Electric Bills for Homes with Tinted and Untinted Windows

Sample 1 Tinted	Sample 2 Untinted
$175	$170
$165	$190
$170	$215
$185	$240
$155	$215
$145	$210
$200	$175
	$220
	$255

Sample 1 median = $170
Sample 2 median = $215

Table 8.10 Wilcoxon T_1 and T_2 Calculations

Sorted Electricity Cost	Cost Rank	Sample 1 Rank	Sample 2 Rank
145	1	1	
155	2	2	
165	3	3	
170	4.5	4.5	
170	4.5		4.5
175	6.5	6.5	
175	6.5		6.5
185	8	8	
190	9		9
200	10	10	
210	11		11
215	12.5		12.5
215	12.5		12.5
220	14		14
240	15		15
255	16		16
		$T_1 = 35$	$T_2 = 101$

6. The expected value of 68 is based on the sum of the ranks for 16 observations being 136 and an expectation of both T_1 and T_2 being equal to 68 if the samples were from identical populations.

Figure 8.2 Part of the Wilcoxon Sum of Ranks Critical Value Table

Wilcoxon Sum of Ranks Test

Lower Critical Values of T_1

	1-tailed α	0.005	0.01	0.025	0.05	0.1
	2-tailed α	0.01	0.02	0.05	0.1	0.2
n_1	n_2					
6	10	27	29	32	35	38
6	11	28	30	34	37	40
6	12	30	32	35	38	42
7	7	32	34	36	39	41
7	8	34	35	38	41	44
7	9	35	37	40	43	46
7	10	37	39	42	45	49
7	11	38	40	44	47	51
7	12	40	42	46	49	54
8	8	43	45	49	51	55
8	9	45	47	51	54	58
8	10	47	49	53	56	60

highly confident that cooling costs are lower in this geographic location when windows have solar tint.

A statistical software package is not necessary for this test. Application of the test by setting up a spreadsheet similar to Table 8.10 is fairly simple, and the critical values table in the appendix is easy to apply. While calculation of T_1 is all that is required to conduct the test, it is good practice to calculate T_2 as well and verify that $T_1 + T_2 = [n(n + 1)]/2$ as a means of checking for ranking errors.

Mann-Whitney U Statistic

The Mann-Whitney U statistic is similar to the Wilcoxon sum of ranks test. Both tests provide the same result and can be used interchangeably, depending on the software or tables available to the analyst. The Mann-Whitney U analysis also begins with ranking the data and calculating T_1. Once this is done, subtract the minimum possible sum of ranks for Sample 1 to calculate a measure of how much the actual sum of ranks exceeds the minimum possible value.

The minimum sum of ranks for the smaller Sample 1 is equal to $[n_1(n_1 + 1)]/2$. Therefore, the Mann-Whitney U statistic is

$$\text{Mann-Whitney U} = T_1 - \frac{n_1(n_1 + 1)}{2}$$

At its extreme minimum value U would be equal to zero, when all of the elements of Sample 1 were ranked lower than all of the elements of Sample 2. U will be small enough or large enough to be significant whenever the two samples are sufficiently different. This idea is best illustrated by example.

Example Problem

Looking again at the window tint data, we expect the electrical cost for the homes having tinted windows to be less than for the homes not having tinted windows. Therefore, relatively few of the Sample 2 observations should be ranked less than the Sample 1 observations.

Table 8.10 contains the sum of ranks for Sample 1, where $T_1 = 35$. Sample 1 contains seven observations, therefore

$$\mathbf{U = T_1 - \frac{n_1(n_1 + 1)}{2} = 35 - \frac{7(7 + 1)}{2} = 35 - 28 = 7}$$

Looking at the critical values for Mann-Whitney U in Appendix E, the critical value when $n_1 = 7$ and $n_2 = 9$ is 13 at a one-tailed α of 0.025. Because $7 < 13$, we can reject the null hypothesis and conclude that cooling costs are significantly lower with solar tinted windows.

This statistic is available in SPSS, along with a p-value derived from conversion of the U statistic to an approximation of the standard normal Z statistic. The p-value is uncorrected for ties. SPSS also includes the sum of ranks, which SPSS calls "Wilcoxon W."[7] As we know, $T_1 = 35$ for these data, conforming to the SPSS Wilcoxon W value. **Figure 8.3** shows the SPSS data entry format and output for this example problem.

Minitab also includes a "Mann-Whitney" test. A closer look at the Minitab output reveals that the actual calculation is the sum of ranks (which is 35) rather than U (which is 7), along with a p-value that is adjusted for ties. Note also that Minitab uses "eta" (Greek symbol η) when it is referring to

7. We use the symbol W for the Wilcoxon signed ranks test statistic later in the chapter. The sum of ranks W reported by SPSS, and included in Figure 8.3, is not the same test statistic.

a median. **Figure 8.4** shows the data entry format for Minitab along with the output of the test on the example data.

Mann-Whitney U Statistic in SPSS and Minitab

To conduct the Mann-Whitney U statistic tests in SPSS, select **Analyze, Nonparametric Tests,** and then **2 Independent Samples.** A window opens in which you can select the data and the Mann-Whitney U statistic.

Selections in Minitab are **Stat, Nonparametrics,** and then **Mann-Whitney.**

Figure 8.3 SPSS Data and Output

SPSS Data Entry Format

Cost	Tinted
175.00	1.00
165.00	1.00
170.00	1.00
185.00	1.00
155.00	1.00
145.00	1.00
200.00	1.00
170.00	0.00
190.00	0.00
215.00	0.00
240.00	0.00
215.00	0.00
210.00	0.00
175.00	0.00
220.00	0.00
255.00	0.00

SPSS Output

Test Statistics [b]

	Cost
Mann-Whitney U	7.000
Wilcoxon W	35.000
Z	-2.599
Asymp. Sig. (2-tailed)	.009
Exact Sig. [2*(1-tailed Sig.)]	.008[a]

a. Not corrected for ties.

b. Grouping Variable: Tinted

Figure 8.4 Minitab Data and Output

Minitab Data Entry Format

Tinted Cost	Untinted Cost
175	170
165	190
170	215
185	240
155	215
145	210
200	175
	220
	255

Minitab Output

```
              N  Median
Tinted Cost   7  170.00
Untinted Cost 9  215.00


W = 35.0
91.0 Percent CI for ETA1-ETA2 is (-65.00,-15.00)
Test of ETA1 = ETA2 vs ETA1 < ETA2 is significant at 0.0055
The test is significant at 0.0054 (adjusted for ties)
```

Practice Problems 8.2

1. Joseph is an appraiser specializing in apartment properties. He has noticed that gross rent multipliers are generally lower for older apartment projects when compared to newer buildings. He attributes this difference, at least in part, to higher expense ratios at older buildings, which are less resource-efficient and more maintenance-intensive and cost more to insure. Joseph devised a protocol for dividing apartments into "old" and "new" categories. Then he collected random samples of expense ratios for apartment projects of similar style and size in each category:

Apartment Project Expense Ratios	
"New"	**"Old"**
0.38	0.42
0.36	0.39
0.41	0.43
0.395	0.40
0.37	0.41
0.41	0.39
	0.42
	0.41

   Calculate the median expense ratio for each class. Using the table in Appendix F or Minitab, apply the sum of ranks test to examine the research hypothesis that the "new" median expense ratio is less than the "old" median expense ratio at 95% confidence. Write your conclusion.

2. Repeat the test in Problem 1 using the Mann-Whitney U statistic and compare the result to the Problem 1 result.

Two-Sample Test of Central Tendency: Paired Samples

As you will recall from Chapter 7, paired samples occur when repeated measures are obtained from the same items or people, or the items or people in one group have been matched to the items or people in another group based on a common characteristic. One common example is testing the skill or knowledge level of each person in a sample prior to exposure to training (treatment) and then testing each person after treatment to estimate the change, if any, in skill or knowledge attributable to training. Another example is comparing item prices in a sample at two different points in time to assess whether prices had changed over the time period.

The parametric paired samples *t*-test in Chapter 7 is used to analyze the paired differences under the assumption that the sampling distribution of the mean difference was normally distributed. Use of the Wilcoxon signed ranks test relaxes the assumption of normality by looking at the ranks of the positive differences. The central ideas are

1. When there has been a positive effect, there will be a large number of material positive differences.
2. When there has been a negative effect, there will be very few material positive differences.
3. When there has been no effect, the number and size of positive and negative differences will be small and offsetting.

The Wilcoxon signed ranks test statistic (W) is calculated as follows:

$$W = \sum_{i=1}^{n+} R_i$$

where $n+$ is the number of positive differences and R_i is the rank for each positive difference. Therefore, the W statistic is affected both by the number of positive differences ($n+$) and their relative magnitude (ranks). Because this is a nonparametric test, the probability distribution of the differences is immaterial.

Example Problem

The data in **Table 8.11** are from a practice problem in Chapter 7, facilitating comparison of the nonparametric result with the parametric result from the prior chapter. The data came from a real estate school that was in the final stages of designing a classroom module dealing with how to read and interpret metes and bounds legal descriptions and had conducted a test of the proposed instruction module's effectiveness. As you will recall from the prior chapter, the test involved random selection of 15 participants from a list of appraisal

Table 8.11 Wilcoxon Signed Rank Calculations

Student	Pre-Trial Score	Post-Trial Score	Difference	R_i	Difference Sign	Positive Ranks
1	74	82	8	7	+	7
2	50	60	10	9	+	9
3	42	40	-2	2	-	
4	68	84	16	12	+	12
5	88	90	2	2	+	2
6	92	90	-2	2	-	
7	66	78	12	10.5	+	10.5
8	100	100	0	*		
9	35	54	19	13	+	13
10	80	84	4	4	+	4
11	78	78	0	*		
12	58	66	8	7	+	7
13	40	52	12	10.5	+	10.5
14	64	72	8	7	+	7
15	38	44	6	5	+	5
						$\Sigma = W = 87$

trainees, administering a pretreatment examination on metes and bounds interpretation, exposure to classroom education including the proposed metes and bounds module (the "treatment"), and a post-treatment examination on metes and bounds interpretation. Pre- and post-treatment examination results are shown in Table 8.11 along with the signs and ranks for the test score differences. The sum of the positive ranks is equal to 87, which is the Wilcoxon signed ranks test statistic value (W).

This table also illustrates some of the rules applicable to the Wilcoxon signed ranks test. First, zero differences are discarded from the analysis. Differences are ranked as if the zero differences did not exist. Second, as with many other nonparametric tests, ties receive the average rank of the positions held in the sorted data. Third, differences are ranked according to their absolute values in this test (negative signs are ignored for ranking). For example, in Table 8.11 an absolute difference of 2 occupies the first, second, and third positions when the unsigned, non-zero differences are sorted in ascending order. Therefore, all three absolute differences of 2 are ranked 2 (the average of 1, 2, and 3). Likewise, a difference of 8 occupies positions 6, 7, and 8 in the ascending ordered array of unsigned, non-zero differences. Therefore, all three differences of 8 receive the average rank of 7. There are two differences of 12 occupying positions 10 and 11. These two differences receive a rank of 10.5.

There are 15 students in the experiment, but two students show zero difference. There are 13 non-zero differences, which becomes the sample size for this signed ranks analysis. Therefore, for the purposes of selecting a critical value, $n = 13$. We can use this information to assess whether or not to reject the one-tailed null hypothesis of median test score differences being less than or equal to zero, given a test statistic $W = 87$.

The statistical hypotheses are

H_0: Median Difference ≤ 0

and

H_a: Median Difference > 0

Figure 8.5 shows a portion of the signed ranks critical values table from Appendix D with the $n =$

Figure 8.5 Part of the Wilcoxon Signed Ranks Critical Values Table

Wilcoxon Signed Ranks Test

Lower and Upper Critical Values Separated by Comma

1-Tailed α	**0.050**	**0.025**	**0.010**	**0.005**
2-Tailed α	**0.100**	**0.050**	**0.020**	**0.010**
n				
12	17,61	13,65	10,68	7,71
13	21,70	17,74	12,79	10,81
14	25,80	21,84	16,89	13,92

n = total number of positive and negative signed ranks, omitting zero differences

13 row highlighted. Our test statistic of $W = 87$ exceeds the upper critical value at all four one-tailed significance levels, leading to rejection of the null hypothesis in support of the alternative hypothesis that test scores did improve.

The Wilcoxon signed ranks table also identifies lower critical values, which are applicable to left-tailed tests (our example test was right-tailed). Additionally, both critical values would be applicable to a two-tailed test where the null hypothesis is that the difference is zero and the alternative hypothesis is that the difference is not equal to zero.

The Wilcoxon Signed Ranks critical value table in Appendix D handles up to twenty non-zero differences. The signed rank test is approximately normally distributed when $n > 20$ and can be converted to a Z statistic for analysis of larger sample critical values using the following equations:

When non-zero differences (n) are greater than 20,

$$Z = \frac{W - \mu}{\sigma}$$

where

$$\mu = \frac{n(n + 1)}{4}$$

and

$$\sigma = \sqrt{\frac{n(n + 1)(2n + 1)}{24}}$$

Figure 8.6 Example Problem Solution Output from SPSS

Ranks

		N	Mean Rank	Sum of Ranks
Post_Test - Pre_Test	Negative Ranks	2[a]	2.00	4.00
	Positive Ranks	11[b]	7.91	87.00
	Ties	2[c]		
	Total	15		

a. Post_Test < Pre_Test

b. Post_Test > Pre_Test

c. Post_Test = Pre_Test

Test Statistics [b]

	Post_Test - Pre_Test
Z	-2.908[a]
Asymp. Sig. (2-tailed)	.004

a. Based on negative ranks.

b. Wilcoxon Signed Ranks Test

keeping in mind that n signifies the total number of non-zero differences in the sample.

This test is not available in Minitab, but it is available in SPSS by selecting **Analyze, Nonparametric tests,** and then **2 Related samples.** The example problem above was solved a second time using SPSS, and the results are displayed in **Figure 8.6.** Notice that SPSS calculates a Z statistic p-value even though the sample size is only 13. The statistical conclusion using SPSS is consistent with the prior analysis.

Practice Problems 8.3

1. A Wilcoxon signed ranks analysis contains 22 non-zero differences and results in $W = 57$. Convert this W statistic to a Z statistic and test the null hypothesis that the median difference is zero against the alternative hypothesis that the median difference is not equal to zero (two-tailed test). Use a 5% significance level.

2. Economic theory and the appraisal principle of substitution suggest that apartment vacancy increases when home purchases become more affordable. Apartment vacancy data were collected on a random sample consisting of nine similar, mid-sized apartment developments in a major U.S. city during early 2003. Vacancy data were collected again on the same apartment complexes in late 2005, during the easy-mortgage-financing bubble. The data are contained in the table below.

Apartment Complex	Late 2005 Vacancy Rate	Early 2003 Vacancy Rate
Property 1	9%	8%
Property 2	7%	7%
Property 3	10%	8%
Property 4	5%	6%
Property 5	9%	7%
Property 6	6%	6%
Property 7	10%	7%
Property 8	6%	6%
Property 9	8%	4%

Can we say with 95% confidence that median vacancy rates were lower in early 2003?

Test of Central Tendency: More Than Two Independent Samples

The parametric test for difference in central tendency for more than two means is the single factor ANOVA analysis we looked at in Chapter 7. A nonparametric corollary is the Kruskal-Wallis *H*-statistic, which tests differences in central tendency for more than two medians based on ranks and does not require an assumption of normality.

Similar to the Wilcoxon sum of ranks test, the Kruskal-Wallis test begins with ranking all of the observations in the data set without regard to which group each item belongs. Ties are again assigned the average of the ranks that the tied observations would occupy if they were not tied. Once the combined ranks have been calculated, rank sums are calculated for each group. If there is no difference in central tendency among the groups, then the sums of ranks for each group should be similar.

Although the calculations of Kruskal-Wallis *H* shown below appear to be complicated, the test is well suited to a spreadsheet application. Once *H* has been calculated, *p*-values are easily obtained because the test statistic approximately follows a chi-square distribution with degrees of freedom being one less than the number of groups.

The calculations begin with the average rank for the combined groups. Let n_c represent the number of observations in the combined groups. Also, assume for simplicity's sake that there are three groups with sample sizes n_1, n_2, and n_3. Therefore,

$$\mathbf{n_c = n_1 + n_2 + n_3}$$

The first calculation required is the average combined rank $\overline{R}_c$, which is the sum of the combined ranks $[n_c(n_c + 1)]/2$ divided by n_c, or

$$\overline{\mathbf{R}}_c = \frac{\mathbf{n_c(n_c + 1)}}{\mathbf{2n_c}}$$

which simplifies to

$$\overline{\mathbf{R}}_c = \frac{\mathbf{(n_c + 1)}}{\mathbf{2}}$$

Next we calculate D, which is the sum of squared deviations between each group's average rank and the combined average rank, weighted by each group's size:

$$D = \Sigma n_i(\overline{R}_i - \overline{R}_c)^2$$

When there are three groups,

$$D = n_1(\overline{R}_1 - \overline{R}_c)^2 + n_2(\overline{R}_2 - \overline{R}_c)^2 + n_3(\overline{R}_3 - \overline{R}_c)^2$$

With these calculations completed, H is easily computed as follows:

$$H = \frac{12D}{n_c(n_c + 1)}$$

which is distributed chi-square with degrees of freedom equal to the number of groups minus one. When there are three groups, H is distributed chi-square with 2 degrees of freedom. In order to reliably conduct this test, there should be at least five observations in each group.

Example Problem

Table 8.12 presents three random samples each of size $n = 12$ of vacancy rates for Class A, Class B, and Class C office space in a major U.S. city (the same data we analyzed in Chapter 7). The Kruskal-Wallis statistic can be used to assess the research hypothesis that median vacancy rate does vary by class–i.e., Class A, Class B, and Class C office space. The Kruskal-Wallis statistic facilitates this assessment by testing the null hypothesis that all three medians are equal against the alternative hypothesis that not all are equal.

Table 8.12 Office Vacancy Rates by Building Class

Class A Vacancy (%)	Class B Vacancy (%)	Class C Vacancy (%)
12.5	10.4	7.8
9.4	9.6	8.5
9.8	11.1	7.5
14.3	11.5	6.5
8.9	9.3	9.0
13.2	13.6	5.8
12.3	11.9	9.8
8.1	9.5	6.5
11.5	10.4	8.9
12.9	12.0	9.2
10.2	8.4	5.7
13.1	12.8	7.8

$$\overline{R}_c = \frac{n_c + 1}{2} = \frac{36 + 1}{2} = 18.5$$

The average rank by class is the sum of ranks in a group from

Table 8.13 divided by group sample size.

$$\overline{R}_A = \frac{295.5}{12} = 24.625$$

$$\overline{R}_B = \frac{274.5}{12} = 22.875$$

$$\overline{R}_C = \frac{96}{12} = 8.0$$

The group size-weighted sum of squared deviations is

$$D = 12(24.625 - 18.5)^2 + 12(22.875 - 18.5)^2 + 12(8 - 18.5)^2$$

$$= 12(6.125)^2 + 12(4.375)^2 + 12(-10.5)^2$$

$$= 450.1875 + 229.6875 + 1{,}323.0$$

$$= 2{,}002.875$$

Finally,

$$H = \frac{12D}{n_c(n_c + 1)} = \frac{12(2{,}002.875)}{36(36 + 1)} = \frac{24{,}034.5}{1{,}332} = 18.04$$

Using "=CHIDIST" in Excel, the *p*-value of chi-square equal to 18.04 with 2 degrees of freedom is 0.0001. Therefore, we reject the null hypothesis that the median vacancy rates are equal. This is the same conclusion we reached using the ANOVA procedure in Chapter 7. However, this alternative result

Table 8.13 Sum of Ranks by Office Building Class

Class A Vacancy (%)	Rank	Class B Vacancy (%)	Rank	Class C Vacancy (%)	Rank
12.5	30	10.4	22.5	7.8	6.5
9.4	16	9.6	18	8.5	10
9.8	19.5	11.1	24	7.5	5
14.3	36	11.5	25.5	6.5	3.5
8.9	11.5	9.3	15	9.0	13
13.2	34	13.6	35	5.8	2
12.3	29	11.9	27	9.8	19.5
8.1	8	9.5	17	6.5	3.5
11.5	25.5	10.4	22.5	8.9	11.5
12.9	32	12.0	28	9.2	14
10.2	21	8.4	9	5.7	1
13.1	33	12.8	31	7.8	6.5
	Σ = 295.5		Σ = 274.5		Σ = 96

Kruskal-Wallis in SPSS and Minitab

The Kruskal-Wallis test is available in SPSS by selecting **Analyze, Nonparametric Tests,** and then **K Independent Samples.** To replicate the office vacancy example problem in SPSS, enter all of the vacancy data as one variable and enter another numerical variable identifying building class. For example, Class A = 1, Class B = 2, and Class C = 3. If you experiment with SPSS using this data, the SPSS output for the preceding example should look like the table below.

SPSS Output for the Kruskal-Wallis Example Problem

Ranks

	1=A, 2=B, 3=C	N	Mean Rank
Vacancy	1	12	24.63
	2	12	22.88
	3	12	8.00
	Total	36	

Test Statistics [a,b]

	Vacancy
Chi-Square	18.058
df	2
Asymp. Sig.	.000

a. Kruskal Wallis Test

b. Grouping Variable: 1=A, 2=B, 3=C

Minitab also includes the Kruskal-Wallis *H* statistic, results of which are shown below. Data are entered into Minitab in the same manner as they are entered in SPSS. To run this test, select **Stat, Nonparametrics,** and then **Kruskal-Wallis.**

The SPSS chi-square statistic and one of the two Minitab chi-square statistics are adjusted for ties. Minitab also reports chi-square unadjusted for ties, which is the chi-square = 18.04 result we obtained earlier using the equation for *H*.*

Minitab Output for the Kruskal-Wallis Example Problem

```
Kruskal-Wallis Test on Vacancy
Class      N  Median  Ave Rank
1 (A)     12  11.900      24.6
2 (B)     12  10.750      22.9
3 (C)     12   7.800       8.0
Overall 36                18.5
H = 18.04 DF = 2 P = 0.000
H = 18.06  DF = 2  P = 0.000(adjusted for ties)
```

* Divide H by $1 - (\Sigma[(\tau_i - 1)\tau_i(\tau_i + 1)]/[n_c(n_c^2 - 1)])$ to correct for ties where τ_i is the number of ties in the *i*th set of ties. This correction has very little effect when there are only a few ties. The correction for ties shown here is from S. A. Glantz, *Primer of Bio-Statistics*, 3rd ed. (New York: McGraw-Hill, 1992).

did not rely on an assumption of normality. Nevertheless, when sample size is sufficiently large or the population is known to be approximately normal, ANOVA is preferred because it is more powerful. That is, it does a better job of detecting subtle, but significant, differences in central tendency.

Practice Problem 8.4

Residential subdivisions often include three classes of lots:

- Lots abutting relatively busy arterial streets
- Standard lots
- Lots abutting or viewing an amenity feature such as green space or a water feature

The data below are representative samples of sales of similar-sized residential lots abutting busy streets, standard lots, and lots abutting green space from a large tract home subdivision. Test the null hypothesis at 90% confidence that median price per square foot is the same for all three lot categories.

Price per Square Foot by Lot Category		
Standard Lot	**Abutting Busy Street**	**Abutting Green Space**
$2.50	$2.00	$2.75
$2.70	$2.35	$2.60
$2.35	$2.40	$2.65
$2.45	$2.35	$2.50
$2.40	$2.20	$2.45
$2.55	$2.25	$2.55
$2.60		
$2.50		

Nonparametric Tests for Normality

Several statistical goodness-of-fit tests are available for assessing normality, which augment the subjective criteria discussed in Chapter 5. Minitab includes three tests–Anderson-Darling, Ryan-Joiner, and Kolmogorov-Smirnov. The Anderson-Darling test is said to be especially useful for small samples. The following example demonstrates application of the Anderson-Darling test using Minitab.

Example Problem

For the purposes of this example only, vacancy rates from Table 8.12 were combined into one data set with 36 observations. Descriptive statistics for the combined vacancy rates were computed in Minitab by selecting **Stat, Basic Statistics**, and then **Display Descriptive Statistics**. The Minitab output is shown in **Figure 8.7**.

The information in Figure 8.7 can be used to test for normality by assessing the distributive criteria from Chapter 5. Another option is to apply the nonparametric normality tests available in Minitab. Each test is accompanied with a normal probability plot. **Figure 8.8** shows the Anderson-Darling test output. A good learning exercise would be to apply the distributive criteria from Chapter 5 to the information in Figure 8.7 and compare the results to the Anderson-Darling test in Figure 8.8.

Figure 8.7 Descriptive Statistics for Combined Vacancy Data

Variable	Total Count	Mean	SE Mean	StDev	Minimum	Q1	Median	Q3	Maximum
Vacancy	36	9.992	0.381	2.288	5.700	8.425	9.700	11.975	14.300

Figure 8.8 Minitab Tests for Normality: Combined Vacancy Data

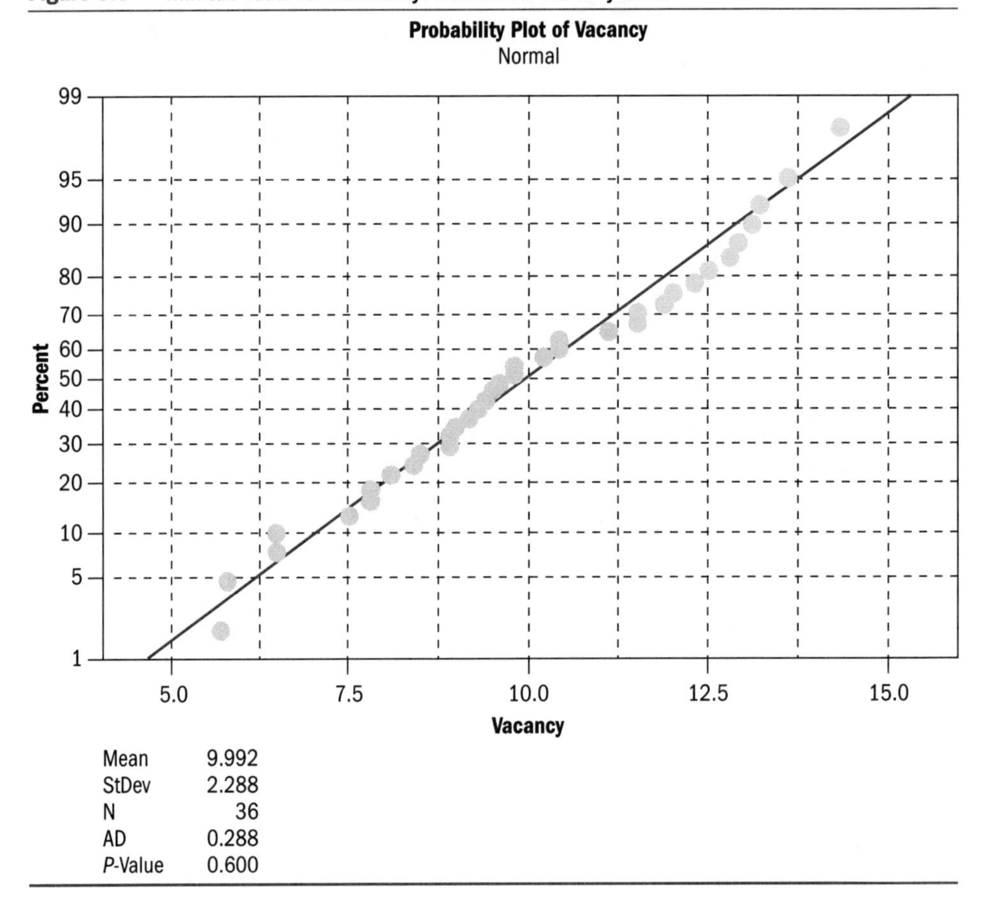

The p-value of 0.60 for the Anderson-Darling test value of 0.288 indicates that we cannot reject the null hypothesis that the combined vacancy data is normally distributed. Similar results are obtained in Minitab with the Ryan-Joiner test value of 0.992 with a reported p-value of > 0.10 and the Kolmogorov-Smirnov test value of 0.076 and a reported p-value of > 0.15.

Nonparametric Tests for Normality in SPSS and Minitab

To test for normality in SPSS, use the Kolmogorov-Smirnov test by selecting **Analyze, Nonparametric Tests,** and then **1-Sample K-S.** Select the **Normal Test Distribution** when the **K-S Test** window opens. In Minitab, select **Stat, Basic Statistics,** and then **Normality Test.**

9 Simple Linear Regression Analysis

> This chapter contains quite a bit of algebra. Take your time with each section, making sure you understand the content of that section before proceeding. Nearly everything we do in this chapter can be done with a calculator and either a *t* statistic table or Excel. The arithmetic is kept as simple as possible.

Simple linear regression is a statistical procedure designed to examine the relationship between two variables. It examines how a dependent variable y is related to an independent variable x. When the relationship between the explanatory variable x and the response variable y is linear, it can be expressed algebraically as

$$\mathbf{y = a + bx}$$

In this equation, when $x = 0$, $y = a$, and the value of the equation's constant a is called the *y-intercept*. When $x = 1$, $y = a + b$, and a 1-unit increase in x results in y increasing by an amount equal to b, which is referred to as the *slope* of the linear equation $y = a + bx$.

The following equation provides a simple numerical example:

$$\mathbf{y = 1 + 2x}$$

We can create a table based on this linear relationship by calculating the value of y for each value of x in the table. **Table 9.1** shows six x values and the corresponding y values computed as $1 + 2x$.

The relationship illustrated in the table is *deterministic* because the value of the variable y can be calculated with certainty when the value of x and the relationship between x and y are known.

Table 9.1 $y = 1 + 2x$

x values	0	1	2	3	4	5
y values	1	3	5	7	9	11

Figure 9.1 Scatter Plot and Line Representing the Deterministic Relationship $y = 1 + 2x$

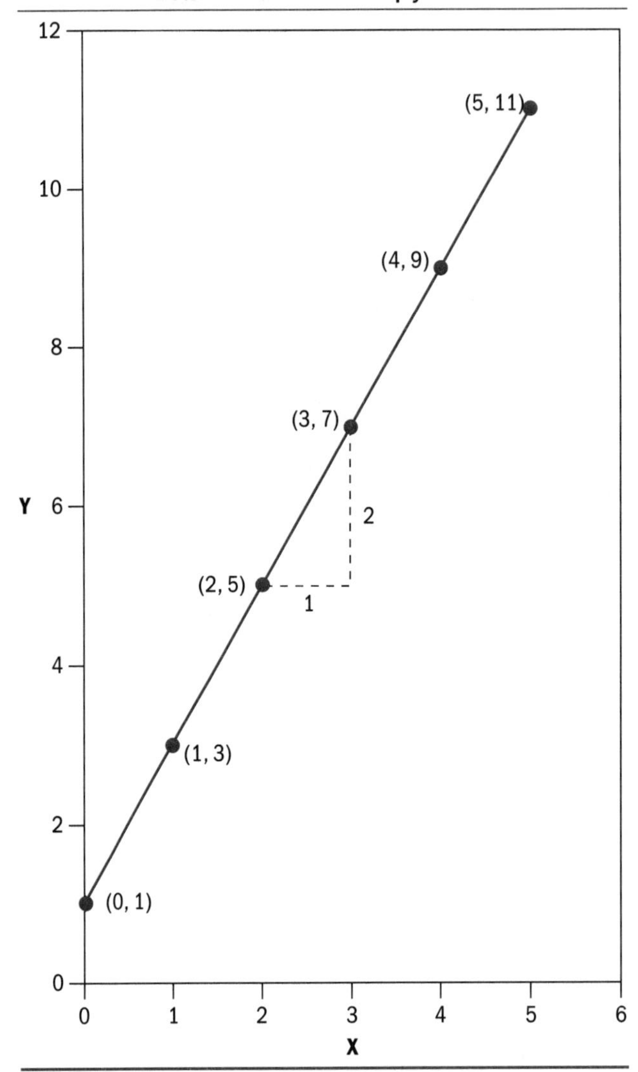

This deterministic relationship is charted in **Figure 9.1**, which shows a scatter plot of each x, y pair from Table 9.1 along with a line that represents the relationship $y = 1 + 2x$. Notice that when x increases horizontally from a value of 2 to a value of 3, there is a corresponding vertical increase in y from a value of 5 to a value of 7. The difference between 7 and 5 of 2 is the change in y given a 1 unit change in x, so 2 is the *slope* of this linear equation (and of the line in Figure 9.1). Notice also that the line crosses the vertical y-axis at 1 (where $x = 0$), so 1 is the y-intercept for this linear equation.

Deterministic relationships are often encountered in the physical sciences and applied mathematics (e.g., chemistry, physics, land surveying, mechanical drawing, engineering). However, deterministic relationships are rarely encountered in economics, business environments, and the social sciences because relationships are not likely to be deterministic when human behavior colors the association between an explanatory variable and a response variable.

Table 9.2 is a slight modification of Table 9.1. The relationship between y and x is no longer deterministic because there is no linear equation that exactly accounts for the value of y given the value of x. This sort of uncertain relationship is characterized by unexplained variance in the response variable and is much more indicative of the kinds of relationships between independent and dependent variables found in the social sciences, where some of the variance in a response variable is likely to be random. (The field of economics uses the word *stochastic* to describe random processes, as shown in the title of Table 9.2.)

Table 9.2 Stochastic Relationship between y and x

x values	0	1	2	3	4	5
y values	1	3.5	5	6.5	9.6	10.4

Figure 9.2 Scatter Plot and Best-Fitting Line for an Uncertain Relationship

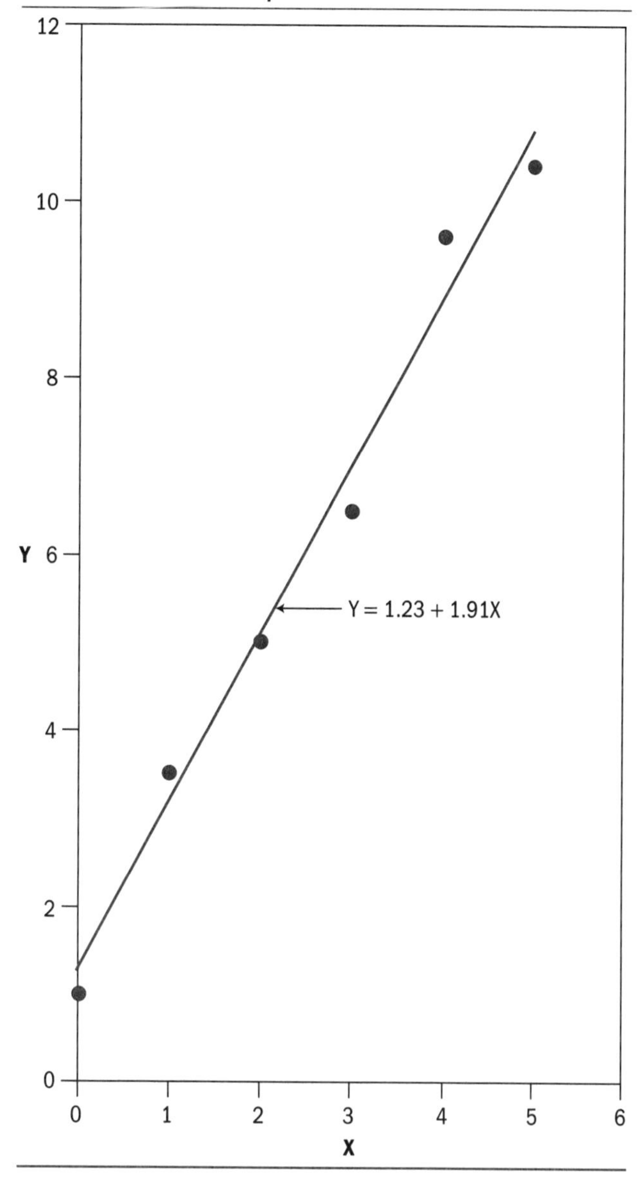

The uncertain nature of this relationship is captured by the chart in **Figure 9.2**. The relationship between y and x is best captured by the line $y = 1.23 + 1.91x$, but as Figure 9.2 shows the line is not a perfect fit to the data. (We will explore how we derived this equation for this line later in the chapter.) According to this equation the expected value of y is 3.14 when x is equal to 1, but the actual value of y in Table 9.2 is 3.5 when x equals 1. Differences between the expected values of y based on the best-fitting linear equation and the actual values of y are called *estimation error*. Estimation error (e) when $x = 1$ is the difference between the actual value of y of 3.5 and the expected (estimated) value of y of 3.14. Therefore, $e_{x=1} = 3.5 - 3.14 = 0.36$.

The line in Figure 9.2 could have been drawn by hand. While it is possible to draw a line between the data points using a ruler that looks like a good fit, someone else might draw a different line that also looks like a good fit, while neither line is in fact the "best" fit. In contrast, the line determined by a linear regression equation (in this case, $y = 1.23 + 1.91x$) is demonstrably the best-fitting line.

Defining and estimating the best fit are what linear regression analysis is all about. The "best" fit can be defined in several ways. One possible definition is the line that minimizes total error. Unfortunately an infinite number of linear equations satisfy this criterion. Positive and negative estimation errors will offset each other when summed, so any line drawn through the point defined by the mean of x and the mean of y will have total error equal to zero.

Another possible definition of the "best" fit is the line that minimizes the absolute value of total error. While this criterion accounts for the offsetting effects of positive and negative error, it may not be the optimal definition of "best" fit because deriving the equation for such a line requires a complex linear programming solution that may not provide a unique equation and is not generally included in statistical software packages.

The third possible definition of "best" fit is the equation that minimizes total squared error. Squaring the error terms converts all the errors into positive values, overcoming the problem of offsetting positive and negative errors. In addition, squaring the error terms enables derivation of a unique answer that is fairly simple to compute. Furthermore, as a practical consideration, "least squares" solutions are included in statistical software packages. This sort of solution is referred to as *ordinary least squares regression*. The ordinary least squares solution derives a unique linear equation that satisfies the following condition:

$$\min[\sum_{i=1}^{n}(y_i - \hat{y}_i)^2]$$

where y_i represents the actual value of y_i in a sample and $\hat{y}_i$ is the expected value of y_i based on the regression equation. In short, the "best-fitting equation" is the equation that minimizes the sum of squared errors (*SSE*).

Deriving the Least Squares Regression Equation

The values of *a* and *b* that minimize the sum of squared errors in a linear equation of the form

$$y_i = a + bx_i + e_i$$

can be derived algebraically as follows:

$$b = \frac{\Sigma x_i y_i - \frac{1}{n}\Sigma x_i \, \Sigma y_i}{\Sigma x_i^2 - \frac{1}{n}(\Sigma x_i)^2}$$

and

$$a = \bar{y} - b\bar{x}$$

Example Problem

Consider the data in **Table 9.3**, which transposes and expands the information from Table 9.2. These data illustrate how easy and straightforward it can be to fit an ordinary least squares equation using simple algebra. Six *x,y* data pairs are included, along with the summations needed to compute *a* and *b* algebraically.

The least squares equation (rounded to 2 decimal places) for the Table 9.3 data is

$$y = 1.23 + 1.91x$$

If you refer back to Figure 9.2, you will see that this is the equa-

Table 9.3 Computations Required to Compute Least Squares Coefficients *a* and *b*

i	x_i	y_i	$x_i y_i$	x_i^2
1	0	1	0	0
2	1	3.5	3.5	1
3	2	5	10	4
4	3	6.5	19.5	9
5	4	9.6	38.4	16
6	5	10.4	52	25
Totals	15	36	123.4	55

$$b = \frac{\Sigma x_i y_i - (1/n)\,\Sigma x_i \, \Sigma y_i}{\Sigma x_1^2 - (1/n)(\Sigma x_i)^2} = \frac{123.4 - (1/6)(15)(36)}{55 - (1/6)(15)^2} = \frac{33.4}{17.5} = 1.90857$$

$$a = \bar{y} - b\bar{x} = 36/6 - 1.90857 \times (15/6) = 6 - 4.77143 = 1.22857$$

tion that corresponds to the best-fitting line. Also, as this example shows, it is good practice to carry several decimal places when calculating a and b, rounding to the number of significant digits appropriate to the analysis only at the end of the calculations.

Practice Problem 9.1

Use the *x*, *y* pairs below to calculate a linear regression equation. Construct a scatter plot and a line showing the linear regression equation you derived.

i	x_i	y_i
1	2	6
2	4	4
3	6	16
4	8	16
5	10	22
6	12	13

Assumptions Underlying the Regression Model

The elements of a simple regression equation are "statistics." The statistics consist of two numbers, a and b, that describe a linear relationship between two sample variables, x and y. If the sample variables are representative of the population, then (as with any other representative statistic) we can infer the true linear relationship in the population from a regression equation derived from a sample.

As with other statistical measures, assumptions must be made about the underlying population in order to draw inferences based on a sample. The underlying assumptions of linear regression analysis are that

- The expected value of regression errors is zero–that is, $E[e_i] = 0$ for all i.
- The variance of regression error is constant.
- Regression errors are normally distributed.
- Regression errors are independent.

The expectation that regression error for each observation is zero means that the regression line passes through the expected values of y_i given x_i for all values of the dependent variable. This implies that the expected y values are linear[1] in their relationship to the x values and that no significant variables are missing from the regression equation. The next two assumptions are self-explanatory, and when the first three assumptions are true the regression errors should mimic a random draw from a normally distributed population. The requirement that regression errors be independent is most applicable to time series data, where a variable's past value may be related systematically to its present value. Consequently, a scatter plot of regression errors and values of the dependent variable y should look random and normally distributed around a mean of zero. When regression errors (residuals) are normally distributed, about 68% of the errors should be within ±1 standard deviation from the mean. (The regression error mean is zero when a regression model includes an intercept term.) In addition, approximately 95% should be within ±2 standard deviations and 99% within ±3 standard deviations from the regression error mean. Many statistics programs will generate standardized residuals (expressed in standard deviation units), which makes it easier to determine the proportions of errors within these standard deviation ranges.

Tests for Normality in Minitab

Minitab includes a test for normality called a Ryan-Joiner test along with two additional tests for normality—Anderson-Darling and Komolgorov-Smirnov. The test is available as an option when a normal probability plot is generated by the program. This test assesses the "straightness" of the regression error normal probability plot and generates test statistic Minitab labels as *RJ*. The null hypothesis for the Ryan-Joiner test is that the regression errors are normally distributed, so low *p*-values result in rejection of the normality assumption.

As we discovered in Chapter 8 these tests are available in Minitab by selecting **Basic Statistics** and then **Normality Test**. Test for normality in SPSS using the Kolmogorov-Smirnov test by selecting **Analyze, Nonparametric Tests**, and then **1-Sample K-S**. Select the **Normal Test Distribution** when the **K-S Test** window opens.

More about Normality

Normality is assumed when making inferences from small samples. When the sample is large, the normality assumption can be relaxed because the sampling distribution of the mean of the regression

1. We will relax this assumption later by fitting non-linear relationships. This involves transforming the *x* variable, *y* variable, or both variables into a form that is linear and then regressing the transformed variables.

coefficient estimator will be approximately normal, based on the Central Limit Theorem. For simple linear regression, a sample of size 30 is considered sufficiently "large." (Chapter 10 will introduce multiple linear regression, which accommodates additional explanatory variables. Larger samples are often required to assure normality when additional explanatory variables are included.[2])

Small departures from normality are not serious concerns.[3] However, large departures are a concern because the credibility of the inference depends on the underlying normality assumption. Departures from normality are difficult to detect because omission of an important variable, misspecification of the regression equation, and nonconstant error variance all affect the appearance of regression errors. Therefore, it is usually best to assess normality after having addressed concerns about the other underlying assumptions.

The most often suggested measure to remediate non-normal regression error, and also to remediate non-constant regression error, is the transformation of the dependent variable Y. Often a simple log transformation is sufficient, where the natural log of Y is regressed on X. Another procedure, suggested by Box and Cox,[4] is to consider a family of transformations on Y, including Y^2, $\sqrt{Y}$, $\ln(Y)$, $1/\sqrt{Y}$, and $1/Y$. This involves running a number of regression models and choosing the transformation resulting in the smallest SSE.

Assumptions and Inference

Given the validity of the four assumptions underlying regression modeling, inferences can be made about the population slope (β), intercept (α), and error (ε) from linear regression solutions for

2. See Terry Dielman, *Applied Regression: Analysis for Business and Economics*, 3rd ed. (Pacific Grove, Calif.: Duxbury/Thomson Learning, 2001), 339. Dielman notes that a sample of size 30, plus 10 to 20 more observations for each additional explanatory variable, is commonly suggested to assure normality.
3. See John Neter, William Wasserman, and Michael H. Kutner, *Applied Linear Statistical Models*, 3rd ed. (Homewood, Ill.: Irwin, 1990), 124.
4. George E. P. Box and D. R. Cox, "An Analysis of Transformations," *Journal of the Royal Statistical Society, Series B (Methodological)* 26, no. 2 (1964): 211-243.

sample slope (b), intercept (a), and error (e). Stated algebraically, we use regression analysis to infer a population relationship of the form

$$\mathbf{y}_i = \alpha + \beta \mathbf{x}_i + \varepsilon_i$$

from a sample relationship of the form

$$\mathbf{y}_i = \mathbf{a} + \mathbf{b}\mathbf{x}_i + \mathbf{e}_i$$

Population parameters α, β, and ε are inferred by solving for sample statistics a, b, and e. This is the exercise we just performed in the previous section of the chapter.

Regression and the Market Value Concept

Consider a hypothetical residential housing population and the relationship between the dependent variable "price" and the independent variable "floor area." Assume for the purposes of this illustration that floor area is the only variable that influences price (i.e., population price is not influenced by any other property characteristic). Within this hypothetical population there exists an underlying price distribution for each floor area. The mean of this underlying price distribution (μ_p) is the expected price given the associated floor area. In appraisal practice, and in linear regression analysis, we make inferences about μ_p by developing an understanding of the relationship between price and floor area. In general, regression models can be used to estimate mean price (most probable price) by studying the underlying relationship between price and relevant elements of comparison. Since market value definitions generally include the phrase *most probable price*, well-specified regression models with price (or rent) as dependent variables are consistent with the appraisal concept of market value (or market rent).

A common misconception is that linear regression applies a stochastic model (i.e., a model accounting for random error) to a data sample to uncover an underlying deterministic population relationship. Actually, linear regression models allow us to make inferences about underlying stochastic relationships in the population because randomness occurs in the populations social scientists study.

Inferences Concerning Population Intercept (α) and Slope (β)

Regression intercepts and slopes a and b are statistics and random variables. (The values computed for a and b will vary with different samples drawn from the same underlying population.) As with any random variable, sampling distributions for a and b are characterized by means and standard deviations. Knowing the sampling distributions for coefficients a and b allows us to make inferences concerning the true population variables α and β.

Sampling Distribution of the Intercept Estimate *a*

Given certain information about a sample distribution, we can make a set of inferences about the population:

1. a is normally distributed or the sample size is large.
2. The expected value of a is the population intercept α.
3. The standard error of a (S_a), which is an estimate of the standard deviation of α, is calculated as

$$S_a = S_e\sqrt{\frac{1}{n} + \frac{\bar{x}^2}{(n-1)S_x^2}}$$

where S_e is the regression standard error (an estimate of the standard deviation of regression error) and the other variables (sample size, sample mean of x, and sample variance of x) are consistent with how they have been defined throughout this book. Regression standard error is calculated as

$$S_e = \sqrt{\frac{\Sigma(y_i - \hat{y}_i)^2}{n-2}} = \sqrt{\frac{SSE}{n-2}}$$

As you will recall $\Sigma(y_i - \hat{y}_i)^2$ is the sum of squared error (*SSE*), which was minimized when calculating the regression coefficients. *SSE* is divided by $n - 2$ because this provides a better estimate of error in the linear relationship found in the underlying population than *SSE* divided by sample size *n*. *SSE* for the data in Table 9.3 is 1.07371, as shown in **Table 9.4**. Therefore,

Table 9.4 *SSE* for Example Problem

i	y_i	Predicted y_i	e_i	e_i^2
1	1.00	1.22857	-0.22857	0.05224
2	3.50	3.13714	0.36286	0.13167
3	5.00	5.04571	-0.04571	0.00209
4	6.50	6.95429	-0.45429	0.20638
5	9.60	8.86286	0.73714	0.54338
6	10.40	10.77143	-0.37143	0.13796
				SSE = 1.07371

$$S_e = \sqrt{\frac{1.07371}{6-2}} = 0.5181$$

and

$$S_a = 0.5181\sqrt{\frac{1}{6} + \frac{(15/6)^2}{(6-1)(1.870829)^2}}$$

$$= 0.5181\sqrt{\frac{1}{6} + \frac{6.25}{17.5}}$$

$$= 0.37497$$

Sampling Distribution of the Slope Estimate *b*

Like the intercept, the slope estimate for the sample is subject to certain conditions:

1. b is normally distributed or the sample size is large.
2. The expected value of b is the population slope β.
3. The standard error of b (S_b), which is an estimate of the standard deviation of β, is calculated as

$$S_b = S_e\sqrt{\frac{1}{(n-1)S_x^2}}$$

S_b for the example problem is

$$S_b = 0.5181\sqrt{\frac{1}{(6-1)(1.870829)^2}} = 0.12385$$

The names for these calculations vary among software programs. *SSE* is called *Residual SS* in Excel, *Residual Error SS* in Minitab, and *Residual Sum of Squares* in SPSS. *Standard error* is called *standard error* in Excel and SPSS, but it is called *SE Coefficient* and *S* in Minitab, depending on the context. With practice, you will become familiar with the terminology that is specific to the software you choose to use.

Three exhibits are included here to assist in identifying the various names for values we have just calculated as they are presented in Excel, Minitab, and SPSS outputs. Excel output is included in **Figure 9.3.** Similar outputs from Minitab and SPSS are presented in **Figures 9.4** and **9.5.**

SSE in Excel, Minitab, and SPSS

To calculate *SSE* in Excel, select **Tools, Data Analysis,** and then **Regression.** In Minitab select **Stat, Regression,** and then **Regression** again. In SPSS select **Analyze, Regression,** and then **Linear.** Data can be entered into Minitab and SPSS by directly importing it from Excel or it can be typed into any two columns.

As you have probably noticed, the software outputs in Figures 9.3, 9.4, and 9.5 contain additional information we have not discussed. For now concentrate on what we have learned so far. We will get to the rest as we advance through this chapter and the next.

Figure 9.3 Excel Regression Output for Table 9.2 Data

Regression Statistics	
Multiple R	0.99168
R Square	0.98344
Adjusted R Square	0.97929
Standard Error	0.51810
Observations	6.00000

ANOVA

	df	*SS*	*MS*	*F*	*Significance F*
Regression	1	63.74629	63.74629	237.47951	0.00010
Residual	4	1.07371	0.26843		
Total	5	64.82000			

	Coefficients	*Standard Error*	*t Stat*	*P-value*
Intercept	1.22857	0.37497	3.27642	0.03060
x	1.90857	0.12385	15.41037	0.00010

Figure 9.4 Minitab Regression Output for Table 9.2 Data

```
The regression equation is
y = 1.23 + 1.91 x

Predictor    Coef   SE Coef      T      P
Constant   1.2286    0.3750   3.28  0.031
x          1.9086    0.1238  15.41  0.000

S = 0.518101   R-Sq = 98.3%   R-Sq(adj) = 97.9%

Analysis of Variance
Source           DF      SS      MS       F      P
Regression        1  63.746  63.746  237.48  0.000
Residual Error    4   1.074   0.268
Total             5  64.820
```

Figure 9.5 SPSS Regression Output for Table 9.2 Data

Model Summary

Model	R	R Square	Adjusted R Square	Std. Error of the Estimate
1	.992[a]	.983	.979	.51810

a. Predictors: (Constant), x

ANOVA[b]

Model		Sum of Squares	df	Mean Square	F	Sig.
1	Regression	63.746	1	63.746	237.480	.000[a]
	Residual	1.074	4	.268		
	Total	64.820	5			

a. Predictors: (Constant), x

b. Dependent Variable: y

Coefficients[a]

Model		Unstandardized Coefficients		Standardized Coefficients	t	Sig.
		B	Std. Error	Beta		
1	(Constant)	1.229	.375		3.276	.031
	x	1.909	.124	.992	15.410	.000

a. Dependent Variable: y

Hypothesis Tests

Simple linear regression usually involves hypothesis tests concerning β. The most common test hypotheses require a two-tailed test where

$H_0: \beta = 0$

and

$H_a: \beta \neq 0$

The *t* statistic for this test is

$$t = \frac{b - 0}{S_b}$$

with $n - 2$ degrees of freedom. For the Table 9.2 example problem,

$$t = \frac{b - 0}{S_b} = \frac{1.90857 - 0}{0.12385} = 15.41$$

The p-value for this t statistic is 0.0001. To verify this, enter "=TDIST(15.41, 4, 2)" in Excel where 15.41 is the t-test statistic, 4 is the number of degrees of freedom, and 2 signifies a two-tailed test.

In the example problem we reject the null hypothesis that $\beta = 0$, and conclude that $\beta \neq 0$. The practical implication of this conclusion is support for the assertion that knowledge of x_i provides information about the expected value of y_i. Therefore, the simple linear regression model's estimated value of y_i based on a known value of x_i is a better estimate of y_i than reliance on the sample mean $\bar{y}$.

One-tailed tests can also be appropriate when the research hypothesis (H_a) is that $\beta > 0$ or $\beta < 0$. Furthermore, it is sometimes appropriate to test a null hypothesis of β being equal to some value other than zero. When this is the case, the non-zero null hypothesis value is substituted for zero when calculating t. For example, a null hypothesis that $\beta \leq 2$ could be examined to support an assertion that $\beta > 2$ by deriving the test statistic for t as follows:

$$t = \frac{b - 2}{S_b}$$

and comparing the result to a one-tailed critical value for t with $n - 2$ degrees of freedom.

SPSS, Minitab, and Excel report t statistics for a null hypothesis of $\beta = 0$ and provide p-values for two-tailed tests. The reported p-value should be halved for a one-tailed test. When the null hypothesis involves a non-zero value, t statistics and p-values must be calculated by hand. The easiest way to do this is to run the standard software test then use the S_b value from the software output in the equation for t to calculate the t statistic for the non-zero null hypothesis.

Confidence Intervals

Creation of a confidence interval provides another way of making inferences about β. A confidence interval on β is constructed as follows:

$$(1 - \alpha) \times 100\% \text{ Confidence Interval on } \beta = b \pm t_{\alpha/2} S_b$$

where α reflects the significance level selected by the analyst.

Assume that 0.05 is an appropriate significance level for the example problem. Therefore, $t = 2.77645$ would be selected for a 95% confidence interval with 4 degrees of freedom. (Enter "=TINV(0.05,4)" in Excel, which automatically divides 0.05 by 2 to calculate *t*.) The 95% confidence interval for the Table 9.2 data is

$$\begin{aligned} \mathbf{1.9086 \pm 2.77645(0.12385)} &= \mathbf{1.9086 \pm 0.34386} \\ &= \mathbf{1.5647 \text{ to } 2.2525} \end{aligned}$$

When this interval does not contain zero, you can be 95% confident that $\beta \neq 0$. Therefore, a confidence interval can serve as a means to conduct a hypothesis test. The center of the confidence interval is the most probable value of β given the sample data, and the confidence interval provides a statement about the degree of precision associated with the estimate of β provided by the regression analysis of the sample data.

Practice Problem 9.2

Use the following hypothetical information on lot size in acres (x) and price in thousands of dollars (y) to derive a linear regression equation. Using $\alpha = 0.10$, test the null hypothesis that β is ≤ 0, and derive a 90% confidence interval for β.

Price ($000)	90	130	230	220	320
Lot Size (Acres)	1	1.2	2	2.4	3

Prediction Using Simple Linear Regression

Predictions concerning the response variable y based on a given value for the explanatory variable x fall into two categories:

- Predictions concerning the *mean of y* given x
- Predictions concerning a single outcome of y given x

Generally speaking, appraisers are more interested in predicting the *mean of y* given x because definitions of market value and market rent focus on most probable price and most probable rent rather than predicting a particular transaction's price or rent. To the extent possible, predictions should be confined to the range of the x variable. Prediction outside of the range of x involves an assumption that the x, y relationship is the same outside of the x variable range as it is inside the x variable range.

Predicting the Mean of y Given x

Figure 9.6 demonstrates the concept of estimating the mean of y conditioned on the value of x. As the chart illustrates, the expected value of y is the center of the sampling distribution of the mean of y around the regression line at the point where the value of x is x_i. The predicted mean of y given x is calculated as follows:

$$\mu_{y|x_i} = \hat{y}_i = a + bx_i$$

Figure 9.6 Estimating Mean of *y* Given *x*

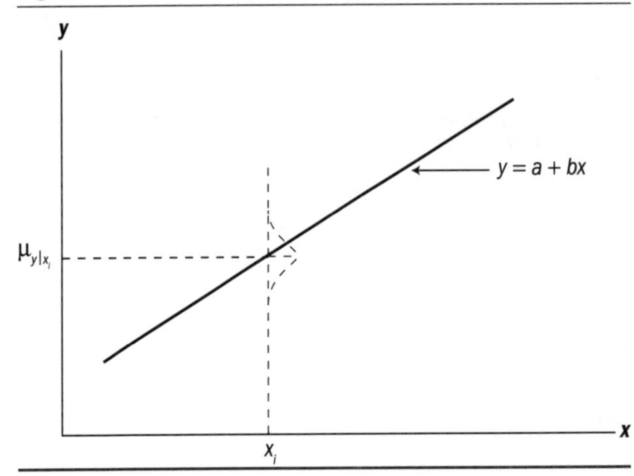

The relationship between x and y is not deterministic. Therefore, there is uncertainty in the estimate of $\mu_{y|x_i}$, as shown in Figure 9.6. This stems from uncertainty concerning the value of α, and the value of β. The standard deviation of the estimated value of $\mu_{y|x_i}$ is called the *standard error of the estimate* of the mean of y given x, or $S_{\mu_{y|x}}$.

The standard error of the estimate of the mean of y given x is calculated as follows:

$$S_{\mu_{y|x}} = S_e\sqrt{\frac{1}{n} + \frac{(x_i - \bar{x})^2}{(n-1)S_x^2}}$$

And, the confidence interval for $\mu_{y|x}$ is

$$\mu_{y|x} \pm t_{\alpha/2} S_{\mu_{y|x}}$$

where t has $n - 2$ degrees of freedom.

Notice from the equation for $S_{\mu_{y|x}}$ that the value of $S_{\mu_{y|x}}$ changes as x changes. The value of $S_{\mu_{y|x}}$ is minimized at $x = \overline{x}$ because $(x - \overline{x})^2 = 0$ when $x = \overline{x}$. Consequently, the confidence interval is the narrowest when $x = \overline{x}$, and becomes wider as $(x - \overline{x})^2$ becomes larger. Varying confidence interval width as x departs from $x - \overline{x}$ is depicted in **Figure 9.7.** As thc exhibit shows, prediction precision deteriorates as values of x deviate from the mean of x.

Figure 9.7 Pattern of Confidence Interval on $\mu_{y|x}$

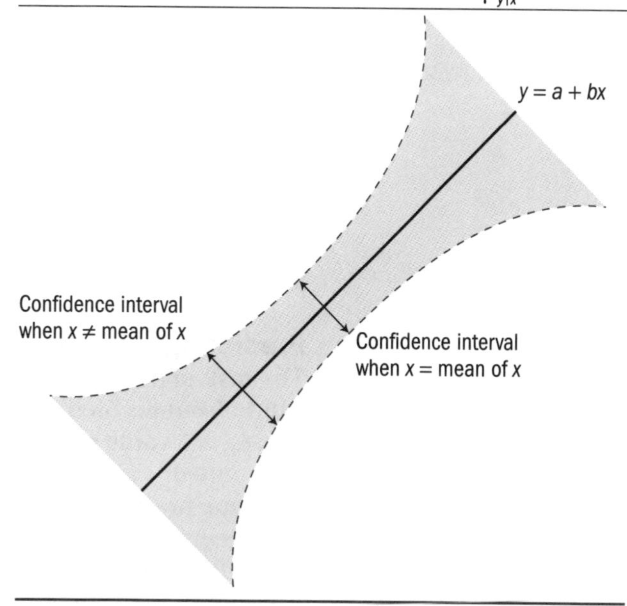

Example Problem

Table 9.5 reiterates the information from Table 9.2 that we have been using as an example throughout the chapter. We will use these data again, this time to derive a 90% confidence interval on $\mu_{y|x=3.5}$.

Based on these data and the previously derived linear regression equation, the predicted mean of y given $x = 3.5$ is

$$\mu_{y|x=3.5} = \hat{y}_{x=3.5} = 1.23 + 1.91 \times 3.5 = 7.92$$

In addition, we previously solved for the regression standard error for this data as

$$S_e = \sqrt{\frac{1.07371}{6-2}} = 0.5181$$

The sample standard deviation of x was

$$S_x = 1.870829$$

and the mean of x was 2.50:

$$\overline{x} = 15 \div 6$$

Table 9.5 Stochastic Relationship between *y* and *x* from Table 9.2

x values	0	1	2	3	4	5
y values	1	3.5	5	6.5	9.6	10.4

Therefore, the standard error of the estimate of the mean of y when $x = 3.5$ is

$$S_{\mu_{y|x=3.5}} = 0.5181\sqrt{\frac{1}{6} + \frac{(3.5 - 2.5)^2}{(6 - 1)(1.870829)^2}}$$

$$= 0.5181\sqrt{\frac{1}{6} + \frac{1}{17.5}}$$

$$= 0.245$$

The value of the t statistic for this estimate is $t_{0.05,4} = 2.132$, and the 90% confidence interval on the mean value of y given $x = 3.5$ is

$$\mu_{y|x} \pm t_{\alpha/2}S_{\mu_{y|x}} = 7.92 \pm 2.132(0.245) = 7.92 \pm 0.52$$

or 7.40 to 8.44.

Predicting a Single Outcome of *y* Given *x*

The task of predicting the value of y given x differs subtly from predicting the *mean of* y given x. Predicting the value of y given x involves estimating the value of y given x for a *single* outcome. Single outcome predictions are less precise because the confidence interval for a single y prediction is wider than the confidence interval for a mean of y prediction. Recall that we encountered the same logic in Chapter 7 when we discovered that the sampling distribution of the mean is less variable than the distribution of the underlying variable.

Applying that logic to a real property appraisal, consider that task of predicting market value (i.e., the most probable price or μ) as opposed to predicting how close a given property's sale price will be to market value. If an appraiser's market value estimate is $250,000, then the expected sale price would be $250,000. Nevertheless, an actual market transaction's price is likely to differ from the expected price, and actual market prices are likely to differ from the expected price substantially more so than the mean of numerous market sales of highly comparable properties. The key idea here is that there will be more variation in individual transaction prices than there will be in mean prices for the same good, all else being equal.

The standard error of a single outcome of y given x is

$$S_{y|x} = S_e\sqrt{1 + \frac{1}{n} + \frac{(x_i - \bar{x})^2}{(n-1)S_x^2}}$$

By looking back at the equation for $S_{\mu_{y|x}}$ we can see that $S_{y|x}$ is greater than $S_{\mu_{y|x}}$ at all values of x as a result of the addition of 1 to the value inside the square root sign. As a result, the confidence interval for an individual y (e.g., individual price) prediction is always wider than the confidence interval for a mean of y (e.g., mean price) prediction.

For the data from the example problem,

$$S_{y|x=3.5} = 0.5181\sqrt{1 + \frac{1}{6} + \frac{(3.5 - 2.5)^2}{(6-1)(1.87084)^2}}$$

$$= 0.5181\sqrt{1 + \frac{1}{6} + \frac{1}{17.5}}$$

$$= 0.573$$

which is more than twice the previously derived standard error of the estimate of the mean of y given x of 0.245.

The 90% confidence interval for $y|x$ is

$$y|x \pm t_{\alpha/2}S_{y|x} = 7.92 \pm 2.132(0.573) = 7.92 \pm 1.22$$

or 6.70 to 9.14, whereas the 90% confidence interval for $\mu_{y|x}$ is 7.40 to 8.44.

Notice that estimates of $y|x$ and $\mu_{y|x}$ are equal at 7.92 because both are defined as the expected value of y when $x = 3.5$. The difference is in the width of the confidence interval. The pattern of the confidence interval for an individual y prediction looks like the pattern shown in Figure 9.7, only wider as shown in **Figure 9.8.**

Figure 9.9 shows the Minitab output for predictions of $\mu_{y|x}$ and $y|x$. The Minitab output is consistent with the algebraic calculations we just completed, with the difference attributed to rounding in our algebraic calculations. In Minitab, *Fit* is the expected value of y, which is 7.909 when $x = 3.5$. *SE Fit* is the standard error of the estimate of $\mu_{y|x}$. The standard error of the estimate of $y|x$

Figure 9.8 Comparison of the Confidence Interval on the Mean with the Prediction Interval on an Individual Outcome

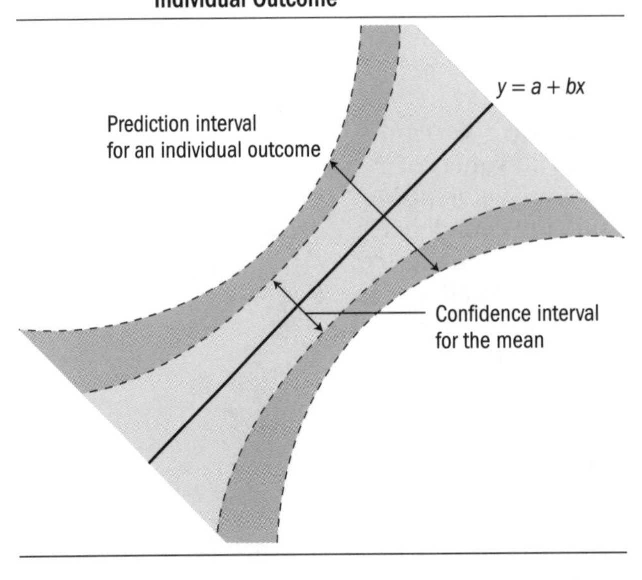

is not given, but it can be calculated algebraically using the values of the two confidence intervals. The *90% CI statistic* is the confidence interval on the mean prediction in Minitab, and *90%PI* is the confidence interval (also called a prediction interval, or PI) on the individual prediction in Minitab. Minitab also lists the values for the new observations used to derive the predictions. In this example, there is only one new predictor, $x = 3.5$.

Figure 9.9 Minitab Confidence Interval Output for Example Data

```
Predicted Values for New Observations

New
Obs    Fit  SE Fit       90% CI            90% PI
  1  7.909   0.245  (7.386, 8.431)  (6.687, 9.130)


Values of Predictors for New Observations

New
Obs     x
  1  3.50
```

Confidence Intervals in Minitab and SPSS

To replicate the output in Figure 9.9, enter the data into Minitab, select **Options** in the **Regression Window**, enter 3.5 in the **Prediction intervals for new observations** box, and then enter 90 in the **Confidence level** box prior to running the regression.

In SPSS, add an additional data entry with $x = 3.5$ and y missing. When you run the regression in SPSS, select **Save** in the **Linear Regression Window.** Then check both **Prediction Intervals** and enter 90 in the **Confidence Interval %** box. After the regression has been run, SPSS will insert confidence intervals for all of the x values into your data, including confidence intervals for $x = 3.5$.

Practice Problem 9.3

The following hypothetical information on lot size in acres (x) and price in thousands of dollars (y) was used in Practice Problem 9.2 in the derivation of a linear regression equation. Using $\alpha = 0.10$, derive an expected price, confidence interval for the mean price, and a confidence interval for an individual transaction price when lot size = 2.3 acres.

Price ($000)	90	130	230	220	320
Lot Size (Acres)	1	1.2	2	2.4	3

Coefficient of Determination (R^2)

The coefficient of determination, abbreviated as R^2, is a measure of the extent to which the regression equation accounts for variation in the dependent variable y. To gain an understanding of R^2 and how it is calculated, start with the assumption that you have a representative sample of the outcome variable y only. In this case the expected value of y is the mean of y, which is referred to as $\bar{y}$. If another variable exists–call it x–that accounts for some of the variation in y, then you can develop a more refined expectation concerning the value of y if you know the value of x and understand how x and y are related. This more refined estimate of y is the conditional mean of y given the value of x, which is referred to as $\mu_{y|x}$.

The coefficient of determination is a measure of the extent to which $\mu_{y|x}$ is an improvement over $\bar{y}$ as an estimate of the expected value of y. Values for R^2 can vary from 0 (x is totally unrelated to, and explains nothing about, y) to 1 (the x, y relationship is deterministic). Therefore, R^2 values are often described as "explanatory power," ranging from "none" (0) to "perfect" (1).

Figure 9.10 shows how total deviation in y is distributed among explained[5] and unexplained components where "explanation" is provided by the regression equation $y = a + bx$. When $x = x_i$, total deviation in y from its mean is the quantity $y_i - \bar{y}$. As shown, part of total deviation is accounted

5. "Explained" does not mean "caused" when referring to regression models. The fact that knowledge of the value of x explains, or accounts for, a portion of the variation in y does not imply that x causes y. Causation is a much stronger statement than explanation.

Figure 9.10 Elements of Deviation in *y*

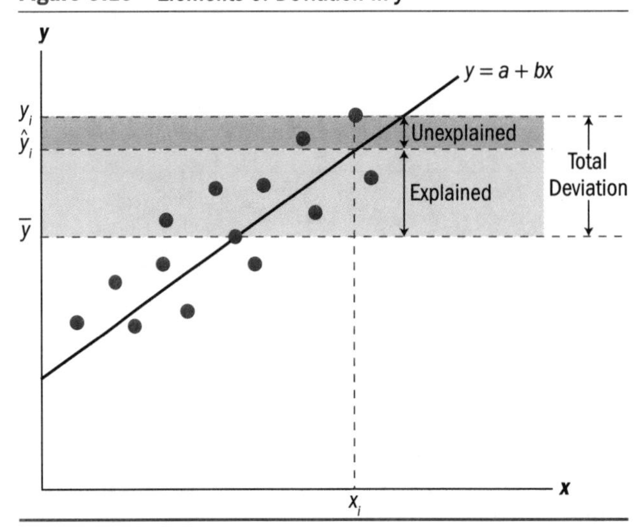

for by the equation $y = a + bx_i$. Therefore, the explained deviation is $\hat{y}_i - \bar{y}$. The remainder is unaccounted for by the regression equation. Therefore, the unexplained deviation is $y_i - \hat{y}_i$.

Considering the numerical values for all of the outcome *y* variables in the date set in aggregate, total deviation is $\Sigma(y_i - \bar{y})$. Total deviation explained by the regression is $\Sigma(\hat{y}_i - \bar{y})$, and total unexplained deviation (regression error) is $\Sigma(y_i - \hat{y}_i)$. For computational purposes, each of these deviations is squared to account for the offsetting effects of positive and negative values. Acronyms generally used in statistics to indicate the squared deviations are

$$\text{SST} = \Sigma(y_i - \bar{y})^2 \quad \text{(total sum of squares)}$$
$$\text{SSR} = \Sigma(\hat{y}_i - \bar{y})^2 \quad \text{(regression sum of squares)}$$
$$\text{SSE} = \Sigma(y_i - \hat{y}_i)^2 \quad \text{(error sum of squares)}$$

and

$$\text{SST} = \text{SSR} + \text{SSE}$$

These squared deviations are used to calculate the coefficient of determination (R^2). The coefficient of determination is the proportion of *SST* accounted for by the linear regression equation. This proportion can be expressed in either of two equivalent ways:

$$R^2 = \frac{\text{SSR}}{\text{SST}}$$

or

$$R^2 = 1 - \frac{\text{SSE}}{\text{SST}}$$

R^2 ranges from 0 to 1 because it is 0 when $SSR = 0$ and it is 1 when $SSE = 0$. Another way of looking at this is when $SSE = 0$, $SSR = SST$, and the ratio of SSR to SST is 1. When $R^2 = 1$ the model is deterministic because there is no estimation error.

The coefficient of determination is universally provided in statistical software least squares outputs. Excel and SPSS report R^2 as proportions (Figures 9.3 and 9.5) and Minitab reports it as a percentage (Figure 9.4).

Practice Problems 9.4

1. A regression model reports $R^2 = 0.820$. In addition, SSE was calculated to be 3.80. What are the values of SSR and SST?
2. Based on the following sample, what proportion of deviation in house price (in thousands of dollars) is explained by living area (SF)? Write the regression equation and calculate SSE, SSR, and SST. Although this can be done by hand, it is more efficient to solve this problem in Excel, SPSS, or Minitab. Assume all of the assumptions underlying linear regression modeling are satisfied.

Price ($000)	200	240	195	180	210	220	200
SF	1,800	2,000	1,840	1,710	1,800	1,980	1,860

Transformations for Curvilinear Relationships

One of the assumptions underlying linear regression models is that the relationship between the dependent variable and the independent variable is linear. When the relationship is curvilinear, a linear model will not satisfy the assumption that $E[e_i] = 0$ for all values of the response variable. Instead, the expected value of regression error will be non-zero when non-linear data are being fit to a linear equation.

The data in **Table 9.6** and **Figure 9.11** illustrate this issue. The relationship between y and x is not linear. The relationship is best approximated by $y = 1/x$, but it is not deterministic. A simple linear regression model can be fit to these data, but the data must be transformed to conform to the assumption of linearity. Since we know the approximate form of the relationship, we can easily transform the data into a linear form.

One suitable transformation is as follows:

If $y = 1/x$, then

$$\ln(y) = \ln(1) - \ln(x)$$

which is linear in the natural logs of x and y. We can therefore fit a linear regression model to the transformed data by substituting $\ln(y)$ for y and $\ln(x)$ for x.

Panel A of Figure 9.11 illustrates the raw, untransformed data fit to a linear equation (solid line) and to the nonlinear $y = 1/x$ equation (dashed line). The dashed line's curvilinear equation provides a superior fit. When a linear equation is fit to the untransformed data, as in the solid line, the expected value of regression error is negative in the middle of the curve (from about $x = 1.5$ to $x = 5.5$) and the expected value of regression error is positive at both ends of the curve. In contrast, Panel B shows how the transformed data fit a linear equation. The natural log transformation satisfies the assumption of linearity, leading to a better fit and less estimation error.

The least squares linear regression model for the Panel B data is $\ln(y) = 0.04034 - 1.03880[\ln(x)]$, which can be converted to a function of x and y by use of exponents where

$$\hat{y} = e^{\ln(y)}$$

When $x = 4$, the expected value of $\ln(y)$ is $0.04034 - 1.03880\ [\ln(4)] = -1.39974$, and $e^{-1.39974} = 0.247$.

Table 9.6 Values of *x* and *y* Variables

x	1	2	3	4	5	6	7
y	0.9	0.6	0.4	0.22	0.18	0.17	0.13
Transformed Values of *x* and *y* Variables							
ln(*x*)	0.0000	0.6931	1.0986	1.3863	1.6094	1.7918	1.9459
ln(*y*)	-0.1054	-0.5108	-0.9163	-1.5141	-1.7148	-1.7720	-2.0402

Figure 9.11 Linear Fit to Untransformed and Transformed Data

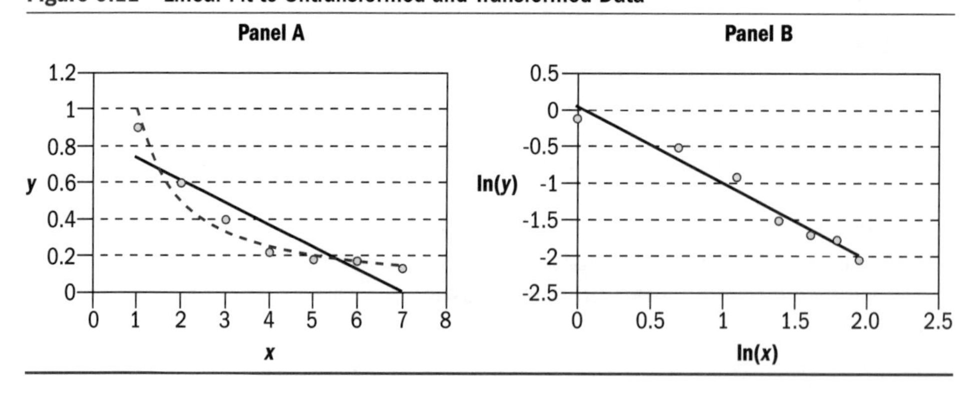

The actual value of y when $x = 4$ is 0.22, and the estimation error is 0.22 – 0.247 = -0.027.

By comparison, when a linear regression model is fit to the untransformed data as in Panel A, the estimation equation is $y = 0.85571 - 0.12107x$, and when $x = 4$ the expected value of $y = 0.85571 - 0.12107(4) = 0.371$. The estimation error is equal to 0.22 – 0.371 = -0.151, which is more than five times the estimation error for the transformed data shown in Panel B.

Another suitable transformation option is to regress y on the reciprocal of x. **Table 9.7** shows both x and the reciprocal of x. When y is regressed on the reciprocal of x, the resulting regression equation is

$$y = 0.03 - 0.9216\left(\frac{1}{x}\right)$$

As **Figure 9.12** shows, although the data are not linear in the relationship of y to x, they appear to be

Table 9.7 Reciprocal Transformation of x

x	1	2	3	4	5	6	7
$1/x$	1	0.50	0.33333	0.25	0.20	0.16667	0.14286
y	0.9	0.6	0.4	0.22	0.18	0.17	0.13

Figure 9.12 Linear Fit to Table 9.7 Reciprocal Transformation

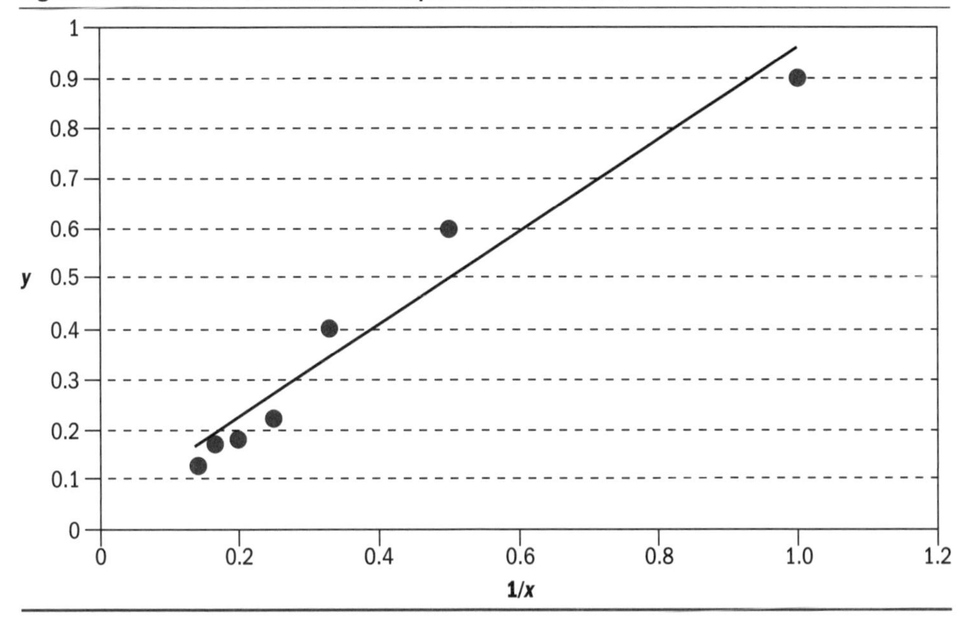

linear in the relationship of y to the reciprocal of x. The regression line fit in Figure 9.12 is similar to the log transformation's fit in Panel B of Figure 9.11.

This simple exercise demonstrates how fitting a linear equation to untransformed, nonlinear data results in several undesirable consequences. Estimates of $\mu_{y|x}$ and $y|x$ are biased.[6] Estimation error is inflated, resulting in excessively large confidence intervals on α, β, $\mu_{y|x}$, and $y|x$. Larger estimation error also affects hypothesis tests relying on the t statistic. The t-test statistics will not be credible because both the numerator and denominator in the equation $t = (b - 0)/S_b$ will be inaccurate.

Usually a visual inspection of a scatter plot will reveal whether or not the relationship between a dependent variable and an independent variable is linear. For example, Panel A of Figure 9.11 shows a clearly nonlinear relationship. Economic and valuation theories and market experience are also important in making a determination of when a nonlinear relationship might be expected. Knowledge of when decreasing and increasing returns to scale are at play is a prime example of this.

Example Problem

Consider the x and y values table and scatter plot in **Figure 9.13**. The relationship shown by this sample is nonlinear and represents decreasing returns to y based on the value (scale) of x. This is the sort of relationship you might expect to see in some residential markets if you plotted lot price (y) against lot size (x).

Figure 9.14 shows the results of curve fit analysis of these data in SPSS, which includes fitting linear, logarithmic, and quadratic curves. Both the logarithmic and quadratic curves fit the data better than the linear equation.

An important cautionary statement is

Curve Fitting in SPSS and Minitab

A **Curve Estimation** algorithm is available in the SPSS menu of options that appear after selecting **Analyze** and then **Regression**. Figure 9.14 was created using curve estimation in SPSS. Minitab will fit a quadratic and cubic curve, but none of the other curve types fit in SPSS. This is not too big of a handicap, however, because quadratic and cubic curves are adaptable to many nonlinear relationships.

6. When the expected error is non-zero, the estimates are referred to as being "biased."

Figure 9.13 Scatter Plot of y and x

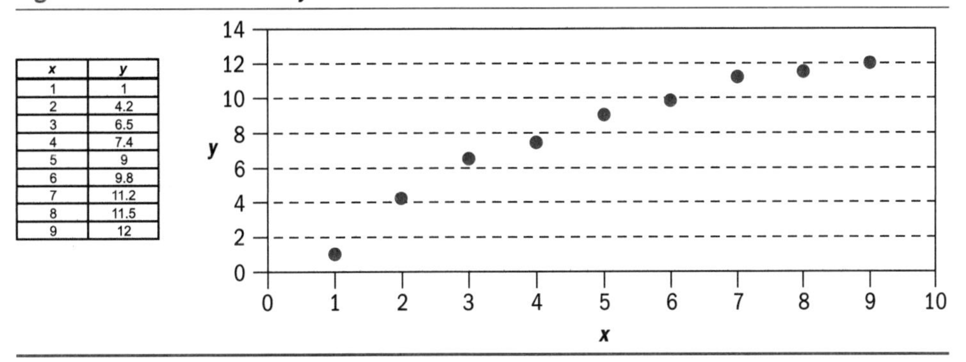

x	y
1	1
2	4.2
3	6.5
4	7.4
5	9
6	9.8
7	11.2
8	11.5
9	12

Figure 9.14 SPSS Curve Fit Illustration

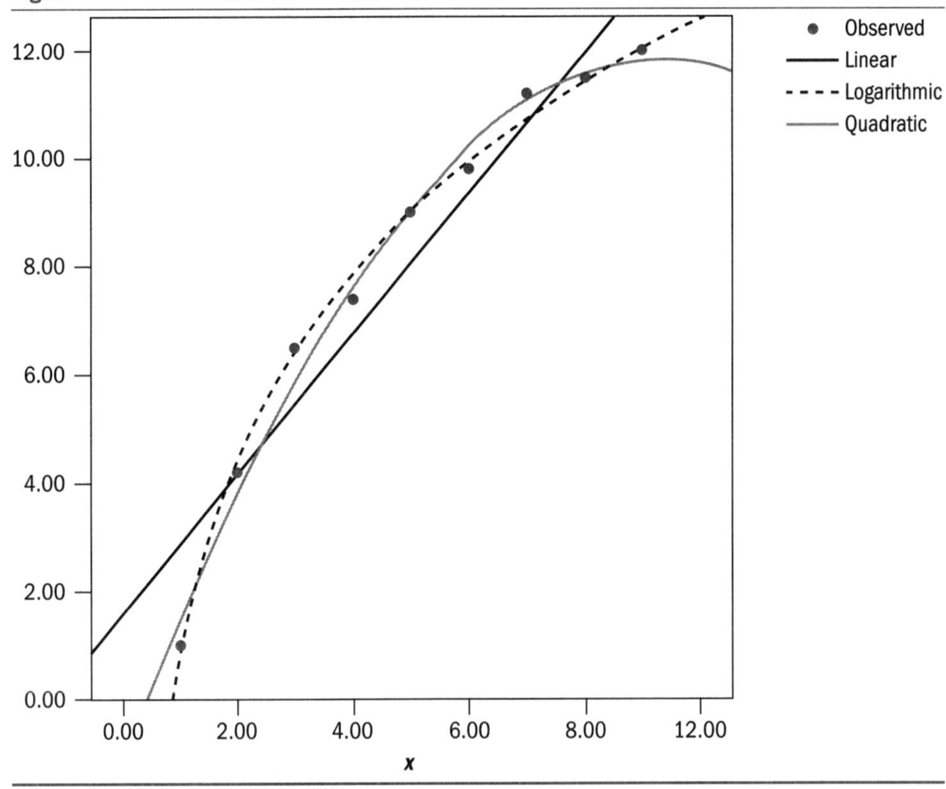

warranted here. The preceding example problem relies on a small, nine-observation data set. Small data sets can be misleading because a small random sample can appear to be nonlinear even though the underlying population is linear. For example, assume the random sample in Figure 9.13 was drawn from the population illustrated in

Figure 9.15. Due to an unfortunate and admittedly unlikely random draw, the shape of the nine-observation sample is inconsistent with the underlying linearity of the population–an example of one possible consequence of sampling error. Under these circumstances, reliance on the smaller sample results in an inappropriate conclusion of nonlinearity.

The lesson here is that small samples can be misleading, and they should be judged against theory and, in keeping with the Competency Rule of the Uniform Standards of Professional Appraisal Practice,[7] the analyst's experience and knowledge of the market. When small sample implications are counterintuitive, more investigation may be necessary to produce a definitive and credible valuation result.

Figure 9.15 Illustration of an Unrepresentative Curvilinear Sample (Dark-Colored Dots) from a Linear Population (Both Colors of Dots)

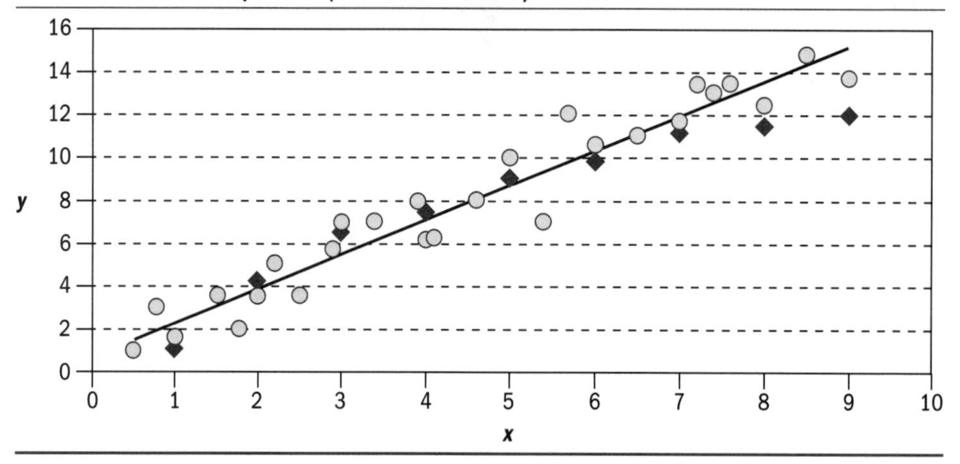

Types of Nonlinear Relationships

Nonlinear curve types can be reciprocal, logarithmic, exponential, or quadratic (polynomial), to name a few. Examples of typically encountered curve shapes are shown in **Figures 9.16, 9.17**, and **9.18**, along with some suggested transformations used to fit them.

7. The Uniform Standards of Professional Appraisal Practice (USPAP) is published and updated regularly by The Appraisal Foundation. Of particular note in this context, Advisory Opinion 18 contains a summary of relevant USPAP references concerning the use of statistical software for data analysis in appraisal, including an in-depth statement concerning competency.

Figure 9.16 Downward-Sloping Curve: Decreasing at a Decreasing Rate

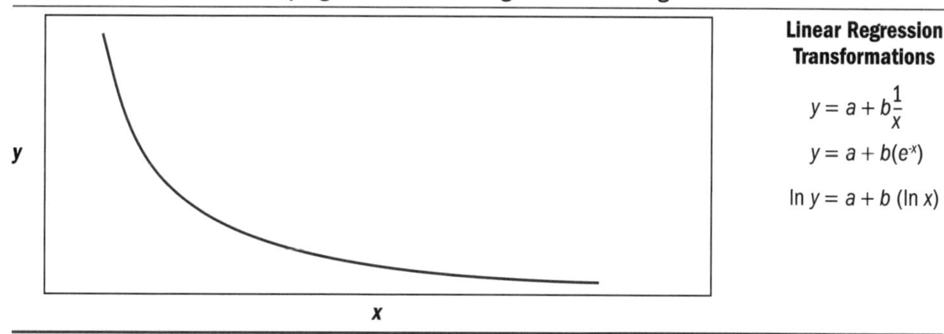

Figure 9.17 Upward-Sloping Curve: Increasing at a Decreasing Rate

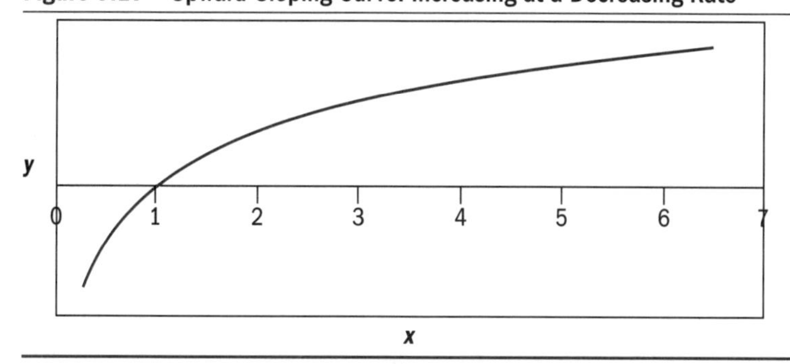

Linear Regression Transformations

$y = a + b(\ln x)$

$y = a + b(\sqrt{x})$

$y = a + b\frac{1}{x}$

Figure 9.18 Upward-Sloping Curve: Increasing at an Increasing Rate

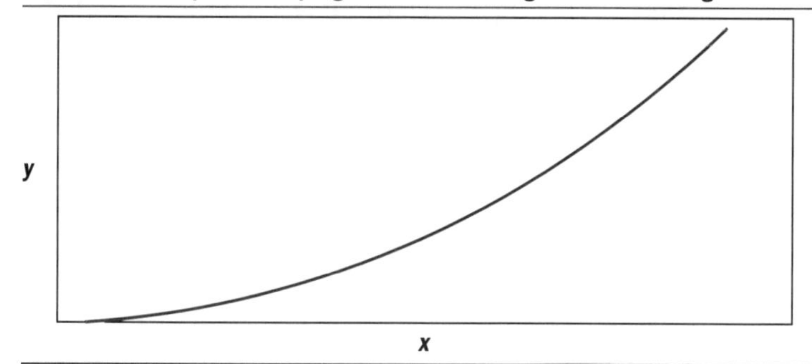

Linear Regression Transformations

$y = a + bx^2$

$y = a + b(e^x)$

A polynomial transformation can accommodate curves that are changing at a varying rate, which may be accelerating or decelerating. These transformations result in two (or more) independent variables, which will become clearer in the next chapter.

A polynomial linear regression transformation relies on the following equation:

$$\mathbf{y = a + b_1x + b_2x^2 + \ldots + b_kx^k}$$

Any kth order polynomial is theoretically possible. Generally speaking, however, many nonlinear relationships can be modeled sufficiently with a second-order polynomial where $y = a + b_1x + b_2x^2$, although the addition of an x^3 term is sometimes beneficial.

Another useful and often-used transformation involves a logarithmic transformation of the dependent variable only, resulting in the regression equation $\ln(y) = x$. This sort of transformation is called a "semilogarithmic" equation. It is also sometimes loosely referred to as a "log-linear model," which can be confusing because this name is also associated with models used to analyze multidimensional contingency tables.

Semilogarithmic models are useful because the coefficients on the continuous independent variables indicate the percentage change in the untransformed dependent variable given a 1-unit change in the independent variable. A small amount of work is involved to interpret the coefficient on a categorical variable as a percentage change, however. When an independent variable is categorical (i.e., coded 1, 0), the percentage change in the dependent variable associated with a dummy variable category coded as 1 is calculated as $100(e^{\text{COEF}} - 1)$. In this calculation, "COEF" refers to the coefficient of the categorical variable.[8]

For example, in a study investigating the effect of foreclosure status on apartment building sale prices, foreclosure status was coded as a dummy variable with a value of 1 if an apartment sale involved a bank seller who obtained the sold property through foreclosure.[9] Otherwise, the foreclosure status variable was coded 0. The coefficient of "foreclosure" was statistically significant and equal to -0.251231. This coefficient indicated a foreclosure sale price difference of -22.2% based on the following calculation:

$$\textbf{PRICE DIFFERENCE} = 100(e^{-0.251231} - 1) = -22.2\%$$

8. For more on this topic, see Robert Halvorsen and Raymond Palmquist, "The Interpretation of Dummy Variables in Semi-logarithmic Equations," *American Economic Review* 70, no. 3 (1980): 474-75.
9. William G. Hardin and Marvin L. Wolverton, "An Introduction to the Analysis of Covariance Model Using an Empirical Test of Foreclosure Status on Sale Price," *Assessment Journal* 6, no. 1 (1999): 50-55.

Practice Problem 9.5

Many research papers indicate that exposure to excessive street noise has a negative impact on home price. The effect appears to diminish with distance from the source of the noise. Using the data below test whether the increase in price per square foot as distance increases is linear or curvilinear. Plot the data, try some transformations, and investigate which model results in the greatest explanatory power. Assume that the prices have been adjusted for all other elements of comparison.

Price per Sq. Ft.	Distance
110	50
134	80
135	100
142	150
140	180
145	220
145	250
146	300
146.5	400
144	500
149	800
150	1,000

Assessing Prediction Accuracy

Two commonly used and easily conducted assessments of prediction accuracy involve data splitting or calculation of the prediction sum of squares known as a *PRESS* statistic.

Data Splitting

When a data set is large enough, *data splitting* provides a valid means of assessing prediction accuracy. With this method the data are randomly divided into two subsets–an estimation subset and a holdout subset. The estimation subset is used to develop several competing models (i.e., functional forms or transformations). The competing regression model results are each used to predict separate dependent variable values for the holdout subset. The model that predicts the holdout dependent variable values the best is the superior model for prediction purposes. Two measures of prediction accuracy are *mean square prediction error* and *mean absolute prediction error*, which are computed as follows:

$$\text{Mean Square Prediction Error} = \sum_{i=1}^{n'} \frac{(y_i - \hat{y}_i)^2}{n'}$$

$$\text{Mean Absolute Prediction Error} = \sum_{i=1}^{n'} \frac{|y_i - \hat{y}_i|}{n'}$$

Table 9.8 Data Splitting Example

Full Sample		Holdout Subset		Estimation Subset	
x	*y*	*x*	*y*	*x*	*y*
1.000	1.000			1.000	1.000
8.500	0.122	8.500	0.122		
2.000	0.600			2.000	0.600
7.500	0.115			7.500	0.115
8.000	0.136			8.000	0.136
7.000	0.129			7.000	0.129
6.000	0.189	6.000	0.189		
3.500	0.265			3.500	0.265
4.500	0.240	4.500	0.240		
6.500	0.154			6.500	0.154
5.000	0.200			5.000	0.200
4.000	0.238			4.000	0.238
9.000	0.111			9.000	0.111
1.500	0.700	1.500	0.700		
2.500	0.400			2.500	0.400
3.000	0.315			3.000	0.315
5.500	0.182			5.500	0.182

Table 9.9 Model 1 and Model 2 Holdout Subset Prediction Errors

		Model 1		Model 2	
x	y	$\hat{y}_i$	$y_i - \hat{y}_i$	$\hat{y}_i$	$y_i - \hat{y}_i$
8.5	0.122	0.008067	0.114	0.113217	0.009
6.0	0.189	0.211335	-0.022	0.162476	0.027
4.5	0.240	0.333296	-0.093	0.218959	0.021
1.5	0.700	0.577217	0.123	0.684197	0.016

where n' is the size of the holdout sample. Holdout predictions can be compared on each of these two measures. The smaller the error, the better the prediction.

For example, look at the data in **Table 9.8.** The full sample includes 17 observations on a dependent variable (y) and an independent variable (x). Four of the data elements were randomly chosen to be held out of the analysis (the holdout subset) and 13 were used for model building (estimation subset).

Two competing models are being examined in this example. Model 1 is a linear model of the form $y = a + bx$, and Model 2 is of the form $\ln(y) = c + d[\ln(x)]$. Regression coefficients for the two models (rounded here for display) are

$$\text{Model 1: } y = \mathbf{0.69915 - 0.08131x}$$

$$\text{Model 2: } \mathbf{\ln(y) = 0.0409 - 1.0371[\ln(x)]}$$

Based on the information presented in **Table 9.9**, holdout subset prediction statistics are

Model 1: Mean Square Prediction Error = 0.0093
Mean Absolute Prediction Error = 0.0881

Model 2: Mean Square Prediction Error = 0.0004
Mean Absolute Prediction Error = 0.0180

Model 2 is a better predictor based on both mean square prediction error and mean absolute prediction error. Therefore, the nonlinear regression equation is the better choice for predicting y given the value of x.

PRESS Statistic

Minitab can be used to calculate a *PRESS* ("prediction sum of squares") statistic and an associated prediction R^2 statistic. To calculate *PRESS*, each observation is omitted in turn from a series of regression models having $n - 1$ observations. This results in an iterative series of n estimation samples of size $n - 1$. The predicted value for the omitted dependent variable is calculated for each of the n iterations, based on the corresponding value of the omitted independent variable. The result is a collection of n observed y values and n predicted y values. The *PRESS* residual is equal to $y_i - \hat{y}_i'$ where $\hat{y}_i'$ is the predicted value of y_i based on the other $n - 1$ observations. The *PRESS* statistic is the sum of squared *PRESS* residuals:

$$\text{PRESS} = \sum_{i=1}^{n} (y_i - \hat{y}_i')^2$$

A corollary is the *prediction* R^2 statistic:

$$R^2_{PRED} = 1 - \frac{\text{PRESS}}{\text{SST}}$$

The competing model having the lowest *PRESS* statistic or the largest R^2_{PRED} statistic is the best predictive model. Be aware that R^2_{PRED} is not interpreted the same way as R^2 and should not be compared to R^2.

PRESS and R^2_{PRED} statistics for the two models previously derived from the full sample data in Table 9.8 are as follows:

$$\text{Model 1: } \textbf{PRESS} = \mathbf{0.4335}$$
$$\mathbf{R^2_{PRED} = 0.555}$$

$$\text{Model 2: } \textbf{PRESS} = \mathbf{0.1330}$$
$$\mathbf{R^2_{PRED} = 0.981}$$

Since Model 2 has the lower *PRESS* statistic and the correspondingly higher R^2_{PRED}, it is the better predictive model. The results are consistent with the prior data splitting analysis.

Once a model has been chosen, it is appropriate to develop a regression equation using the full data set and then to use the full sample result as the basis for analysis. Running the example's preferred nonlinear model on the full sample of 17 observations yields the following estimation equation:

$$\ln(y) = 0.04236 - 1.0242[\ln(x)]$$

$R^2 = 0.985$, and the t statistic on the coefficient of $\ln(x)$ is highly significant at -31.51.

Note that this prediction accuracy example problem uses simple data. Data splitting and *PRESS* are not necessary to reach a conclusion that a nonlinear model will predict better. A scatter plot of the relationship between y and x reveals an obvious nonlinear relationship. (Try creating the scatter plot to observe the relationship.) Use of simple data allows us to concentrate on the methodology without the confounding effects complex data would cause. Be aware that model choice is not so simple in most real-world settings where the dependent variable is usually related to multiple predictor variables and the function form of the relationship among the variables is not easily identified. Nevertheless, the prediction accuracy assessment skills demonstrated here are especially valuable in complex real-world settings.

Prediction Accuracy in Minitab

PRESS and **Predicted R-Square** statistics can be generated in Minitab by clicking the **Options** button in the **Regression** window. These statistics are not available in Excel or SPSS.

Simple Linear Regression for Time Trends

Data gathered on a single item (firm, city, industry, person, etc.) over a time sequence is called *time series data.* In contrast, data gathered on numerous items at a given point in time is called *cross-sectional data*, and the term *panel data* describes the collection of data on numerous items over a time sequence. Most real property value estimation issues deal with cross-sectional data, created by selecting sale or rent comparables from (or adjusting them to) a common point in time.

Occasionally it is beneficial to be able to construct a linear-trend, time series model to either forecast values of a dependent variable into the future or gain a greater understanding of the structure of change over a period of time. When this is the case, linear regression analysis can be used to model the trend[10] and to forecast, or extrapolate, the trend into the future. As you might expect, forecasts far into the future are less reliable than forecasts into the less distant future.

Linear-trend, time series forecasts assume that the past pattern will hold into the future, which may or may not be a valid assumption. When a model fits past time series data well, it could be a mistake to assume that the same model will be able to make accurate forecasts. A valid forecast requires an accurate model of past trends *plus* an assessment of the assumptions concerning continuation of the pattern into the future and reasonable expectations regarding the length of a credible time frame for the future forecast.

A time trend may be linear or curvilinear, and the dependent variable can be any measure, such as population, housing starts, number of sales, product output, or price. The independent variable is always a measure of time in constant units–weeks, months, quarters, years, and so on. When the time trend follows a linear pattern, a regression model looks like the linear models we have been studying, except for the substitution of t (time) for x. Therefore, a linear trend model is of the form

$$\mathbf{y_i = a + bt_i + e_i}$$

The forecast for some future time period t' is

$$\mathbf{y' = a + bt'}$$

A time trend can also follow a curvilinear path, and when this is the case the linear transformations discussed earlier in the chapter may be applicable. One particular form of nonlinear time trend associated with population growth or financial returns is a compound rate of change where

$$\mathbf{y = a(1 + b)^t}$$

10. As used here, a *trend* is defined as a tendency for a dependent variable to move in a predictable pattern over time.

This nonlinear relationship can be regressed linearly by taking the logarithms of both sides of the equation, whereby

$$\ln(y) = \ln(a) + t \ln(1 + b)$$

and the linear regression model is

$$\ln(y) = c + dt$$

In this model c is an estimator for $\ln(a)$ and d is an estimator for $\ln(1 + b)$. Therefore, $a = e^c$ and $1 + b = e^d$.

Example Problem

Suppose you had observations on y for $t = 0$ to $t = 12$ and you regressed $\ln(y) = c + dt$ to obtain the following solution:

$$\ln(y) = 0.693 + 0.0488(t)$$

There are two ways to solve for the expected value of y when $t = 4$:

1. $$\ln(y) = 0.693 + 0.0488(4) = 0.8882$$
   $$y = e^{0.8882} = 2.43$$

2. $$\ln(a) = 0.693, \ a = e^{0.693} = 2.0$$
   $$\ln(1 + b) = 0.0488, \ 1 + b = e^{0.0488} = 1.05$$
   $$y = 2(1.05)^4 = 2(1.215) = 2.43$$

The first solution yields a direct estimate of y. The second solution yields the same estimate for y, and also provides estimates for the intercept ($y = 2$ when $t = 0$) and the compound rate of increase ($b - 1 = 0.05$). The time trend indicated by the regression result is illustrated in **Figure 9.19**. As the exhibit shows, if you were forecasting the value of y for the next time period ($t = 13$), the expected value of y would be $3.77 = 2 \times (1.05)^{13}$.

Figure 9.19 Compound Rate of Change

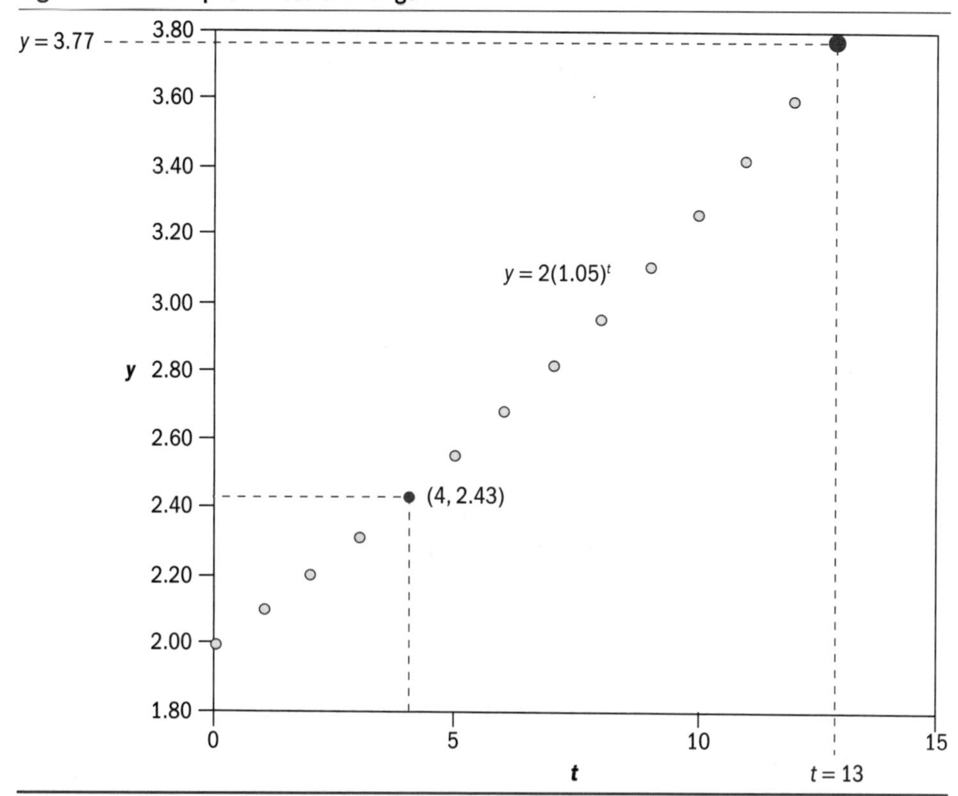

Practice Problem 9.6

Assume that population change follows a compound annual growth model. As shown in the table below, you have population data for the past 10 years. Derive a model that best fits population growth. Estimate the compound annual growth rate. What is the forecast population for two more years in the future (t = 10 and t = 11), assuming the past population trend extends into the future?

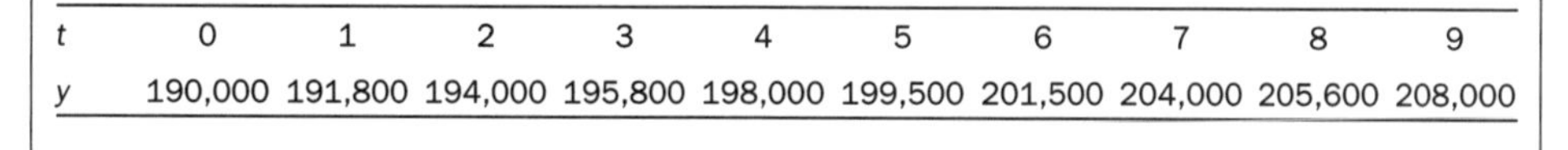

t	0	1	2	3	4	5	6	7	8	9
y	190,000	191,800	194,000	195,800	198,000	199,500	201,500	204,000	205,600	208,000

Analyzing Regression Errors

A simple scatter plot of regression errors plotted against the values of the dependent variable y can reveal a great deal about how well a regression model accounts for systematic variation in y. In addition, investigation of regression errors can identify departures from assumptions implicit in linear regression analysis concerning the underlying population. Ideally, the regression error will be random and exhibit no discernable pattern throughout the range of y values, as in **Figure 9.20**. When this is true, the expected value of regression error is zero, regression errors are independent, and the variance of regression error is constant across the values of y.

Variation from the ideal random distribution of regression error shown in Figure 9.20 can indicate departures from the assumptions underlying regression modeling. Error analysis is also useful for identifying observations that may be incompatible or erroneous. Panels A, B, C, and D of **Figure 9.21** show patterns of error distribution that signal a need for additional analysis.

Figure 9.20 Ideal Regression Error Pattern

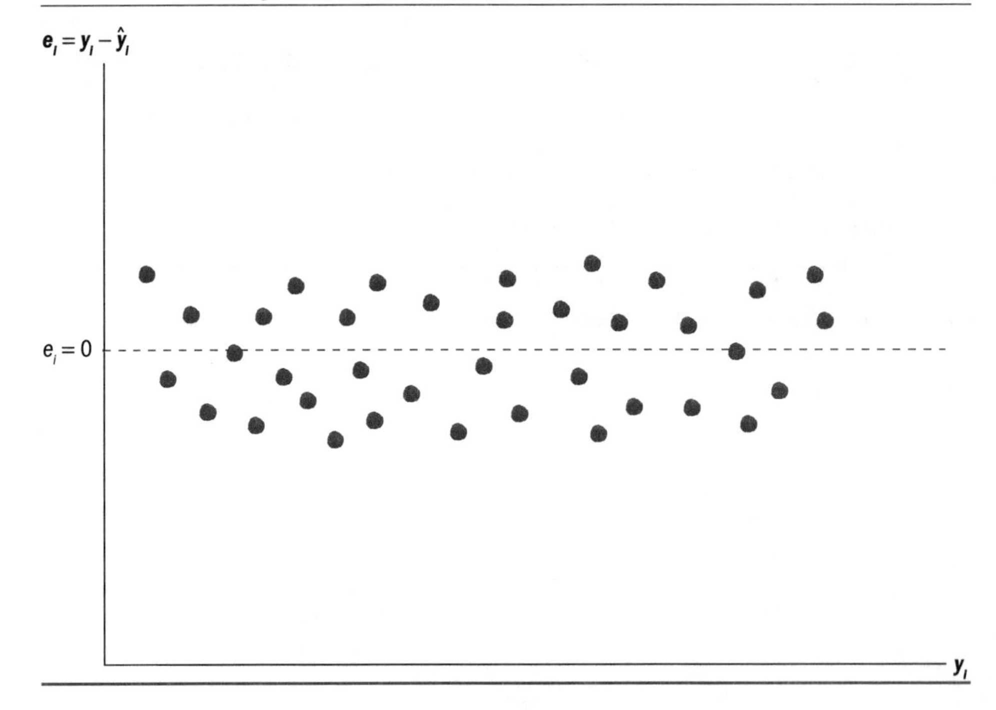

Figure 9.21 Unexpected Error Distributions

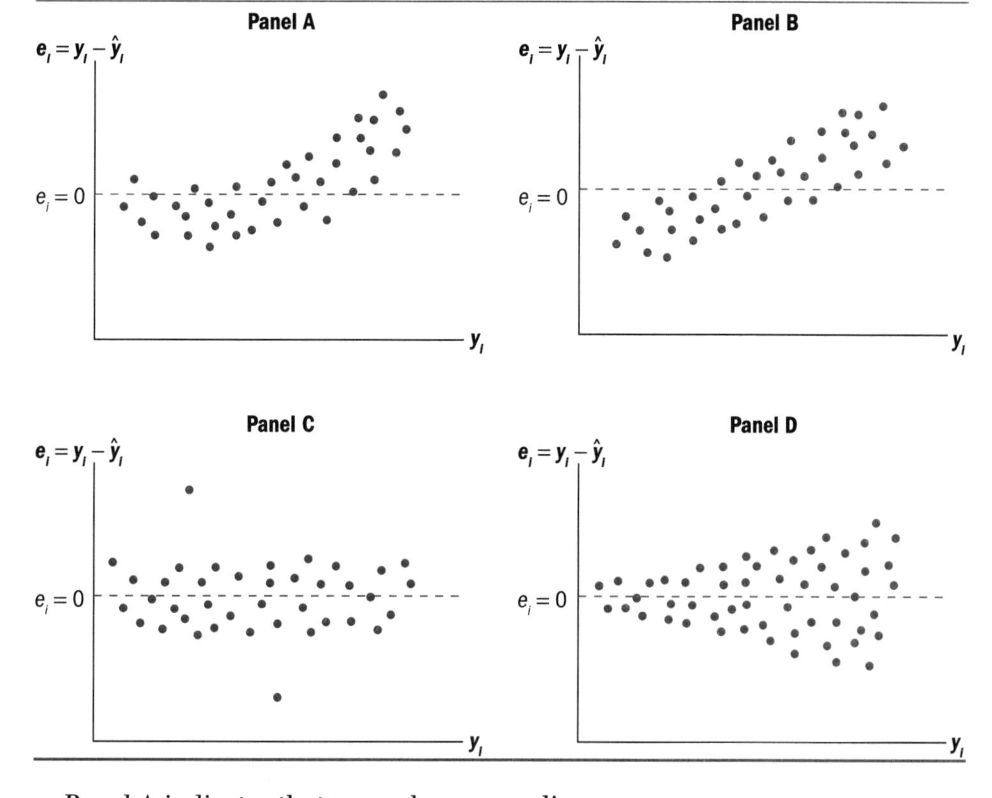

Panel A indicates that *y* may have a nonlinear relationship to the independent variable. When this error pattern is evident, examine a scatter plot of *x* and *y* to determine if the relationship between them is linear in appearance. If not, decide on which transformations of *x* or *y* may be appropriate and test the transformations to see if they improve the regression equation fit. If the relationship between *x* and *y* appears to be linear, a curvilinear residual pattern is probably indicative of an additional, but omitted, variable that accounts for systematic and nonlinear error in the estimates of *y*. (Adding additional variables will be discussed in the next chapter.)

Panel B is indicative of an omitted linear independent variable. Simple linear regression may not be capturing all of the systematic variation in *y*. Inclusion of one or more additional independent variables may be necessary to alter the error pattern in Panel B to the desired pattern illustrated in Figure 9.20. As Panel B shows, when this occurs the expected values of regression error are not zero for all values of the response variable. Violation of this underlying regression modeling assumption is also evident in Panel A.

Panel C includes two outliers–observations that exhibit unusually large errors. Outliers may be data points from a different population than the one being studied. They can also result from data entry errors–e.g., the values for x or y may have been miscoded for these observations. Additionally, they can indicate that the population includes occasional outcomes that vary greatly from expected outcomes. A good protocol for handling outliers is to check for data entry errors first. If there are no data entry errors, then check for unaccounted-for characteristics of the observation that may be the source of the large regression error. If there is no valid reason for removing it, an outlier should not be dropped simply because it is an outlier.

Panel D illustrates non-constant error variance (which is called *heteroskedasticity*). Regression modeling assumes constant error variance (which is called *homoskedasticity*). When regression errors are heteroskedastic, the coefficient estimates (a and b) may no longer minimize the sum of squared error, and the estimates of the standard errors of the coefficients (S_a and S_b) are unreliable. Therefore, conclusions concerning the statistical significance of the regression coefficients may be misleading.

To understand why this is so, let's take a closer look at Panel D, as shown in more detail in **Figure 9.22**. When errors are not constant across the range of values of the dependent variable, we are unsure what is being reflected in the regression model's standard error calculation. Is it indicative of the relatively small error variance at position A or the relatively large error variance at position C? Or, is it a compromise as shown at position B? What is clear is that the model fit depends on the values of the response variable, and is most likely a consequence of correlation between regression error and the values of the explanatory variable used to estimate the response variable, y.

Figure 9.22 Non-constant Regression Error

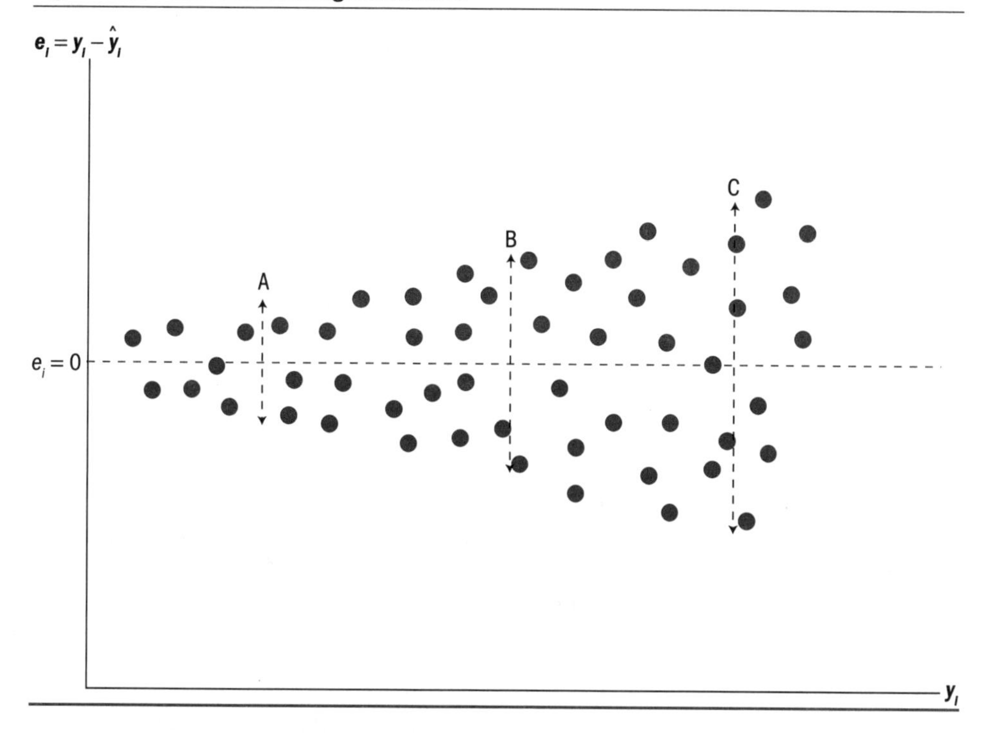

Dealing with Heteroskedasticity

A simple and sometimes effective treatment for heteroskedasticity is to replace *y* with the natural logarithm of *y*. This solution often works because the distribution of ln(*y*) is less variable than the distribution of *y*. A major drawback to this heteroskedasticity treatment is that the natural logarithm of *y* is undefined for $y \leq 0$.

Another option is to apply a solution referred to as *weighted least squares*. This solution is effective when the change in error variance is correlated with values of the independent variable, which may be the case with some real property data. For example, market prices for high-priced homes can vary greatly in comparison with variation in market prices for mid-priced homes. This can be attributed to the relatively better availability of market information in the mid-price range or a result of shallower markets for high-priced homes. When a large range of home prices is included in a regression model and living area and home price are highly correlated, error variance could be positively correlated with square footage.

Recall that a simple linear regression equation is of the form

$$y_i = a + bx_i + e_i$$

When e is systematically related to x, the regression equation can be modified through division by x, i.e., weighting each ith observation by the value of x_i. The modified weighted least squares regression equation is of the form

$$\frac{y_i}{x_i} = \frac{a}{x_i} + b + \frac{e_i}{x_i} = b + a\left(\frac{1}{x_i}\right) + e'_i$$

where $e'_i = e_i / x_i$. Notice that b is the intercept and a is the slope in the modified model. Division by various powers of x (i.e., x^{POWER}) can be attempted to determine which exponent works best. One drawback to a weighted least squares model is that the model is undefined when $x = 0$.

A number of statistical routines test for non-constant error variance. A fairly simple test, attributed to Szroeter,[11] that is easily adaptable

Weighting in SPSS

SPSS includes a weighted least squares option in its regression menu when you purchase the regression add-on package. Choose **Analyze, Regression**, and **Weight Estimation.** The **Weight Estimation** window will ask you to select the dependent variable, the independent variable (or variables), and the weight variable. You can also select a range of "powers" and the size of the intervals separating the test powers. The output will include a list of the powers and associated "log-likelihood" values. The lowest log-likelihood value indicates the best fit. A regression equation is also included in the output employing the power that fits best along with estimates of regression coefficients and t statistics derived from the weighted least squares model. In addition, SPSS allows you to save the weights as a new variable, which can be used in further analysis of the data. Excel does not address this issue, and the weight algorithm in Minitab is not as versatile or user-friendly.

Some statistical programs included more advanced features for dealing with heteroskedasticity. For example, *Stata* and *EViews* are statistical packages that allow the user to invoke a procedure that calculates "robust standard errors." Standard errors calculated in this manner are robust against loss of reliability due to non-constant error variance. Robust standard errors allow the analyst to test variable significance knowing that t statistics are credible.

11. See W. Griffiths and K. Surekha, "A Monte Carlo Evaluation of the Power of Some Tests for Heteroscedasticity," *Journal of Econometrics* 31, no. 2 (1986): 219-231.

to a spreadsheet is demonstrated in a regression analysis textbook by Dielman.[12] Use of this test is summarized and demonstrated below, beginning with hypothesis statements:

H_0: Constant error variance

H_a: Non-constant error variance

Decision Rule: Reject the null hypothesis of constant error variance if $Q \geq$ the critical value of Z consistent with the choice of α.

Test Statistic:

$$Q = \left(\frac{6n}{n^2 - 1}\right)^{1/2}\left(h - \frac{n+1}{2}\right)$$

where

$$h = \frac{\Sigma i e_i^2}{\Sigma e_i^2}$$

Data are arranged in ascending order sorted by the values of x to conduct this test, and i is the rank of each error term associated with the ordered array of x. Q follows a standard normal distribution. So the critical value of Z is determined by the significance level associated with α.

Example Problem

The goal of this section is to present the topic unclouded by real-world complications and consequently provide an opportunity for learning by keeping the example simple enough to allow for easy replication of the results for those readers inclined to do so.

Table 9.10 contains a set of x and y values illustrating heteroskedasticity. These data are used to demonstrate the preceding Q statistic test for heteroskedasticity and the weighted least squares procedure. A simple linear regression of y on x shown in **Figure 9.23** yields the equation $y = 5.30 + 0.708x$. Also, $S_b = 0.105$ and the t statistic on b is 6.74. $SST =$ 421.96, $SSR = 268.46$, $SSE = 153.50$, and $R^2 = 0.636$.

Regression errors are plotted against the values of y in **Figure 9.24**. It is obvious that the regression

12. Terry Dielman, *Applied Regression Analysis for Business and Economics*, 3rd ed. (Pacific Grove, Calif.: Duxbury, 2001), 318.

Table 9.10 *x* **and** *y* **Values**

x	y
1.00	6.90
2.00	5.20
2.50	8.25
3.00	5.80
3.20	8.88
4.00	6.40
4.80	10.32
5.00	7.00
5.00	10.50
6.00	7.60
6.30	11.67
7.00	8.20
7.90	13.11
8.00	8.80
8.20	13.38
9.00	9.40
9.10	14.19
10.00	10.00
10.50	15.45
11.00	10.60
11.60	16.44
12.00	11.20
12.30	17.07
13.00	11.80
14.00	18.60
15.00	13.00
16.00	20.40
17.00	14.20

Figure 9.23 Ordinary Least Squares Regression Output (Excel)

Regression Statistics	
Multiple R	0.798
R Square	0.636
Adjusted R Square	0.622
Standard Error	2.430
Observations	28

ANOVA

	df	*SS*	*MS*	*F*	*Significance F*
Regression	1	268.455	268.455	45.470	0.000
Residual	26	153.502	5.904		
Total	27	421.957			

	Coefficients	*Standard Error*	*t Stat*	*P-value*
Intercept	5.300	0.992	5.344	0.000
x	0.708	0.105	6.743	0.000

errors are not constant and increase as y increases. The correlation between x and the absolute value of the errors is 0.96. So, the size of the error is strongly related to the values of x.

Table 9.11 illustrates how the data should be organized to calculate Q when testing for heteroskedasticity. Errors associated with ascending values of x are squared and then multiplied by their rank. The bottom of the table includes the sums required to calculate h:

$$h = \frac{\Sigma ie_i^2}{\Sigma e_i^2} = \frac{2890.633}{153.5025} = 18.83$$

Since $n = 28$,

$$Q = \left(\frac{6n}{n^2 - 1}\right)^{1/2}\left(h - \frac{n+1}{2}\right) = \left(\frac{168}{783}\right)^{1/2}\left(18.83 - \frac{29}{2}\right) = 2.01$$

Since $Z_{0.05} = 1.96$ and $2.01 \geq 1.96$, we can reject the null hypothesis of constant error variance with 95% confidence and conclude that the data are heteroskedastic.

Figure 9.25 shows output from the SPSS weighted least squares procedure applied to the same Table 9.10 data. It identifies the correction for heteroskedasticity in this example being to weight the regression by x^1 (x to the power of 1). This is the same as dividing by x_i because $x^1 = x$. **Table 9.12** compares the example problem's ordinary least squares regression of y on x from Figure 9.23 to weighted least squares regression of y on x from Figure 9.25.

We know that the least squares regression result is heteroskedastic, so the weighted least squares (WLS) coefficients, t statistics, and coefficient of determination are more credible. With WLS the estimation equation changes slightly from

$$y = 5.30 + 0.708x$$

to

$$y = 5.45 + 0.690x$$

Although the estimation equation is not dramatically different, it is a better fit to the data because

Figure 9.24 Scatter Plot of Regression Errors and Values of *y*

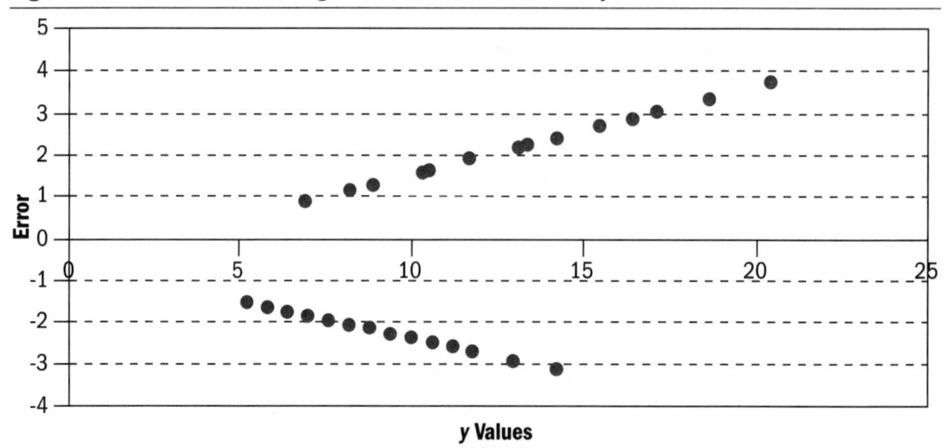

Table 9.11 Data Format and Analysis for Heteroskedasticity Test

Ordered Obs (*i*)	*x*	*e*	e^2	ie^2
1	1.00	0.892015	0.795692	0.795692
2	2.00	-1.51601	2.298286	4.596572
3	2.50	1.179977	1.392347	4.17704
4	3.00	-1.624035	2.63749	10.54996
5	3.20	1.31436	1.727541	8.637707
6	4.00	-1.732061	3.000034	18.0002
7	4.80	1.621519	2.629324	18.40527
8	5.00	-1.840086	3.385916	27.08733
9	5.00	1.659914	2.755315	24.79783
10	6.00	-1.948111	3.795138	37.95138
11	6.30	1.909481	3.646118	40.1073
12	7.00	-2.056137	4.227698	50.73238
13	7.90	2.216641	4.913495	63.87544
14	8.00	-2.164162	4.683597	65.57036
15	8.20	2.274233	5.172135	77.58203
16	9.00	-2.272187	5.162835	82.60537
17	9.10	2.44701	5.987858	101.7936
18	10.00	-2.380213	5.665413	101.9774
19	10.50	2.715775	7.375432	140.1332
20	11.00	-2.488238	6.191329	123.8266
21	11.60	2.926947	8.567017	179.9074
22	12.00	-2.596263	6.740584	148.2928
23	12.30	3.061329	9.371735	215.5499
24	13.00	-2.704289	7.313178	175.5163
25	14.00	3.387686	11.47642	286.9104
26	15.00	-2.920339	8.528383	221.738
27	16.00	3.771635	14.22523	384.0813
28	17.00	-3.13639	9.836943	275.4344
Totals			153.5025	2890.633

Figure 9.25 SPSS Weighted Least Squares Output

Log-Likelihood Values[b]

Power	-2.000	-77.577
	-1.800	-75.896
	-1.600	-74.259
	-1.400	-72.672
	-1.200	-71.140
	-1.000	-69.670
	-.800	-68.269
	-.600	-66.948
	-.400	-65.718
	-.200	-64.592
	.000	-63.589
	.200	-62.727
	.400	-62.030
	.600	-61.524
	.800	-61.234
	1.000	-61.187[a]
	1.200	-61.406
	1.400	-61.906
	1.600	-62.694
	1.800	-63.767
	2.000	-65.112

ANOVA

	Sum of Squares	df	Mean Square	F	Sig.
Regression	41.693	1	41.693	58.284	.000
Residual	18.599	26	.715		
Total	60.291	27			

Coefficients

	Unstandardized Coefficients		Standardized Coefficients			
	B	Std. Error	Beta	Std. Error	t	Sig.
(Constant)	5.452	.599			9.105	.000
x	.690	.090	.832	.109	7.634	.000

a. The corresponding power is selected for further analysis because it maximizes the log-likelihood function.

b. Dependent variable: y, source variable: x

Table 9.12 Comparison of Ordinary Least Squares and Weighted Least Squares Results

	Ordinary Least Squares	Weighted Least Squares
S_b	0.105	0.090
t statistic on *b*	6.74	7.63
Estimate of *b*	0.708	0.690
SSE	153.5	18.599
SST	421.96	60.291
R^2	0.636	0.692

the least squares equation did not completely minimize the sum of squared error.

Analysis of heteroskedasticity is typically not as simple as this when working with real-world data. The pattern of error variability is usually not as distinct, and the choice of the independent variable associated with error variance is not always obvious. When a model contains several independent variables, which is the topic of the next chapter, it may be necessary to look for systematic relationships between regression errors and numerous independent variables.

Real-World Case Study: Part 5

Price as a Function of Improved Living Area

Generally speaking, townhome sale prices are directly related to improved living area. We expect, all else being equal, that larger units will sell at higher prices. The **scatter plot below** of 129 sale prices and improved living areas from the case study data is consistent with this expectation. The correlation between price and living area is approximately 0.72.

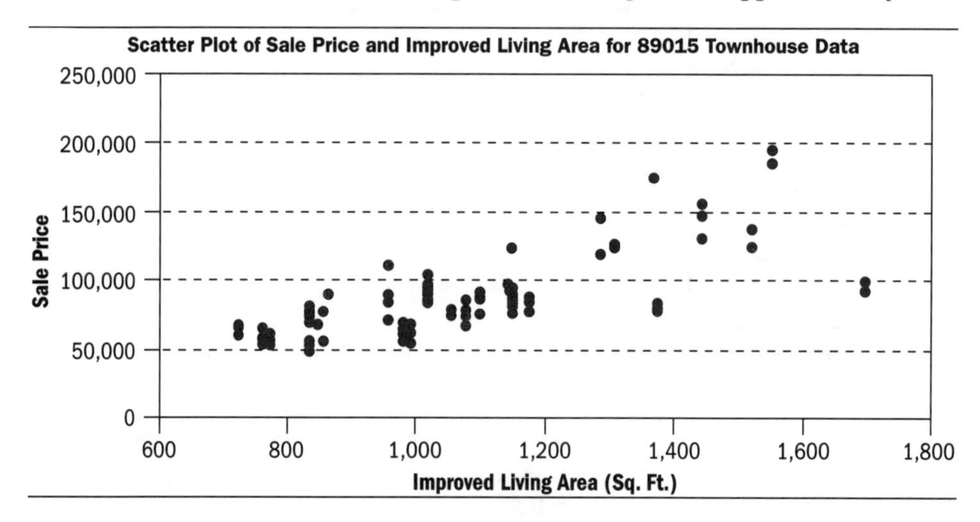

The relationship between price and living area appears to be linear, so a simple linear regression of sale price on improved living area is a reasonable means of assessing how much of the variation in price is "explained" by living area. The **table below** presents the Excel regression output for these data.

Excel Regression Output for Townhouse Sale Price

SUMMARY OUTPUT

Regression Statistics	
Multiple R	0.72046939
R Square	0.51907614
Adjusted R Square	0.51528934
Standard Error	17587.3887
Observations	129

ANOVA

	df	*SS*	*MS*	*F*	*Significance F*
Regression	1	42399544654	42399544654	137.0751	6.31463E-22
Residual	127	39283162771	309316242.3		
Total	128	81682707425			

	Coefficients	*Standard Error*	*t Stat*	*P-value*	*Lower 95%*	*Upper 95%*
Intercept	-12523.7418	8437.598187	-1.484278055	0.140213	-29220.22528	4172.741617
Living Area (SF)	91.0136858	7.773694493	11.70790618	6.31E-22	75.63094787	106.3964237

This simple linear regression model suggests that variation in improved living area is associated with 51.9% of the variation in sale price for these townhomes. Expected price is identified by the following regression equation:

Price = -$12,524 + $91.01(Improved Living Area)

The model indicates that price changes by $91.01 for each one square foot change in improved living area across the range of living area contained in the data. The relationship is significant. $S_b = 7.77$ and the *t* statistic is

$$t = \frac{91.01 - 0}{7.77} = 11.71$$

which you can read directly from the Excel output.

The 95% confidence interval on the change in price associated with variation in living area is $75.63 to $106.40. At first glance, the model seems to do a good job of forecasting sale price. However, appraisal training and experience suggest that influences on townhome prices are likely to be more complex than the relationship captured in this simple linear equation.

One way to determine if there may be other factors at play in explaining townhome price is to look at a scatter plot of regression errors and sale prices as in the illustration below. The error scatter plot looks similar to Panel B of Figure 9.21, which appeared earlier in the chapter. This suggests that a systematic linear relationship exists between regression error and sale price even after accounting for the influence of living area on price. Although the results are not shown in detail here, linear regression of the error terms on price reveals that 48% of the regression error variance is accounted for by the dependent variable price. The systematic relationship between regression error and price is illustrated by the regression line in the **scatter plot on the following page**.

Simple linear regression employing living area as the only explanatory variable appears to be insufficient for building a credible townhouse price model, and additional systematic error can

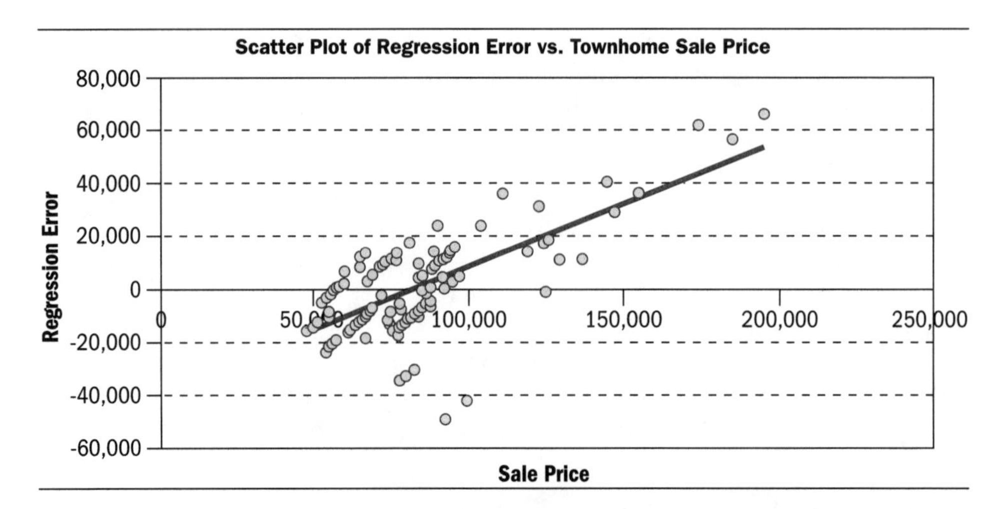

be accounted for by inclusion of omitted variables. Multiple regression analysis–the subject of Chapter 10–allows variables to be added to a model to improve explanatory power. Possible significant omitted variables (i.e., elements of comparison) for these data might include date of sale, room count, lot size, property age, parking arrangement, and other factors. We refine and improve our townhouse price model in the next chapter.

Multiple Linear Regression Analysis

From the Simple Linear Regression Model to the Multiple Linear Regression Model

Chapter 9 introduced the concept of the linear equation and how to derive a linear regression estimate of the underlying linear relationship. We used sample data to infer a linear relationship between variables y and x of the form

$$\mathbf{y} = \alpha + \beta\mathbf{x} + \varepsilon$$

where ε indicates that there is a random element to the response variable y in the population being studied. In addition, we discovered that a single x variable is sometimes incapable of explaining all of the systematic (nonrandom) variation in y. Multiple linear regression overcomes this shortcoming by expanding the simple, one-independent-variable model to include two or more explanatory variables. The least complex multiple linear regression equation is of the form

$$\mathbf{y} = \mathbf{a} + \mathbf{b}_1\mathbf{x}_1 + \mathbf{b}_2\mathbf{x}_2 + \mathbf{e}$$

which is used to infer an underlying population relationship between a dependent variable y and two explanatory variables x_1 and x_2. As with simple linear regression, a is an estimate

> Multiple regression modeling creates new complexities in the analysis of data. Nevertheless, the underlying principles are unchanged from simple linear regression. For this reason Chapter 10 does not repeat a number of important topics such as assumptions underlying regression modeling and inference, the sampling distribution of the mean of *y*, and curve fitting. Consequently, Chapters 9 and 10 are best read together and in sequence.

of the y-intercept (the value of y when both x_1 and x_2 are equal to zero), b_1 estimates the slope of the relationship between x_1 and y, and b_2 estimates the slope of the relationship between x_2 and y. The population equation being inferred via a two independent variable multiple linear regression analysis is therefore

$$\mathbf{y} = \alpha + \beta_1 \mathbf{x}_1 + \beta_2 \mathbf{x}_2 + \varepsilon$$

Multiple linear regression equations can accommodate more than two independent variables, so the population linear equation can be written more generally as

$$\mathbf{y} = \alpha + \beta_1 \mathbf{x}_1 + \beta_2 \mathbf{x}_2 + \ldots + \beta_k \mathbf{x}_k + \varepsilon$$

where k indicates the number of explanatory variables in the model. Inclusion of more than one explanatory variable increases the calculation complexity of linear regression analysis. Calculating model coefficients a and b_1 through b_k by hand requires linear algebra, which is a cumbersome process beyond the mathematical prerequisites for this book, so we will rely on the computer to handle that task. More important for our purposes than working the algebra are a variety of higher-level decisions about the statistical model:

- How many and which independent variables to include in a model
- The functional form of the relationships between the response variable and the various explanatory variables
- How to account for interrelationships between the independent variables that confound the interpretation of the explanatory variable coefficients

The goal of multiple linear regression is the same as the goal of simple linear regression–to derive a model that provides the best fit for the sample data. Again, the definition of "best" fit is the model that minimizes the sum of squared errors. As you will recall from Chapter 9, this can be written as

$$\mathbf{min} = [\sum_{i=1}^{n} (\mathbf{y}_i - \hat{\mathbf{y}}_i)^2] = \mathbf{min[SSE]}$$

where n represents sample size and i identifies each observation of the dependent variable y. The "hat" over the y_i variable signifies the value of the dependent variable estimated by the model when $x_1 = x_{1i}$, $x_2 = x_{2i}$, and $x_k = x_{ki}$. In the equation above, the differences between the actual y values in the sample and estimated y values from the regression model reflect the error term (e) in the regression equation. The minimized SSE estimate provides a measure of randomness in the underlying population relationship (ε), assuming a well-specified model and adequate data.

Inference, confidence intervals for the regression coefficients, prediction, and prediction confidence intervals are all similar to simple linear regression. However, there are some subtle differences in what these entail in multiple linear regression modeling, which are illustrated within an example problem based on the data in **Table 10.1**. The example problem is quite long because it covers a lot of material.

Example Problem

Table 10.1 contains information on an actual sample of 59 residential lots from a mid-sized southwestern U.S. city. The data include sale price, lot size in acres, and a view score. All of the lots are located in the foothills of a large, highly visible mountain range. City views vary among the lots, and city views are important determinants of lot value in this market. View scores provide a means to differentiate among city views. The view score is a measure of city view panorama in degrees of a semicircle, which can vary from no city view of 0 degrees to a full, 180-degree city view extending uninterrupted across the city-facing portion of the lot.

The hypothesized relationship among the Table 10.1 variables is

$$\textbf{PRICE} = \alpha + \beta_1\textbf{(ACRES)} + \beta_2\textbf{(VIEW)} + \varepsilon$$

A multiple linear regression model derived from the sample results in the following equation:

$$\textbf{PRICE} = \$89{,}347 + \$20{,}438\ \textbf{(ACRES)} + \$1{,}121\textbf{(VIEW)}$$

Table 10.1 59 Sales of Residential View Lots

Price	Size (AC)	View	Price	Size (AC)	View
160,000	1.579	30	135,000	0.632	20
126,000	1.269	30	90,000	0.500	0
125,000	1.141	15	146,250	0.848	25
135,000	0.827	15	124,815	0.830	20
133,500	0.843	20	130,000	0.830	20
90,000	1.073	30	345,000	1.766	150
130,000	1.185	30	89,474	0.830	0
325,500	2.192	100	185,000	1.208	90
155,000	1.295	30	156,000	0.839	40
125,000	0.950	15	159,000	0.829	30
150,000	1.295	30	220,200	1.022	90
180,000	1.642	20	153,000	0.831	30
168,500	1.579	30	156,000	0.830	30
153,500	1.837	50	320,000	3.340	150
160,000	1.154	60	205,000	1.640	100
185,000	1.645	60	150,000	0.800	40
100,000	1.372	10	160,000	1.010	50
87,650	1.000	0	215,000	1.069	100
125,000	0.636	20	142,500	1.440	20
167,000	0.734	60	300,000	1.027	150
170,000	1.837	50	125,000	2.650	20
185,000	1.050	15	142,500	1.250	100
150,000	1.100	30	110,000	1.230	10
140,000	1.220	25	235,000	0.895	120
140,000	1.100	10	164,000	1.230	10
178,000	1.070	30	290,000	1.670	150
156,750	1.070	20	243,750	1.100	130
149,000	1.130	10	340,000	2.950	160
105,000	1.310	0	275,000	1.280	150
115,000	0.680	15			

Figure 10.1 shows SPSS output from the multiple regression analysis, which best illustrates all of the information relevant to the example problem. (The problem can also be solved using Excel or Minitab.) To interpret this output, we will look at all the regression statistics that we learned about in Chapter 9 as well as some regression statistics not discussed there.

As with simple linear regression, the coefficient of determination (R^2) in multiple linear regression analysis is a measure of how much variation in the response variable *PRICE* is explained by the example's two independent, or explanatory, variables

ACRES and *VIEW*. For the equation derived from this sample, $R^2 = 0.836$.

The SPSS output also includes the statistic R, which is referred to as *Multiple R* in Excel but is not reported in Minitab. The statistic R is actually a Pearson correlation coefficient measuring the correlation between the predicted values of the dependent variable and the actual values of the dependent variable in the sample. Even though Minitab does not report R in its regression output, it can be easily calculated as the square root of R^2. In this case, $R = 0.915$.

The output in Figure 10.1 also includes the *Adjusted* R^2 statistic, which is abbreviated as R^2_{ADJ}. In this example, $R^2_{ADJ} = 0.831$, which is slightly less than R^2. The importance of R^2_{ADJ} stems from the fact that addition of *any* random variable to a

Figure 10.1 SPSS Output for Example Problem

Model Summary

Model	R	R Square	Adjusted R Square	Std. Error of the Estimate
1	.915[a]	.836	.831	26082.57214

a. Predictors: (Constant), View, Acres

ANOVA[b]

Model		Sum of Squares	df	Mean Square	F	Sig.
1	Regression	1.9E+011	2	9.744E+010	143.228	.000[a]
	Residual	3.8E+010	56	680300569.7		
	Total	2.3E+011	58			

a. Predictors: (Constant), View, Acres

b. Dependent Variable: Price

Coefficients[a]

Model		Unstandardized Coefficients		Standardized Coefficients		
		B	Std. Error	Beta	t	Sig.
1	(Constant)	89346.783	8652.239		10.326	.000
	Acres	20438.379	7246.856	.172	2.820	.007
	View	1120.602	83.267	.822	13.458	.000

a. Dependent Variable: Price

regression model will increase R^2, whether or not the new variable is truly relevant to systematic variation in the dependent variable. However, R^2_{ADJ} will increase with the introduction of an additional variable only when the increase is in excess of the expected increase in R^2 that would occur with the introduction of an irrelevant variable.

Recall from Chapter 9 that R^2 is the proportion of SST represented by SSR, which can be expressed in the example problem as

$$R^2 = 1 - \frac{SSE}{SST} = \frac{3.81E10}{2.33E11} = 1 - \frac{3.81}{23.3} = 1 - 0.164 = 0.836$$

Adjusted R^2 is calculated differently, accounting for the number of explanatory variables (k) in the model:

$$R^2_{ADJ} = 1 - \frac{\frac{SSE}{n-k-1}}{\frac{SST}{n-1}} = 1 - \frac{\frac{3.81}{56}}{\frac{23.3}{58}} = 1 - \frac{0.068}{0.402} = 0.831$$

When a variable is added to the model, the divisor of SSE $(n - k - 1)$ decreases in size relative to the divisor of SST $(n - 1)$. The decrease in SSE from the addition of a new variable (i.e., the increase in SSR) must offset division by a smaller number relative to the divisor of SST. Otherwise, R^2_{ADJ} will remain constant or decrease when a new variable is added. When a variable is added to a model and R^2_{ADJ} does not increase, the new variable is explaining no more than would be explained by adding any totally irrelevant random variable. For this reason, some analysts prefer to report R^2_{ADJ} instead of R^2 as a measure of explained variance absent the effect of the number of independent variables. In addition, the examination of change in R^2_{ADJ} as variables are added to a model is a way of assessing whether or not an added variable is actually explanatory. Finally, R^2_{ADJ} will always be less than R^2, and R^2_{ADJ} can be negative when model fit is so poor that it explains less than a model including an equal number of totally random and irrelevant explanatory variables.

Inference in multiple linear regression is similar to inference in simple linear regression. One added inferential measure is the model F statistic.[1] The F statistic tests the null hypothesis that all of the independent variable coefficients are zero versus the alternative hypothesis that at least one is not zero. In other words, it assesses whether or not the *model* provides a better measure of the expected value of y than the mean of y.

Three inferences are relevant to the example problem:

The first inference involves the model F statistic:

H_0: $\beta_{ACRES} = \beta_{VIEW} = 0$

H_a: At least one $\neq 0$

$F = 143.23$, p-value $= 0.000$

H_0 is rejected, and we conclude that at least one of the independent variables is significant.

The second inference involves the t statistic for the lot size variable:

H_0: $\beta_{ACRES} = 0$

H_a: $\beta_{ACRES} \neq 0$

$t = 2.82$, p-value $= 0.007$

H_0 is rejected, and we conclude that lot size is a significant independent variable.

The third inference involves the t statistic for the view variable:

H_0: $\beta_{VIEW} = 0$

H_a: $\beta_{VIEW} \neq 0$

$t = 13.46$, p-value $= 0.000$

H_0 is rejected, and we conclude that view is a significant independent variable.

Confidence intervals can be developed for the independent variable coefficients in the same way they were developed in Chapter 9. The only difference is the degrees of freedom applicable to the confidence interval t statistic. In simple regression degrees of freedom for the t statistic is equal to $n - 2$. In multiple regression, degrees of freedom is equal to $n - k - 1$. (Note that $n - k - 1 = n - 2$ in the simple regression case where k is always equal to

1. A model F statistic is also reported in simple linear regression software output, but it tests the same hypothesis as the t statistic when there is a single explanatory variable.

1.) Degrees of freedom for the example problem is 59 – 2 – 1 = 56. For a 90% confidence interval, $t_{0.05,\,56}$ = 1.673, and the confidence intervals are

90% confidence interval on β_{ACRES} = \$20,438 ± 1.673(\$7,247)

or \$8,314 to \$32,562, and

90% confidence interval on β_{VIEW} = \$1,121 ± 1.673(\$83)

or \$982 to \$1,260.

As in simple linear regression, a multiple linear regression model can be used for prediction. For example, the prediction for *PRICE* when *ACRES* equals 1.2 and *VIEW* equals 80 is

PRICE = \$89,347 + \$20,438.38(1.2) + \$1,120.60(80)

= 203,521

Again, as in simple linear regression, \$203,521 is the expected mean price as well as the expected price for a single outcome. The calculation of prediction confidence intervals in multiple linear regression by hand is too mathematically cumbersome to show here. However, confidence intervals are available for the mean and for a single outcome in both SPSS and Minitab. Derivation of these intervals in SPSS and Minitab is identical to the process shown in Chapter 9, and it would be a good idea to review this topic from the prior chapter and to replicate the results below. (Excel does not provide prediction confidence intervals in its regression procedure.) The 90% confidence intervals for *PRICE* when *ACRES* = 1.2 and *VIEW* = 80 are

90% confidence interval for $\mu_{y|x_1,\,x_2}$ = \$196,195 to \$210,847

and

90% confidence interval for $y|x_1, x_2$ = \$159,286 to \$247,756

One important feature of using multiple regression analysis in appraisal is the ability to include an expression of precision along with a point estimate value opinion. If the example data were being used to develop an opinion of market value, the value opinion of $203,500 (rounded) could be accompanied with a statement to the effect that $203,500 is an estimate of most probable price and we are 90% confident that the expected price (i.e., mean price, most probable price, market value) lies within a range of $196,000 to $211,000. This statement dramatically alters what is provided in a traditional appraisal, which provides a point estimate value opinion without any statement concerning the precision of the estimate.

As stated in Chapter 9 but worth repeating here, predictions concerning the response variable *y* based on given values for the explanatory variables fall into two categories: predictions concerning the *mean of y* and predictions concerning a single outcome of *y*. Generally speaking, appraisers are more interested in predicting the *mean of y* given specific values of the explanatory variables because definitions of market value and market rent focus on most probable price and most probable rent rather than predicting a particular transaction price or rent. Therefore, confidence intervals for the mean are more appropriate than prediction intervals for a single outcome[2] in appraisal applications involving market value or market rent.

The last topic illustrated by this example problem is the *Standardized Beta Coefficient* provided in the SPSS output included in Figure 10.1.[3] The beta coefficient for each variable is another way to

2. Some analysts reserve the phrase "prediction interval" for use with a confidence interval for a single outcome and refer to a confidence interval for the mean outcome as "confidence interval." This can be confusing because the phrase "confidence interval" has additional meaning in statistics, and the confidence interval for predicting the mean is in fact a prediction interval. Use of more precise language helps avoid this sort of confusion.
3. This information is not provided in Excel or in Minitab. Beta coefficients are calculated by regressing standardized *y* values on standardized *x* values. Beta coefficients can be generated fairly easily in Excel or Minitab by standardizing all of the data inputs prior to running the regression. For example, a standardized x_i variable value would be $(x_i - \bar{x})/S_x$ and a standardized y_i value would be $(y_i - \bar{y})/S_y$.

measure the effect of the x variable on the y variable. In the example problem, the beta coefficient on *ACRES* is 0.172, which means that a change of 1 standard deviation in *ACRES* results in a change of 0.172 standard deviation in *PRICE*. Likewise, a change of 1 standard deviation in *VIEW* results in a change of 0.822 standard deviation in *PRICE*. Therefore, *PRICE* is more sensitive to variation in *VIEW* than to variation in *ACRES* in the market modeled by this equation.

What does this mean in an appraisal context? The regression coefficients on *ACRES* and *VIEW* indicate the contributions of these variables to price when all else is held constant. In an appraisal context, the coefficients are the adjustment factors used to adjust comparable properties to the subject property–assuming that the data used to derive the regression model are representative of the subject property and its market. Regression

Practice Problem 10.1

The data set below is a sample of 20 condominium sales in a market where the elevation of the unit above ground level is an important determinant of market value. Assume normality for the purpose of this exercise, and develop and interpret an equation to estimate *PRICE* as a function of living area (*SF*) and elevation above ground (*FLOOR*). Derive and interpret the model F statistic, t statistics on the two coefficients, R, R^2, and R^2_{ADJ}. Setting α at 0.05, calculate confidence intervals on the coefficients for *SF* and *FLOOR*. Develop a prediction of the mean price when $SF = 1{,}240$ and $FLOOR = 11$. What is the 95% confidence interval for this mean price prediction? Derive and interpret the beta coefficients on *SF* and *FLOOR*.

Price	SF	Floor
519,000	1,200	4
539,000	1,350	3
572,000	1,100	7
455,000	1,180	2
520,000	1,240	2
618,000	1,200	9
520,000	1,390	1
550,000	1,220	5
669,000	1,160	10
795,000	1,450	12
540,000	1,325	4
612,000	1,400	6
645,000	1,310	8
492,000	1,160	3
650,000	1,220	9
620,000	1,350	7
529,000	1,180	5
670,000	1,320	8
710,000	1,250	12
479,000	1,100	3

models facilitate derivation of numerous adjustment factors from large amounts of data, which would be overwhelming and possibly inaccurate if an appraiser were to rely solely on paired sales. The case study analysis at the end of this chapter provides a good illustration of this sort of problem.

Unusual Observations

Outliers

In Chapter 9, we discussed outliers briefly, identifying them visually as the observations with unusually large regression errors in Panel C of Figure 9.21. Keep in mind that outliers are not necessarily bad. Outliers provide valuable information when they accurately reflect the extent of uncertainty present in the underlying population. Outliers can also be misleading if they misrepresent the amount of randomness in the underlying population. Misleading outliers can be a consequence of data entry errors (the observation's x values, y values, or both may not have been accurate entered into the data set), use of unrepresentative data, or failure to correctly specify the regression model. An example of unrepresentative data would be one observation of Class C office building rent in a sample of Class A office rent data. Incorrect model specification could involve omission of a variable that accounts for some of the outlier error or failure to correctly model the functional form of the relationship.

Outlier detection takes advantage of what we know about the normal distribution. Although any value is theoretically possible when a distribution is normal, approximately 95% of the regression errors should lie between -2 and +2 standard deviations from the mean. Therefore, it is beneficial to revisit observations with estimation errors that exceed 2 standard deviations from the mean, assessing data entry accuracy, representativeness, and any omitted characteristic of the data or other model specification errors that may account for the large departure from the mean.

A good way to identify outlying errors is to convert nominal error to standard error. This is done the same way you standardize any variable–subtract mean error from estimated error for each

> **Standard Error in Excel, SPSS, and Minitab**
>
> Excel will calculate and list standard errors for each observation when you check the **Standardized Residuals** box in the **Regression** window. In SPSS you can elect to save standard errors to your data set by clicking the **Save** button in the **Linear Regression** window and checking **Standardized.** In Minitab click the **Results** button in the **Regression** window and check **In addition, the full table of fits and residuals.** The program output will list standard residuals under the column called "St Resid," flagging residuals in excess of 2 standard deviations. The program also provides a list limited to "unusual observations" as one of its regular output options, with standard residuals greater than 2 flagged.
>
> SPSS offers another version of standard residual called a "studentized residual." Studentized residuals are computed by accounting for nonconstant regression error, providing a means to make violation of regression assumptions more apparent in some situations.

observation and divide by the standard deviation of error. Since mean error is forced to zero in a linear regression model, this term drops out of the calculation, leaving each observation's estimated error divided by the standard deviation of error:

$$\textbf{Standard Error} = \frac{e_i}{S_e}$$

Once this has been done, you can easily identify large standard errors (Minitab automatically does this for you) and reassess the observations, looking for data entry errors, unaccounted-for characteristics of the observation, representativeness of the observation, and the like.

Leverage

Leverage identifies unusual sets of independent variables by measuring the distance an observation's independent variables are from the independent variable centroid. The centroid of the x variables is the set of means of all of the explanatory variables $(\bar{x}_1, \bar{x}_2, \ldots, \bar{x}_k)$.

For example, a 6-bedroom, 7-bath house with a living area of 6,000 square feet would be identified as having high leverage if most of the independent variable observations in the data set were between 1,500 and 2,500 square feet with 2 or 3 bedrooms and 2 to 4 bathrooms. An observation with leverage may or may not unduly influence the regression equation.

Figure 10.2 plots independent variable observations x_1 and x_2, which could be used to develop a multiple linear regression equation of the form $y = a + b_1x_1 + b_2x_2 + e$. The independent variable centroid $(\bar{x}_1, \bar{x}_2)$ is identified by the unfilled circle and the data points are identified by filled circles. Data pair $(x_{1,i}, x_{2,i})$ is more distant from the independent variable centroid than the remaining independent variable pairs, therefore it has the most leverage. Leverage (h) is calculated in SPSS and in Minitab (but not in Excel). The criterion suggested to identify unusual observations based on leverage is

$$h > \frac{2(k+1)}{n}$$

where k is the number of independent variables in the model and n is the sample size. As with outliers, unusual observations resulting from leverage should be reexamined to ensure that there are no data entry errors and that the observations are representative of the population being studied.

Leverage in Minitab and SPSS

The h value is the criterion used by Minitab to flag observations with high leverage. SPSS allows you to add leverage values for each observation to the data set for further analysis by checking the **Leverage values** box in the **Save** window prior to running a regression analysis. SPSS does not flag unusual leverage, leaving the calculation to the analyst.

Figure 10.2 Leverage with Two Independent Variables

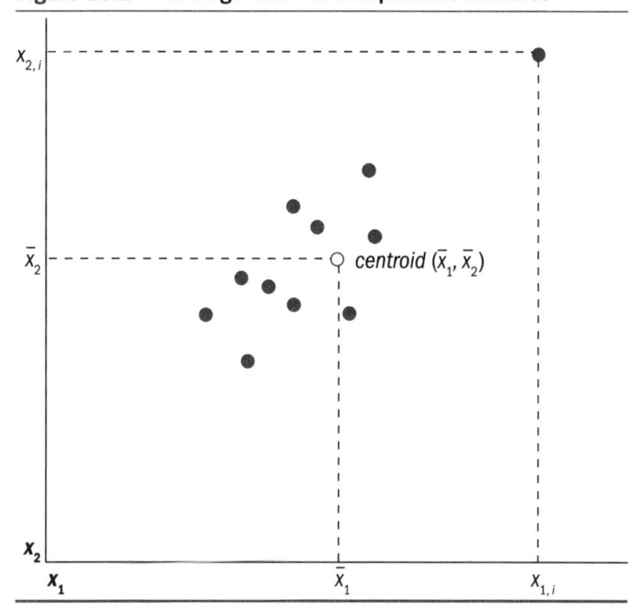

Outliers and Leverage Combined

Outliers are extreme y values that are poorly explained by the regression model and result in large regression errors. Leverage identifies sets of extreme x values that have the potential to unduly influence the regression equation. *DfFit* is a statistic available in SPSS that combines leverage and outliers into a single measure intended to identify observations with excessive influence on the regression equation. *DfFit* for each observation i is the change in the predicted value of the dependent variable when observation i is omitted from the calculation. A more easily applied statistic in SPSS is *Standardized DfFit*,

Standardized *DfFit* in SPSS

SPSS will calculate *Standardized DfFit* for each observation and add the value to the data set for further analysis when you check the **Standardized DfFit** box in the **Save** window.

which allows you to identify *DfFit* values that are more than 2 standard deviations from mean *DfFit.*

As with other unusual observation measures, a high *Standardized DfFit* value identifies observations that should be reexamined. High *Standardized DfFit* values, identification as an outlier, and high leverage in and of themselves are insufficient justification for removing an observation from an analysis. These tools are designed to identify unusual observations in terms of how the observations conform to the sample. If upon reexamination unusual observations prove to be representative of the population you are studying, then they most likely provide valuable information and should not be discarded.

Practice Problem 10.2

The data below contain 24 observations on townhouse sale price and two independent variables, living area in square feet and property age. Regress *PRICE* on *SF* and *AGE*, calculate standardized residuals, and identify outliers using the absolute standardized residual greater than 2 criterion. If you have software capable of providing other unusual observation statistics, calculate and analyze leverage and *Standardized DfFit*.

PRICE	*SF*	*AGE*
85,000	1,019	7
80,500	1,151	9
71,500	836	9
89,900	1,019	3
91,900	1,696	18
74,500	1,101	20
54,000	992	22
82,500	1,151	9
52,900	772	33
83,950	1,151	9
83,900	1,151	9
78,000	1,056	18
63,000	984	21
56,900	984	21
91,000	1,019	3
84,000	1,019	8
89,900	1,019	3
61,500	984	21
84,500	1,151	9
54,750	984	21
56,000	772	33
84,500	1,151	9
85,000	1,151	9
82,500	1,376	20

Heteroskedasticity and Multicollinearity

Heteroskedasticity

In Chapter 9, we looked at heteroskedasticity, and now we're ready to address the topic in the context of multiple independent variables. The pictorial analysis of heteroskedasticity is no different in multiple linear regression than it was in simple linear regression–it entails detection of changing error variance with changing the values of the dependent variable through examination of a scatter plot. If the model fit is reasonably good, errors can be charted against either y_i or $\hat{y}_i$. If the fit is not good, then it is

Generating Error (Residual) Plots in SPSS and Minitab

SPSS and Minitab provide a number of residual plot options. Excel only provides preprogrammed error plots against the values of the independent variables. Other charts are easy to create in Excel, however, by use of the **Residuals** and **Standardized Residuals** outputs available in the **Regression** window.

Minitab offers a fairly complete menu of residual plots, which are available by selecting the **Graphs** button in the **Regression** window. The Minitab **Graphs** window is shown below. It provides for numerous residual plots that are useful for assessing heteroskedasticity and normality (e.g., histogram and normal plot of residuals). Selecting **Residuals versus fits** as shown generates a plot of residuals versus $\hat{y}_i$. Also, adding the variable "Price" in the **Residuals versus the variables** dialog box as shown generates a plot of residuals against y_i. (In this illustration, price is the response variable for the data.) You can also plot standardized residuals to facilitate outlier identification. **Deleted** residuals in the Minitab window are standardized errors computed for each observation by leaving the observation out of the model and predicting the outcome using the remaining variables. This prevents each observation from contributing to predicting itself. Deleted residuals follow a t distribution with $n - k - 1$ degrees of freedom, which provides a different outlier cutoff point for 95% of the observation errors compared to the "±2 rule" used for standardized residuals.The **Residuals versus order** option is of benefit only when the order of data entry is important.

Minitab Regression "Graphs" Window

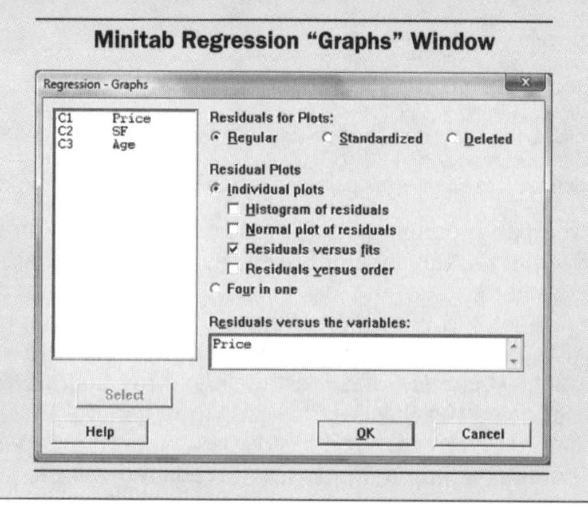

probably better to chart errors against $\hat{y}_i$. When in doubt, create two scatter plots, and if either chart looks suspect, then further analysis is warranted.

Identifying the Source of Nonconstant Error

When you suspect model heteroskedasticity, you must identify which of the x variables is the primary source of the systematic change in error variance before you can calculate Q (see Chapter 9) and test the hypothesis of no heteroskedasticity using the Z distribution. In some circumstances this knowledge is intuitive. For example, the culprit is usually improved living area in residential price models.

Absent strong intuition and experience regarding the source of error variance, independent variables can be eliminated from consideration by

The linear regression plots window in SPSS is shown below.

SPSS Linear Regression "Plots" Window

Linear Regression: Plots
DEPENDNT
*ZPRED
*ZRESID
*DRESID
*ADJPRED
*SRESID
*SDRESID
Scatter 1 of 1
Previous
Next
Y:
X:
Continue
Cancel
Help
Standardized Residual Plots
Histogram
Normal probability plot
Produce all partial plots

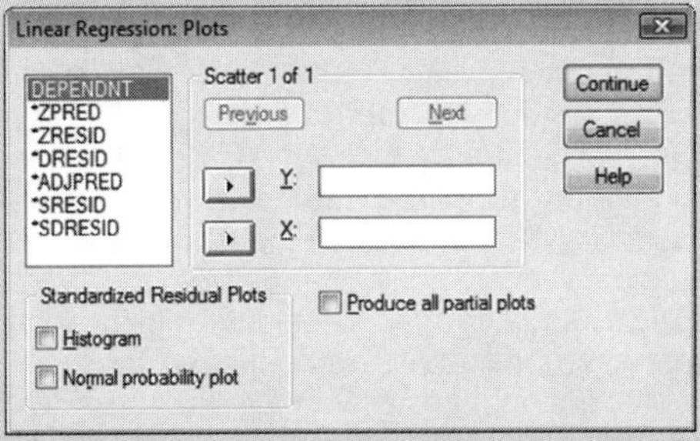

Its use is not as straightforward as the Minitab window. The menu on the left of the **Linear Regression: Plots** window lists the variables you can select for error examination. The user controls the axis chosen for selected variables by placement in the **Y:** or **X:** box to the right of the variable menu. Available variables are

DEPENDNT	dependent variable y
ZPRED	standardized predicted values of y
ZRESID	standardized regression errors
DRESID	residuals with ith case omitted
ADJPRED	predicted value of y with ith observation omitted
SRESID	studentized residuals
SDRESID	studentized residuals with ith observation omitted

As with Minitab, you can produce plots for assessing normality in SPSS by checking **Histogram** and **Normal probability plot** boxes. **Partial plots** are scatter plots of the dependent variable against each of the independent variables. Although SPSS provides a number of charting options in this window, it does not provide the option of generating a plot of actual (nominal) regression error against y or $\hat{y}$. In order to examine these two charts you must save the **Unstandardized residuals** and the **Unstandardized predicted values** to your data set using the **Save** window and then use the SPSS **Graphs** utility to create a scatter plot of unstandardized residuals against the dependent variable or unstandardized residuals against unstandardized predicted values.

plotting the absolute values of regression errors against each of the x variables to identify x variables that are uncorrelated with the size of regression error. The remaining x variables can then be regressed on the absolute values of errors to identify the variable that best explains error variance.[4] The explanatory variable that explains the greatest amount of error variance can then be incorporated into the analysis of Q (to array errors) and, if warranted by the Q statistic, used as a weighting variable in a weighted least squares model.[5]

Example Problem

The data on 24 condominium sales from Practice Problem 10.2 are used for this example. A scatter plot of regression errors in a regression of price on living area and age (**Figure 10.3**) shows that regression error tends to be relatively large for lower priced units and relatively small for higher priced units. Absolute values of error are derived (we are concerned about the size of error here, not the direction) and regressed on age and living area. The

Figure 10.3 Regression Error versus the Dependent Variable Price

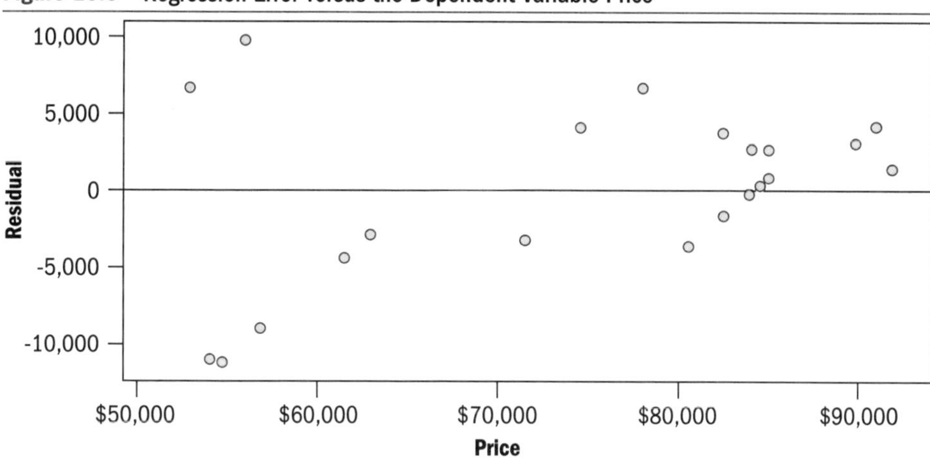

4. Measuring correlation between each of the x variables and the absolute values of regression error will also accomplish this.
5. Some statistical programs included more advanced features for dealing with heteroskedasticity. For example, *Stata* and *EViews* are statistical packages that allow the user to invoke a procedure that calculates "robust standard errors." Standard errors calculated in this manner are resistant to any loss of reliability due to non-constant error variance. This allows the analyst to test variable significance knowing that *t* statistics are credible.

R^2 for absolute error as a function of age is 0.429 and the correlation coefficient for absolute error versus age is 0.655. Living area is not as highly correlated with absolute error, so the nonconstant regression error illustrated in Figure 10.3 appears to be explained mostly by the age variable.

Figure 10.4 shows the relationship between the absolute values of regression error and age. Absolute error appears to lie within a range of approximately \$0 to \$4,000 when age is less than 10, and lie within an approximate range of \$1,000 to \$11,000 when age is greater than 15. There is no data for the 10 to 15 age range. Since older units are expected to sell for lower prices, all else equal, this relationship between error and age is consistent with the observation of greater error variance in lower-priced units illustrated in Figure 10.3.

Based on this information, regression error should be arrayed in ascending order by property age and tested for heteroskedasticity as we did in Table 9.11. As shown in **Table 10.2**, when this is done, $Q = 3.17$ and we can reject the null hypothesis that error variance is constant.

Multicollinearity

In an ideal world regression coefficients on each explanatory variable would fully and independently reflect the relationship between the explanatory variable and the response variable. This ideal is seldom achieved, however.

Consider housing price models. Important explanatory variables usually include improved living area, number of bedrooms, number of bathrooms, and perhaps lot size, among others. Larger homes tend to have more bedrooms, more bathrooms, and often occupy larger lots. These explanatory variables can be highly correlated with each other. Together they capture a latent construct we could call "home size." Individually they tend to "speak for each other" in a linear regression model. Strong correlation among independent variables and the tendency for groups of variables to share explanatory power is referred to as *multicollinearity* in regression analysis.

When variables are highly correlated, the regression coefficients on the multicollinear

Figure 10.4 Absolute Value of Regression Error versus Age

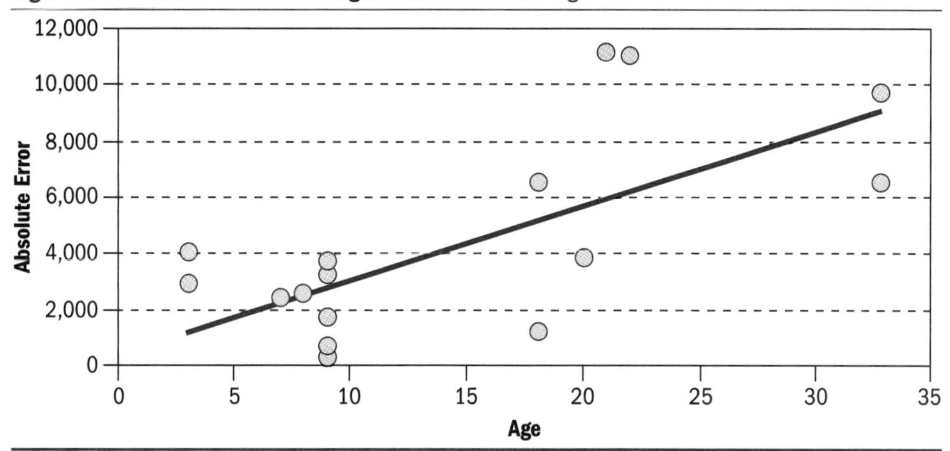

Table 10.2 Calculation of Heteroskedasticity Statistic Q for the Example Data

Ordered Obs (i)	x	e	e^2	ie^2
1	3	2,978.6	8,872,058	8,872,058
2	3	4,078.6	16,634,978	33,269,956
3	3	2,978.6	8,872,058	26,616,174
4	7	2,504.0	6,270,016	25,080,064
5	8	2,610.4	6,814,188	34,070,941
6	9	-3,735.3	13,952,466	83,714,797
7	9	-3,304.2	10,917,738	76,424,163
8	9	-1,735.3	3,011,266	24,090,129
9	9	-285.3	81,396	732,565
10	9	-335.3	112,426	1,124,261
11	9	264.7	70,066	770,727
12	9	264.7	70,066	840,793
13	9	764.7	584,766	7,601,959
14	18	1,304.5	1,701,720	23,824,084
15	18	6,566.2	43,114,982	646,724,737
16	20	3,931.6	15,457,479	247,319,657
17	20	3,698.1	13,675,944	232,491,041
18	21	-2,959.0	8,755,681	157,602,258
19	21	-9,059.0	82,065,481	1,559,244,139
20	21	-4,459.0	19,882,681	397,653,620
21	21	-11,209.0	125,641,681	2,638,475,301
22	22	-11,092.2	123,036,901	2,706,811,818
23	33	6,564.6	43,093,973	991,161,383
24	33	9,664.6	93,404,493	2,241,707,836
		Totals	646,094,504	12,166,224,460

$$h = \frac{\sum_i i\hat{e}_i^2}{\sum_i \hat{e}_i^2} = 18.83041 \qquad Q = \left(\frac{6n}{n^2 - 1}\right)^{1/2}\left(h - \frac{n+1}{2}\right) = 3.16796$$

variables may be counterintuitive and are often unreliable. The coefficients are characterized as being "unreliable" because dropping one of the highly correlated variables can cause a dramatic change in one or more of the coefficients of the other correlated variables remaining in the model as the remaining variables each account for some of the variance explained by the dropped variable. The multicollinearity quandary is twofold:

- Leaving all of the variables in the model results in shared explanatory power. This results in the possibility of the appearance of a lack of significance for variables that are actually significant, and the possibility that the signs associated with some of the variable coefficients may be counterintuitive.
- Omitting some of the correlated variables results in changes in the coefficients on the remaining correlated variables and loss of predictive power when all of the omitted variable's explanatory power has not been accounted for by the variables left in the model.

Testing for Multicollinearity

Several tests for multicollinearity exist, some more formal than others. A variety of ways to detect multicollinearity, including references that are good sources for more in-depth discussions of this topic, are as follows:

- Pearson correlation coefficients among pairs of independent variables are high. Dielman[6] suggests $r > 0.50$, and Hair, et al.[7] suggest $r \geq 0.90$ as criteria for "high."
- Important independent variables are insignificant in the model.
- The model is significant (i.e., a significant F statistic) but has no, or very few, significant t statistics.
- Signs on some coefficients are the opposite of what is expected based on theory or prior experience.

---

6. Terry Dielman, *Applied Regression: Analysis for Business and Economics*, 3rd ed. (Pacific Grove, Calif.: Duxbury/Thomson Learning, 2001).
7. Joseph F. Hair, Ralph E. Anderson, Ronald C. Tatham, and William C. Black, *Multivariate Data Analysis with Readings*, 3rd ed. (New York: Macmillan, 1992).

- Variable coefficients change substantially when a variable is dropped or added to the model.
- A variance inflation factor (*VIF*) on any variable is greater than 10. The *VIF* statistic is based on the coefficient of determination (R^2_j) between one independent variable *j* and the other independent variables in the model, as follows:

$$\mathrm{VIF}_j = \frac{1}{(1 - \mathrm{R}^2_j)}$$

- Mean *VIF* value is considerably greater than 1.
- *VIF* is high in relation to the overall fit of the model, i.e., $VIF > (1 - R^2)^{-1}$, where R^2 is the model coefficient of determination when all of the variables are included.[8] When a *VIF* value exceeds this threshold, the independent variable is related more closely to other independent variables than it is to the dependent variable.

Variance Inflation Factors in SPSS and Minitab

*VIF*s are available for each independent variable in SPSS output by clicking the **Statistics** button in the **Linear regression** window and then checking the **Collinearity diagnostics** box. In Minitab click the **Options** button in the **Regression** window and then select the **Variance inflation factors** box. This statistic is not available in Excel, although it can be calculated using Excel's spreadsheet capabilities and analysis tools.

Correcting for Multicollinearity

Some of the corrections for multicollinearity are beyond the scope of this book, including principal component analysis and ridge regression.[9] Others include adding observations or removing one or some of the correlated variables, which introduces the problem of losing information provided by the

8. Rudolph J. Freund and William J. Wilson, *Regression Analysis: Statistical Modeling of a Response Variable* (San Diego, Calif.: Academic Press, 1998).
9. See Hair, et al., for principal component analysis. For ridge regression, see John Neter, William Wasserman, and Michael H. Kutner, *Applied Linear Statistical Models*, 3rd ed. (Homewood, Ill.: Irwin, 1990); Eruico J. Ferreira and G. Stacy Sirmans, "Ridge Regression in Real Estate Analysis," *The Appraisal Journal* 56, no. 3 (July 1988): 311-319; Alan K. Reichert, James S. Moore, and Chien-Ching Cho, "Stabilizing Statistical Models using Ridge Regression," *The Real Estate Appraiser and Analyst* 51, no. 3 (Fall 1985): 17-22; and John E. Anderson, "Ridge Regression: A New Regression Technique Useful to Appraisers and Assessors," *The Real Estate Appraiser and Analyst* 45, no. 6 (November/December 1979): 48-51.

omitted variables. Adding data does not always work, however, and this solution is usually either costly or impossible.

Fortunately for appraisers, the presence of multicollinearity has no effect on how well a model predicts $\mu_{y \mid x_1 \ldots x_k}$ or on goodness of model fit. Therefore, a well-specified model can be used to derive market value and market rent opinions despite the presence of multicollinearity. One caution, however, is to avoid the temptation to exclude variables that appear to be insignificant when multicollinearity is present. Multicollinearity may be masking the predictive contributions of these variables. If market experience or theory suggests that a multicollinear variable is an important price or rent determinant, then the variable should stay in the model even though it appears to be insignificant.

Example Problem

Table 10.3 contains data on 41 home sales, including *PRICE*, *LIVING AREA*, *BEDROOMS*, and *LOT AREA*. Sale price is positively correlated with each independent variable, as follows:

$$\mathbf{r}_{\text{PRICE, LIVING AREA}} = \mathbf{0.610}$$

$$\mathbf{r}_{\text{PRICE, BEDROOMS}} = \mathbf{0.282}$$

$$\mathbf{r}_{\text{PRICE, LOT AREA}} = \mathbf{0.184}$$

Based on these correlations and what we know about housing prices, we expect a regression model of PRICE on LIVING AREA, BEDROOMS, and LOT SIZE to include positive coefficients for each of these variables.

The multiple linear regression model for the Table 10.3 *PRICE* data is

$$\begin{aligned}\textbf{PRICE} = \$177{,}251 &+ \$112.90(\textbf{LIVING AREA})\\ &+ \$29{,}447(\textbf{BEDROOMS})\\ &- \$26.78(\textbf{LOT AREA})\end{aligned}$$

The model F statistic is highly significant, but the t statistic for *BEDROOMS* is not significant and the t statistic for *LOT AREA* is moderately significant. In addition, the sign on the *LOT AREA* variable is the opposite of what we expect, suggesting that lot value *decreases* by approximately $27 per square

Table 10.3 Data on 41 Home Sales

Sale Price	Living Area (SF)	Bedrooms	Lot Area (SF)
270,000	1,700	4	9,240
344,000	2,296	4	8,910
316,000	2,302	4	8,910
312,000	2,752	5	10,620
320,000	2,752	5	10,620
206,000	1,676	3	7,500
320,000	2,302	4	8,910
352,000	2,200	4	7,650
268,000	2,160	3	7,203
230,000	1,968	4	7,500
310,000	2,302	4	8,910
340,000	2,038	3	7,650
322,000	2,302	4	8,910
286,000	1,672	4	9,240
359,600	2,038	4	7,650
367,600	3,392	5	10,620
357,600	2,038	4	7,650
220,000	1,716	4	8,427
298,000	2,202	5	10,620
216,000	1,984	3	7,500
330,000	2,302	4	8,910
211,600	1,544	3	7,203
335,800	2,302	4	8,910
335,600	2,302	4	8,910
312,000	2,112	5	10,620
320,000	2,302	4	8,910
476,000	2,574	3	6,411
304,000	2,302	4	8,910
252,000	1,968	3	7,500
227,600	1,968	3	7,500
364,000	2,038	3	7,650
336,000	2,038	3	7,650
359,600	2,038	3	7,650
246,000	1,968	3	7,500
338,000	2,302	4	8,910
219,000	1,968	3	7,500
224,000	1,544	3	7,203
338,000	2,302	4	8,910
340,000	2,302	4	8,910
330,000	2,752	5	10,620
288,000	1,920	4	9,240

foot as lots get bigger for homes in this price, size, and lot size range. The multiple linear regression output for these data is shown in **Figure 10.5**.

The regression model illustrates several of the characteristics associated with multicollinearity:

- The model is statistically significant (*p*-value = 0.0001).
- Important variables appear to be insignificant or moderately significant (*BEDROOMS* and *LOT AREA*).
- One variable has a counterintuitive sign (*LOT AREA*).

Table 10.4 shows correlations among the model variables. All three of the independent variables are significantly correlated, and the correlation between *BEDROOMS* and *LOT AREA* is particularly high at 0.913. In addition, variance inflation factors on the three variables are

LIVING AREA	*VIF* = 1.6
BEDROOMS	*VIF* = 6.4
LOT AREA	*VIF* = 6.1

Figure 10.5 Excel Regression Output

Regression Statistics	
Multiple R	0.6566
R Square	0.4312
Adjusted R Square	0.3850
Standard Error	44336.2088
Observations	41

ANOVA

	df	*SS*	*MS*	*F*	*Significance F*
Regression	3	55128062160	18376020720	9.3483	0.0001
Residual	37	72730878328	1965699414		
Total	40	1.27859E+11			

	Coefficients	*Standard Error*	*t Stat*	*P-value*
Intercept	177251.08	60669.643	2.922	0.006
Living Area (SF)	112.90	24.788	4.555	0.000
Bedrooms	29446.67	26097.158	1.128	0.266
Lot Area (SF)	-26.78	15.112	-1.772	0.085

Table 10.4 Model Variable Correlations

	Sale Price	Living Area (SF)	Bedrooms	Lot Area (SF)
Sale Price	1			
Living Area (SF)	0.610	1		
Bedrooms	0.282	0.598	1	
Lot Area (SF)	0.184	0.569	0.913	1

Although no variance inflation factor is greater than 10, mean *VIF* is 4.70, which is considerably greater than 1, and is also indicative of multicollinearity.

We are left with two choices–either reduce the model to *PRICE* as a function of the most significant variable, which is *LIVING AREA*, or work with the full, but multicollinear, model. The full model does a better job of predicting *PRICE*, so it is the obvious choice for predictive modeling. The other option, reducing the model by dropping *BEDROOMS* and *LOT AREA*, is not as attractive for two reasons–it is an inferior predictive model, and the coefficient on *LIVING AREA* will be biased as it accounts for additional systematic variation in *PRICE* through correlations with the dropped variables. Another possible option is to gather more sales data, but this is generally not a viable alternative.

One note of caution: although multicollinearity does not affect predictive power, attempting to interpret the meaning of the coefficients on the model's highly correlated variables may lead to misleading conclusions. In this example, it would be good professional practice to discuss the extent of correlation among the independent variables and to disclose that the coefficients should not be interpreted as traditional appraisal comparison grid adjustments for these elements of comparison.

Indicator Variables

Categorization

Indicator variables provide a means of including categorical variables in a linear regression model. Indicator variables do what the name implies–they *indicate* whether or not an observation belongs to a stated category. Typically, members of the category are coded 1 and nonmembers are coded 0. Potential applications for developing real estate value opinions are many. For example, consider the following categories as potential indicator variables in the analysis of a set of properties:

Category	Coded 1	Coded 0
Swimming Pool	Pool	No pool
Brick exterior	Brick	All other exteriors
Class B office space	Class B	Class A, C, or D
Golf frontage	Golf frontage	No golf course frontage
Vacant	Vacant	Not vacant

The indicator variables on this list are binary (yes, no) and are limited to two mutually exclusive and collectively exhaustive categories (category member and nonmember). One indicator variable coded 0,1 is sufficient to account for the two categories–member and nonmember.

In many situations it is important to be able to account for more than two categories. Class of office space provides a good example of this. If office space in a given market is classified as either A, B, C, or D, it would be beneficial to be able to distinguish among them when analyzing rent or sale data. These four categories can be captured through creation of three related binary indicator variables (typically called "dummy" variables) as follows:

Category	Class A Dummy	Class B Dummy	Class C Dummy
Class A	1	0	0
Class B	0	1	0
Class C	0	0	1
Class D	0	0	0

Three indicator variables–Class A, Class B, and Class C–are sufficient for identification of membership in one of four office building class categories. In addition, these three dummy variables satisfy two important criteria for creation of related indicator variable sets:

- The categories should be mutually exclusive.
- The categories should be collectively exhaustive.

The first criterion means that the name of the category and decisions regarding membership should not be ambiguous, and it should not be possible for one observation to belong to more than one category. The second criterion means that the categories accommodate all possible classifications (i.e., if there were "Class E" office space, the preceding categories would not be collectively exhaustive because the Class D and E buildings would be classified together).

Complete categorization can be accomplished by creating one less dummy variable than the number of categories. The "missing" category is accounted for when all of the other categories are coded 0.

Interpreting Indicator Variables

Consider a regression model of the form

$$y = a + bx + cD + e$$

where x is a continuous numerical variable and D is a binary (0,1) dummy variable. When $D = 0$, the model reduces to a simple linear model:

$$y = a + bx + e$$

When $D = 1$, the model reduces to a different simple linear model with a new intercept:

$$y = (a + c) + bx + e$$

Therefore, the model with the binary indicator variable accounts for two different sub-population equations having the same slope β and two different intercepts $\alpha_1 = a$ and $\alpha_2 = (a + c)$. **Figure 10.6** illustrates interpretation of the two different intercepts when $D = 0$ and $D = 1$. Two estimates of population regression equations are picture in this line graph, both having the same slope but different intercepts. Although the $D = 1$ line with intercept $a + c$ is pictured as being located above the $D = 0$ line in the exhibit, it could just as well be below the $D = 0$ line if the coefficient c happened to be negative.

Just as one indicator variable can produce two different equations in Figure 10.6, two indicator variables would result in three different equations, and so forth. This is why the number of indicator variables required to produce a complete model is always one less than the number of subpopulation equations.

Another way to model a phenomenon such as this is to divide the data

Figure 10.6 Illustration of Meaning of Dummy Variable Coefficients

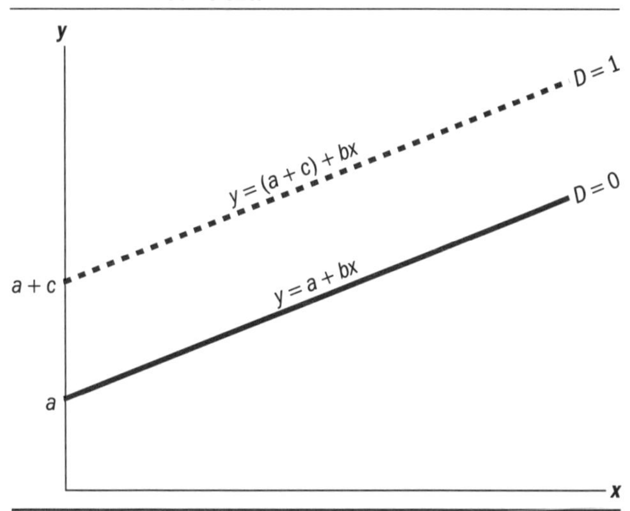

into two different categories and develop a linear model for each category. However, if the undivided data set provides more information concerning how the continuous variable x relates to the outcome variable y than two separate models, this information is lost when two separate models are created.

Example Problem

Table 10.5 contains 23 sales of residential condominium units including the variables *PRICE*, improved living area *SF*, height above ground *FLOOR*, and an indicator variable *POOL* coded 1 if the condominium property includes a pool and 0 otherwise. Analysis of these data is accomplished by creating a multiple linear regression model of the form:

$$\mathbf{PRICE = a + b_1(SF) + b_2(FLOOR) + c(POOL)}$$

As **Figure 10.7** shows, the model is highly significant ($F = 384.8$) and each independent variable is significant (low *p*-values on the *t* statistics). The

Table 10.5 Condominium Sale Data

PRICE	*SF*	*FLOOR*	*POOL*
519,000	1,200	4	0
539,000	1,350	3	0
571,000	1,100	7	1
455,000	1,180	2	0
511,000	1,240	2	1
618,000	1,200	9	0
520,000	1,390	1	0
550,000	1,220	5	0
560,000	1,160	6	1
669,000	1,160	10	0
795,000	1,450	12	1
576,000	1,325	4	1
612,000	1,400	6	0
645,000	1,310	8	0
492,000	1,160	3	0
619,000	1,250	7	1
650,000	1,220	9	0
641,000	1,350	7	1
529,000	1,180	5	0
670,000	1,320	8	0
615,000	1,400	5	1
715,000	1,250	12	0
479,000	1,100	3	0

Figure 10.7 Excel Regression Output for Table 10.5 Condominium Sale Data

ANOVA

	df	*SS*	*MS*	*F*	*Significance F*
Regression	3	1.455E+11	4.852E+10	384.81	0.0000
Residual	19	2.395E+09	1.261E+08		
Total	22	1.479E+11			

	Coefficients	*Standard Error*	*t Stat*	*P-value*
Intercept	74,106.55	30,309.00	2.45	0.0244
SF	296.48	24.37	12.17	0.0000
Floor	22,999.70	776.39	29.62	0.0000
Pool	12,351.00	5,022.60	2.46	0.0237

inferred population multiple linear regression model from these data is

PRICE = $74,107 + $296.48(SF)
+ $23,000(FLOOR)
+ $12,351(POOL)

The multiple linear regression model confirms, based on the significant t statistic for *POOL*, that two subcategories of condominiums exist in this market–condominium properties with and without pools. Equations for modeling these subcategories are

With pool: **PRICE = $86,458 + $296.48(SF) + $23,000(FLOOR)**

Without pool: **PRICE = $74,107 + $296.48(SF) + $23,000(FLOOR)**

If the t statistic and corresponding p-value for *POOL* had exceeded the significance level considered appropriate for the analysis, then a question would have arisen as to whether the pool variable was important enough to include in a price-estimation model. As we will discover later in the chapter, the decision to leave in or omit a variable involves more than the significance of its t statistic.

Interaction Variables

Interaction variables are created by multiplying two independent variables and are used when the effect of one of the two interacting independent variables depends on the value of the other independent variable. Interaction variables facilitate construction of a single linear regression model when categorical variables may affect either the intercept, the slope, or both the intercept and the slope of one or more of the numerical variables in a model. They can also be used to model interactions between two or more continuous variables.

Consider a regression model of the form:

$$\mathbf{y = a + bx + c_1D + c_2Dx + e}$$

where x is a continuous numerical variable and D is a binary (0,1) dummy variable. When $D = 0$, the model reduces to the simple linear equation:

$$\mathbf{y = a + bx + e}$$

When $D = 1$, the model reduces to a different simple linear equation with a new intercept and a new slope:

$$\mathbf{y = (a + c_1) + (b + c_2)x + e}$$

The model with the binary indicator variable can account for two different sub-population equations having two different intercepts, $\alpha_1 = a$ and $\alpha_2 = (a + c_1)$, and two different slopes, $\beta_1 = b$ and $\beta_2 = (b + c_2)$. **Figure 10.8** illustrates the differing slopes and intercepts when $D = 0$ and $D = 1$. The $D = 1$ line with intercept $a + c_1$ and slope $b + c_2$ is pictured as being located above, and having a steeper slope than, the $D = 0$ line in the exhibit. It could just as well be below the $D = 0$ line or have a less steep slope if the coefficient c_1 or the coefficient c_2 happened to be negative.

Figure 10.8 Illustration of Meaning of Dummy and Interaction Variable Coefficients

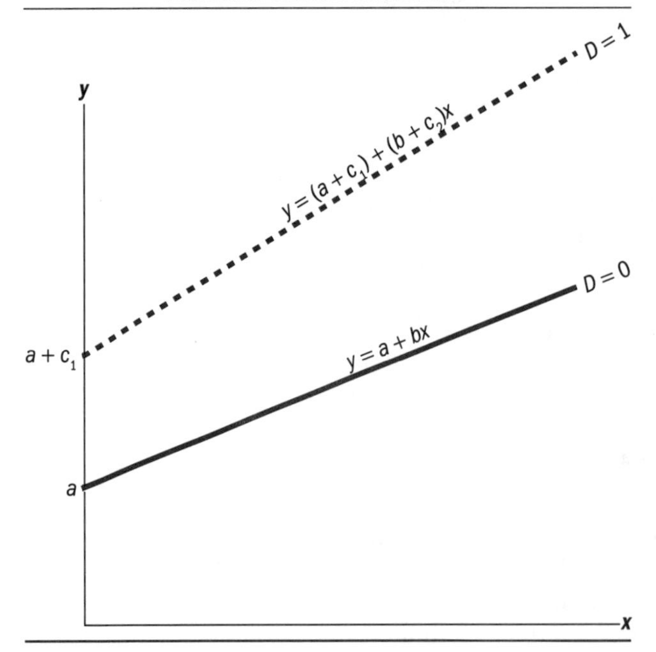

Example Problem

Table 10.6 contains the same 23 condominium sales from Table 10.5. The variables include *PRICE*, improved living area *SF*, height above ground *FLOOR*, and an indicator variable *POOL* coded 1 if the condominium property has a pool and 0 otherwise. Pools are located outdoors at ground level in this market due to its mild climate. Table 10.6 includes an additional interaction variable labeled *POOL* × *FLOOR*. The interaction variable is included to test the theory that the price effect of condominium unit height above ground is influenced by the existence of a ground-level pool. Price could be affected if purchasers of higher-floor units in projects with pools anticipate being inconvenienced by inferior pool access compared with lower-floor units.

Analysis of these data is accomplished by creating a linear regression model of the form

$$\mathbf{PRICE = a + b_1(SF) + b_2(FLOOR) + c_1(POOL) + c_2(POOL \times FLOOR)}$$

Table 10.6 Condominium Sale Data

PRICE	*SF*	*FLOOR*	*POOL*	*POOL* × *FLOOR*
519,000	1,200	4	0	0
539,000	1,350	3	0	0
571,000	1,100	7	1	7
455,000	1,180	2	0	0
511,000	1,240	2	1	2
618,000	1,200	9	0	0
520,000	1,390	1	0	0
550,000	1,220	5	0	0
560,000	1,160	6	1	6
669,000	1,160	10	0	0
795,000	1,450	12	1	12
576,000	1,325	4	1	4
612,000	1,400	6	0	0
645,000	1,310	8	0	0
492,000	1,160	3	0	0
619,000	1,250	7	1	7
650,000	1,220	9	0	0
641,000	1,350	7	1	7
529,000	1,180	5	0	0
670,000	1,320	8	0	0
615,000	1,400	5	1	5
715,000	1,250	12	0	0
479,000	1,100	3	0	0

When *POOL* = 0, the price estimation model is the multiple linear regression equation

$$\text{PRICE} = a + b_1(\text{SF}) + b_2(\text{FLOOR})$$

When *POOL* = 1, the price estimation model is the multiple linear regression equation

$$\text{PRICE} = (a + c_1) + b_1(\text{SF}) + (b_2 + c_2)(\text{FLOOR})$$

which reflects the overall impact of a pool through an altered intercept and a change in the floor price coefficient in condominium properties having pools.

Based on **Figure 10.9**, the linear regression equation including the *POOL* × *FLOOR* interaction variable is

$$\begin{aligned}\text{PRICE} = \$68{,}397 &+ \$299.69(\text{SF}) \\ &+ \$23{,}293(\text{FLOOR}) \\ &+ \$18{,}832(\text{POOL}) \\ &- \$1{,}076(\text{POOL} \times \text{FLOOR})\end{aligned}$$

Based on this multiple linear regression model, the two subcategory price equations are

With pool: $$\text{PRICE} = \$87{,}229 + \$299.69(\text{SF}) + \$22{,}217(\text{FLOOR})$$

Without pool: $$\text{PRICE} = \$68{,}397 + \$299.69(\text{SF}) + \$23{,}293(\text{FLOOR})$$

Notice that the *POOL* variable no longer appears to be significant at either the 10% or 5% level. Also, the *POOL* × *FLOOR* interaction has a small *t* statistic and a *p*-value of only 0.5545. In

Figure 10.9 Excel Regression Output for Table 10.6 Condominium Sale Data

ANOVA

	df	*SS*	*MS*	*F*	*Significance F*
Regression	4	1.456E+11	3.640E+10	279.02	0.0000
Residual	18	2.348E+09	1.305E+08		
Total	22	1.479E+11			

	Coefficients	*Standard Error*	*t Stat*	*P-value*
Intercept	68,397.48	32,254.65	2.12	0.0481
SF	299.69	25.35	11.82	0.0000
Floor	23,292.74	927.59	25.11	0.0000
Pool	18,831.77	11,911.38	1.58	0.1313
Pool x Floor	-1,076.32	1,787.03	-0.60	0.5545

comparison the t statistic on the previous *POOL* indicator variable, shown in Figure 10.7, was significant at the 5% level (p-value = 0.0237).

This new result does not mean that the existence of a pool has lost its significance as a price-determining characteristic of condominiums in this market. Rather, creation of the interaction variable $POOL \times FLOOR$ has introduced multicollinearity into the model through the *POOL* and $POOL \times FLOOR$ variables. The correlation coefficient $r_{POOL,\, POOL \times FLOOR} = 0.88$, and the two correlated variables are sharing explanatory power.

If you were to compare R^2_{ADJ} for the two models with and without the interaction variable, you would discover that R^2_{ADJ} decreases slightly when the interaction variable is added. You should recall that this means addition of the interaction variable explains no more than adding a totally irrelevant random variable and does not improve the predictive power of the model. Failure to improve R^2_{ADJ} combined with the low t statistic imply that the model without the interaction variable is probably a better choice.

Model Building Issues

We looked at some model building issues in Chapter 9 such as fitting nonlinear relationships and assessing prediction accuracy. Adding multiple independent variables to the model forces us to think about new variable selection procedures, including comparison of two models using the partial F test.

Choosing appropriate variables for a regression model involves more than applying the statistical tests and procedures designed to identify the variables that result in the best fit. The human element is equally important in identifying and producing the best model. Furthermore, the size of the data set puts constraints on how many variables that data set will accommodate. A good rule of thumb is to include at least 10 to 15 observations per independent variable. That is, $n/k \geq 10$ to 15. Hair, et al. suggest an absolute minimum of $n/k \geq 4$ and Neter, et al. suggest a minimum of $n/k \geq 6$ to 10. When the ratio of n to k is too low, model fit and prediction statistics can be misleading.

Therefore, caution is advised whenever $n/k < 10$. Dielman[10] notes that 30 observations plus 10 to 20 per additional independent variable is also often suggested as a rule of thumb.

Note that all of these sample size rules of thumb are suggestions, not requirements. If the number of observations in the data set is just one more than the number of explanatory variables, you can still derive a regression equation using the data. However, it will not be possible to evaluate and account for the model's underlying assumptions with such a small amount of data. In addition, your model may be relying on few observations having a given characteristic. Therefore, the data may not be representative of the underlying phenomenon you are studying. Ultimately you will have to determine an appropriate sample size by balancing the costs, benefits, and practicality of including more data.

A large amount of data can help you uncover relationships that may not be apparent in a small data set. Being able to narrow an analysis down to expose minor effects on price or rent may be beneficial in some instances, while it may represent "overkill" in others. For example, you may be able to gather enough data to isolate a $200 adjustment factor in a residential market where homes sell for $300,000 on average, but it may not be practical or beneficial to gather the large amount of data necessary to do so.

The knowledge, experience, and judgment of the analyst contribute greatly to the credibility of the model ultimately chosen as being the best. Knowledge–including underlying economic and valuation theory and an understanding of the market–is a crucial element of deriving expectations concerning the important variables and the correct functional form of the relationship between the explanatory variables and the outcome variable. In addition, the valuation professional and analyst defines the problem and ensures that the statistical tests relate well to the problem being addressed. The human element also includes the identification of reliable data sources, data confirmation, and data entry, which is largely a function of professional experience.

---

10. Terry Dielman, *Applied Regression: Analysis for Business and Economics*, 3rd ed. (Pacific Grove, Calif.: Duxbury/Thomson Learning, 2001).

Additionally, the analyst must decide whether or not the assumptions underlying regression analysis are valid. Real-world data are rarely if ever perfectly normal, linear, and homoskedastic, so model building usually involves a judgment of whether the validity of the regression assumptions is close enough to ideal to produce credible results. This may involve developing several models with differing variable transformations for comparison.

The quantitative variable selection routines and techniques that follow do not always select the best combination of variables. The final model should be a result of applying judgment to the quantitative procedures, which is something computers cannot do. Two automated variable selection routines are commonly available in statistical software programs–all possible regression models and stepwise regression. Minitab includes both, and SPSS includes stepwise regression.

Recall that Table 10.1 contains data on 59 view lots located in a mid-sized city in the southwestern U.S. The data set in Table 10.1 was not complete. Several indicator variables from the example problem using these data were withheld because the topic of indicator variables had not been introduced. **Table 10.7** includes the indicator variables to illustrate the "all possible regression models" and "stepwise" procedures.

All Possible Regression Models

The all possible regression models procedure runs the entire set of possible regression models given the number of possible combinations of independent variables that can be included, and all the possible regression models are then ranked based on one or more criteria. The researcher can look over the rankings and choose models for further consideration. This routine is called *best subsets regression* in Minitab. It is accessed by selecting **Stat, Regression,** and then **Best Subsets.**

Best subsets regression output includes four model selection criteria:

- R^2
- R^2_{ADJ}
- S_e
- C_p

Table 10.7 View Lot Price Data

Price	Size (AC)	View	Developability Constraint	Adjacent +	Adjacent –
160,000	1.579	30	0	1	0
126,000	1.269	30	0	0	1
125,000	1.141	15	0	0	1
135,000	0.827	15	0	0	1
133,500	0.843	20	0	0	1
90,000	1.073	30	1	0	0
130,000	1.185	30	0	0	0
325,500	2.192	100	0	0	0
155,000	1.295	30	0	1	0
125,000	0.950	15	0	0	1
150,000	1.295	30	0	1	0
180,000	1.642	20	0	1	0
168,500	1.579	30	0	1	0
153,500	1.837	50	0	1	0
160,000	1.154	60	0	0	0
185,000	1.645	60	0	0	1
100,000	1.372	10	0	0	0
87,650	1.000	0	0	0	0
125,000	0.636	20	0	0	0
167,000	0.734	60	0	0	1
170,000	1.837	50	0	1	0
185,000	1.050	15	0	1	0
150,000	1.100	30	0	0	1
140,000	1.220	25	0	0	1
140,000	1.100	10	0	0	1
178,000	1.070	30	0	0	0
156,750	1.070	20	0	0	0
149,000	1.130	10	0	0	0
105,000	1.310	0	0	0	0
115,000	0.680	15	0	0	0
135,000	0.632	20	0	1	0
90,000	0.500	0	0	0	0
146,250	0.848	25	0	0	0
124,815	0.830	20	0	0	0
130,000	0.830	20	0	0	0
345,000	1.766	150	0	0	0
89,474	0.830	0	1	0	1
185,000	1.208	90	0	0	1
156,000	0.839	40	0	0	0
159,000	0.829	30	0	0	0
220,200	1.022	90	0	0	0
153,000	0.831	30	0	0	0
156,000	0.830	30	0	0	0
320,000	3.340	150	0	0	0
205,000	1.640	100	0	0	0
150,000	0.800	40	0	0	0
160,000	1.010	50	1	0	0
215,000	1.069	100	0	0	0
142,500	1.440	20	0	0	1
300,000	1.027	150	0	0	0
125,000	2.650	20	1	1	0
142,500	1.250	100	1	0	0
110,000	1.230	10	0	0	0
235,000	0.895	120	0	0	0
164,000	1.230	10	0	0	0
290,000	1.670	150	1	0	0
243,750	1.100	130	0	0	0
340,000	2.950	160	0	0	0
275,000	1.280	150	1	0	0

You are familiar with R-squared, adjusted R-squared, and regression standard error by now, but we have not discussed Mallow's C_p. C_p assesses goodness of fit by comparing the SSE of the variable subset to mean squared error with all of the variables included in the model. When there is no bias in the model, the expected value of C_p is $k + 1$. When using this criterion to assist in best model choice, low C_p values and those near $k + 1$ should be considered.

Figure 10.10 shows the results for a best subsets regression of the Table 10.7 view lot data. Variable names are written vertically in Minitab and include *ACRES*, *VIEW*, *CONSTRAINT*, *ADJ+*, and *ADJ–*. The *CONSTRAINT* variable is a categorical variable identifying lots having topographic developability constraints suggestive of unusually high foundation and site preparation costs. *ADJ+* is a categorical variable coded as 1 when a lot is adjacent to national forest land. *ADJ–* is a categorical variable indicating a lot that is adjacent to a subdivision with homes that are generally lower in price than the comparable custom homes typically built on view lots in this market.

The four model selection criteria statistics are located in the columns labeled "R-Sq," "R-Sq(adj)," "Mallows C-p," and "S." The boxed-in two variable model is the *ACRES* and *VIEW* equation estimated earlier in the chapter and presented in Figure 10.1. Models having the lowest C_p value in each number-of-variables category are flagged with an

Figure 10.10 Minitab Output for Best Subsets Regression

Response is Price

Vars	R - Sq	R - Sq(adj)	Mallows C - p	S	Acres	View	Constraint	ADJ+	ADJ-
*1	81.3	81.0	21.8	27628		X			
1	30.8	29.5	229.9	53198	X				
1	5.3	3.7	334.5	62202					X
*2	84.5	83.9	10.9	25426		X	X		
2	83.6	83.1	14.3	26083	X	X			
2	82.2	81.5	20.4	27239		X		X	
*3	86.9	86.2	2.8	23535	X	X	X		
3	85.3	84.5	9.4	24930		X	X	X	
3	84.7	83.8	12.1	25494		X	X		X
*4	87.1	86.1	4.2	23627	X	X	X		X
4	87.0	86.1	4.3	23646	X	X	X	X	
4	85.4	84.3	11.2	25133		X	X	X	X

asterisk in Figure 10.10. The output was limited to the three best one-, two-, three-, and four-variable models, which is user-selected in Minitab using the **Options** button. While it is possible to compute all possible regression models in Excel and derive Mallow's C_p for each of them, the time and effort required can be extensive. For example, there are 31 possible different simple and multiple linear regression models when there are five independent variables as in this example, computed as

$$\binom{5}{1} + \binom{5}{2} + \binom{5}{3} + \binom{5}{4} + \binom{5}{5} = 5 + 10 + 10 + 5 + 1 = 31$$

(i.e, five one-variable models, 10 two-variable models, 10 three-variable models, five four-variable models, and one five-variable model. Refer to Chapter 2 to refresh your understanding of calculating combinations.)

The model[11] with the highest R^2 is

PRICE = f(ACRES, VIEW, CONSTRAINT, ADJ-)

with an $R^2 = 0.871$ (87.1%). This is not the model with the highest R^2_{ADJ}, however. For this model, Mallow's C_p is 4.2, which is less than $k + 1 = 4 + 1 = 5$. So there is no indication of bias[12] based on Mallow's C_p. The model having the lowest C_p statistic is

PRICE = f(ACRES, VIEW, CONSTRAINT)

with a $C_p = 2.8$, which is less than $k + 1 = 3 + 1 = 4$. This model also has the highest R^2_{ADJ} of 0.862 (86.2%). In comparison, the best one-variable and two-variable models have C_p statistics that far exceed $k + 1$ and much lower R^2 and R^2_{ADJ} statistics. With regard to regression standard error (S_e), the best three-variable model is the lowest at 23,535 and the best four-variable model is the second-lowest at 23,627.

11. The lower case *f* is shorthand for "function of." It means that the dependent variable is a function of the independent variables located inside the parentheses.
12. Mallow's C_p values that are less than $k + 1$ are attributed to random sampling error, not bias. See Neter, Wasserman, and Kutner (448).

Based on the best subsets regression output from Minitab, the modeling choice is reduced to two models, both of which are different from, and better than, the model presented earlier in Figure 10.1:

PRICE = f(ACRES, VIEW, CONSTRAINT, ADJ-)

or

PRICE = f(ACRES, VIEW, CONSTRAINT)

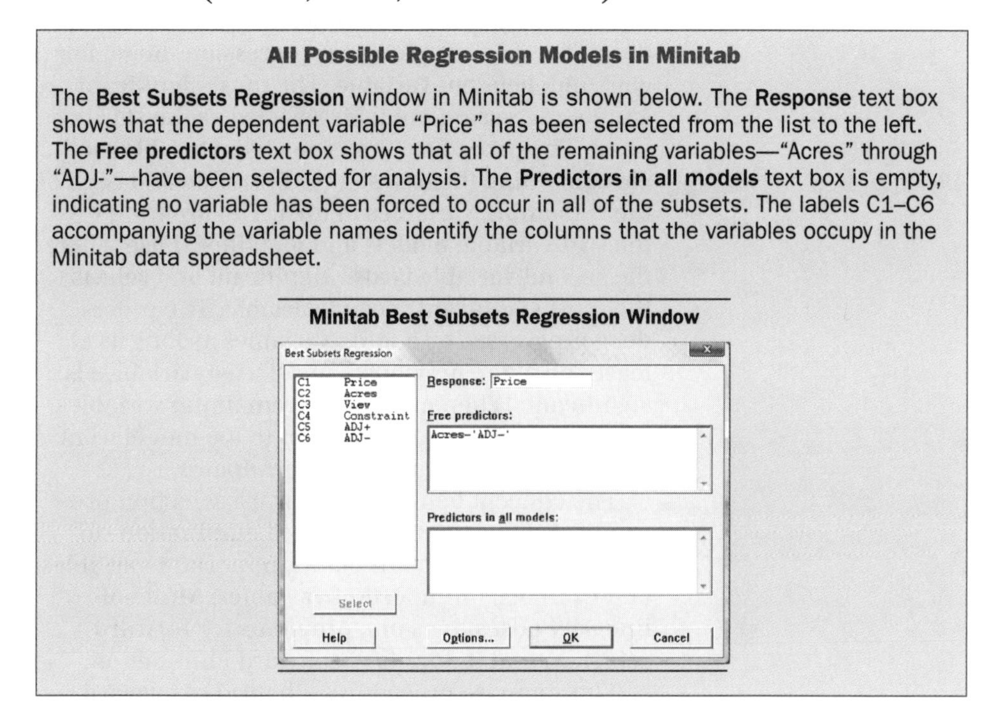

All Possible Regression Models in Minitab

The **Best Subsets Regression** window in Minitab is shown below. The **Response** text box shows that the dependent variable "Price" has been selected from the list to the left. The **Free predictors** text box shows that all of the remaining variables—"Acres" through "ADJ-"—have been selected for analysis. The **Predictors in all models** text box is empty, indicating no variable has been forced to occur in all of the subsets. The labels C1–C6 accompanying the variable names identify the columns that the variables occupy in the Minitab data spreadsheet.

Minitab Best Subsets Regression Window

Stepwise Regression

Stepwise regression can be accomplished by employing a forward selection algorithm, a backward elimination algorithm, or a hybrid of the two simply called "stepwise."

The backward elimination procedure begins with a multiple linear regression model including all of the candidate variables. It then identifies the least significant variable and tests the null hypothesis that its coefficient is zero. If this hypothesis cannot be rejected, the variable is eliminated and the model is rerun with one less variable. The least significant variable from this next model is identified and tested. If a second variable is removed, the process is repeated. The process continues until the null hypothesis of a coefficient value of zero is rejected for all of the surviving variables.

The concept behind the backward elimination procedure is to remove variables that are unimportant and leave in important variables. When multicollinearity is a problem, a seemingly unimportant but highly correlated variable may have been eliminated that the analyst would have preferred to leave in the model.

The forward selection procedure starts with the derivation of a simple linear regression model for each independent variable. The most-significant simple regression model is selected as the starting point. The procedure then runs two-variable models containing the first selected variable and each of the remaining variables in turn. The program tests the two-variable models and identifies those where the second variable is also significant and selects the most significant second variable. The procedure progresses by adding variables as long as at least one of the previously unselected variables is significant. When none of the remaining variables make a significant contribution to the model from the prior step, the procedure terminates.

The concept behind the forward selection procedure is the same as backward elimination–to produce a model that includes important variables and excludes unimportant variables. Multicollinearity poses the same problems for forward selection that it does for backward elimination.

The stepwise procedure–a hybrid of forward selection and backward elimination–starts with forward selection's choice of the most significant simple linear regression model. It then selects the most significant variable from the remaining variables and adds it to the model. This procedure also performs a backward elimination test at each step. Due to correlation among the variables, it is possible that a variable could have been selected moving forward through the data and then eliminated later on when it loses significance in the presence of added variables. The forward-and-backward procedure continues until no significant variable remains to add to the model and none of the model variables qualify for elimination. The stepwise

Stepwise Regression in SPSS

The **Linear Regression** window in SPSS is shown below with "stepwise" selected in the **Method** button menu. "Price" is shown in the **Dependent** text box as the dependent variable for the regression model. The **Independent(s)** text box lists all of the independent variables to include in the stepwise procedure, although they are not all visible due to the small size of the text box. Notice also that the listed variables represent "Block 1 of 1." Data can be divided into blocks with some blocks employing the "enter" method and others employing the "stepwise" method. This facilitates forcing certain variables to be included in all models by employing the enter method. The data set variables are listed in the text box on the left side of the window. The **Options** button opens a window that allows the user to select the level of significance to use for variable entry and elimination.

SPSS Linear Regression Window with Stepwise Selected

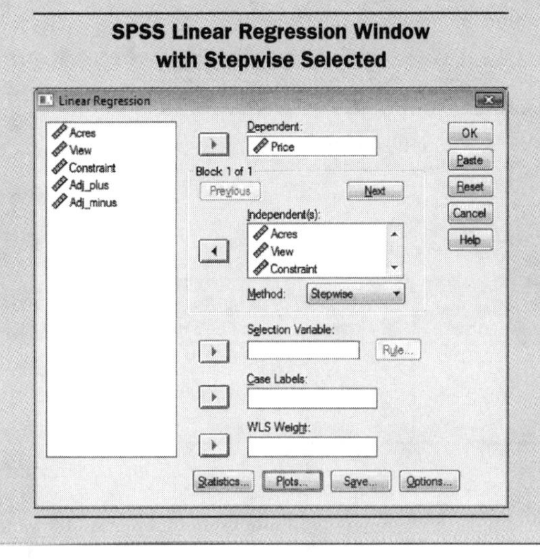

hybrid is more resistant to multicollinearity than either of the other two procedures, but it is not immune from the effects of correlated variables.

The view lot price data from Table 10.7 were analyzed in SPSS using backward elimination, forward selection, and stepwise methods with entrance significance set at 0.05 and elimination significance set at 0.10. All three methods yielded the same result for this fairly simple data set:

PRICE = f(ACRES, VIEW, CONSTRAINT)

which is identical to one of the candidate models from the all possible regression models procedure. The path to the final model was identical in the forward selection and stepwise procedures–*VIEW* entered first, *CONSTRAINT* entered second, and *ACRES* entered last. The path to the final model in the backward elimination procedure was to eliminate *ADJ+* (the least significant independent variable) first and then *ADJ–*, leaving the other three variables.

Figure 10.11 shows a portion of the SPSS output for the forward selection procedure, along with a model summary with all five variables included in the regression equation. The steps and the variables entered at each step are listed as notes a, b, and c below the lower **Model Summary** box. Information provided for each step includes R, R^2, R^2_{ADJ}, and S_e. The ultimate forward selection model (Model 3) has the highest adjusted R-squared and the smallest regression standard error, S_e, of the three. In addition, it has a higher adjusted R-squared and smaller regression standard error than the full five-variable model shown in the upper **Model Summary** box.

Figure 10.11 SPSS Stepwise and Full Model Output Summaries

Model Summary

Model	R	R Square	Adjusted R Square	Std. Error of the Estimate
1	.933[a]	.871	.859	23795.06416

a. Predictors: (Constant), Adj+, Constraint, Acres, Adj-, View

Model Summary

Model	R	R Square	Adjusted R Square	Std. Error of the Estimate
1	.902[a]	.813	.810	27627.86657
2	.919[b]	.845	.839	25426.28823
3	.932[c]	.869	.862	23535.22652

a. Predictors: (Constant), View

b. Predictors: (Constant), View, Constraint

c. Predictors: (Constant), View, Constraint, Acres

Partial *F* Statistic

The partial *F* statistic facilitates the comparison of two regression models, one referred to as the *full* model and the other as the *reduced* model. The reduced model variables are a subset of the full model variables. This test is applicable when at least two less variables have been deleted from the full model to create the reduced model. A partial *F* test is unnecessary when the reduced model contains only one less variable because the *t* statistic in the full model for the eliminated variable provides the same *p*-value as the partial *F* statistic.

So far the all possible regressions and stepwise model building procedures have provided evidence that either one or two of the adjacency indicator variables is insignificant. The stepwise procedures dropped both *ADJ+* and *ADJ–*, whereas the all possible regressions procedure indicated that *ADJ–*

might be a candidate for inclusion in a final model. The partial F statistic provides more information about the model that may help the analyst decide whether or not to drop both of these variables. The partial F statistic is calculated as follows:

$$F = \frac{\frac{SSE_R - SSE_F}{k - k'}}{\frac{SSE_F}{n - k - 1}}$$

where k is the number of independent variables in the full model and k' is the number of independent variables in the reduced model. The partial F statistic has $k - k'$ numerator degrees of freedom and $n - k - 1$ denominator degrees of freedom.

To conduct the partial F test, first run the full data regression model and record SSE_F, which is 30,008,869,165 for the view lot data. Next, run the reduced data regression model and record SSE_R, which is 30,464,878,815 for the view lot data. The difference, $SSE_R - SSE_F = 456{,}009{,}650$. Therefore, the partial F statistic for this analysis is

$$F = \frac{\frac{SSE_R - SSE_F}{k - k'}}{\frac{SSE_F}{n - k - 1}} = \frac{\frac{456{,}009{,}650}{5 - 3}}{\frac{30{,}008{,}869{,}165}{59 - 5 - 1}} = 0.403$$

The critical value of F at $\alpha = 0.05$ with 2 numerator degrees of freedom and 53 denominator degrees of freedom is 3.17.[13] The hypothesis test and conclusions are as follows:

H_0: $\beta_{ADJ+} = \beta_{ADJ-} = 0$

H_a: At least one $\neq 0$

Conclusion: Since $0.403 < 3.17$, we cannot reject the null hypothesis that the coefficients on both omitted adjacency variables are equal to zero, which means both of these variables may be candidates for removal from the model

The partial F statistic is also useful for testing the significance of an indicator variable set. For

13. In Excel, enter =FINV(0.05, 2, 53).

example, many residential real estate markets are thought to be seasonal. A three-variable indicator variable set can be used to reflect seasonality by, for example, coding *Spring*, *Summer*, and *Fall* as 1,0 dummy variables with "winter" reflected in the model when *Spring*, *Summer*, and *Fall* are all coded 0. A partial F test on a full model and a reduced model with the *Spring*, *Summer*, and *Fall* dummy variables omitted can be used to test a seasonality hypothesis by treating the three indicator variable set as capturing the possibility of a seasonal influence on the dependent variable.

PRESS Statistic and Prediction R^2

The last measures of model fit are the *PRESS* statistic and prediction R^2 (R^2_{PRED}), which we discussed in Chapter 9. These two statistics are provided in Minitab and are shown in **Figure 10.12.** The prediction sum of squares (*PRESS*) for the previously discussed three-variable model is slightly less than for the four-variable model including *ADJ-*. Additionally R^2_{PRED} is slightly higher for the three-variable model. These statistics indicate that the three-variable model is a slightly better predictor of price than the four-variable model.

Figure 10.12 Minitab *PRESS* and Prediction R^2 Output

3 VARIABLES	PRESS = 36107199914	R-Sq(pred) = 84.50%
4 VARIABLES	PRESS = 36576179258	R-Sq(pred) = 84.30%

Final Model Selection

In the final analysis of the model we are building here, which is much like final reconciliation in an appraisal report, a number of factors are weighed and point toward the selection of the three-variable model:

- The three-variable model is a slightly better predictor and has a good fit based on Mallow's C_p.
- The increase in the coefficient of determination by addition of *ADJ–* to the model is no better than we would expect from adding a totally irrelevant variable given the decrease in R^2_{ADJ}.
- In addition, the partial F test failed to reject the null hypothesis that the coefficients on both of the adjacency variables is zero, and the three-variable model had the lowest S_e.

- The *PRESS* and R^2_{PRED} values support the three-variable model choice.

Based on this reasoning, the set of independent variables that best explains price for the view lot data, given the information we have at hand, is

ACRES, *VIEW*, and *CONSTRAINT*

The multiple linear regression equation for this relationship is

$$\begin{aligned}\mathbf{PRICE} = \mathbf{\$90{,}767} &+ \mathbf{\$21{,}055(ACRES)} \\ &+ \mathbf{\$1{,}163\ (VIEW)} \\ &- \mathbf{\$35{,}781(CONSTRAINT)}\end{aligned}$$

Can the model be improved? Perhaps it can by testing for linearity in the relationships between *PRICE* and independent variables *ACRES* and *VIEW*. Also, issues of constancy of error variance have not been tested. In addition, there may be a significant interaction between *CONSTRAINT* and either *ACRES* or *VIEW*. These are all areas to explore as ways to hone your analytical skills.

Practice Problem 10.3

Add $ACRES^2$ and $VIEW^2$ variables to the view lot data. Create models to determine if a nonlinear model provides a better fit for either of these variables. In addition, look at interactions between developability constraints and the continuous variables lot size and view. Are the interactions significant? Finally, look at a plot of error versus predicted price for the model you ultimately select and assess whether or not a test for heteroskedasticity is warranted.

Real-World Case Study: Part 6

Developing and Choosing a Multiple Linear Regression Model

The case study data set includes 129 observations on a variety of numerical and categorical variables. Numerical variables include sale price, sale date, number of baths, number of bedrooms, living area, number of fireplaces, number of garage stalls, lot area, year built, and days on the market prior to the sale date. Categorical variables include the season of the year for the sale date and an occupancy description.

Although the data set contains a great deal of information, it is not formatted to facilitate multiple linear regression analysis. First we should put the data into an amenable format, which at a minimum requires converting categorical variables to indicator variables. In addition, the *Year Built* variable should be transformed into an *Age* variable as of the valuation date (i.e., the first half of 2004),[14] and the sale date variable will be transformed into two different indicator variable sets plus a single numerical variable for an eventual best date format comparison.

The reformatted townhouse data set contains the following continuous variables:

- *Sale Price*
- *Time* (date of sale measured semiannually)
- *Baths*
- *Bedrooms*
- *Living Area*
- *Fireplaces*
- *Garage Stalls*
- *Lot Area*
- *Age*
- *Days on Market*

Descriptive statistics for the continuous and numerical count variables are shown in the **following table**:

14. An exact valuation date would be selected for an actual appraisal. Any date in the first half of 2004 works as a valuation date for these data as formatted and analyzed here.

Minitab Descriptive Statistics Output for Continuous Variables

Variable	Total Count	Mean	StDev	Minimum	Median	Maximum
Sale Price	129	84,585	25,262	48,000	83,500	195,000
Time (6 Mos)	129	-3.194	1.969	-6	-4	0
Baths Total	129	2.0155	0.5726	1	2	3
Bedrooms	129	2.1085	0.5190	1	2	4
Living Area (SF)	129	1,067	200	724	1,019	1,696
Fireplaces	129	0.4806	0.5016	0	0	1
Garage	129	0.1240	0.3954	0	0	2
Lot Area (SF)	129	2,789.5	417.7	1,742	2,809	3,978
Age	129	13.612	8.537	1	9	33
Days on Market	129	51.48	57.61	0	34	321

Note that time, as an indication of sale date and time-dependent market conditions, is in semiannual increments and negatively signed, so *Time* = -3, for example, is the third time period prior to the current time period for the data. In other words, the first half of 2004 is the valuation date and is coded as 0, with the value of the *Time* variable decreasing by 1 for each six-month period further back (e.g., *Time* = -3 represents the second half of 2002).

Comparisons can be made between means and medians to identify excessive skewness (look at *Age* and *Days on Market*). Also, minimum and maximum values provide a range that can be used to evaluate how well the data represent the market being studied. Finally, the number of *Fireplaces* variable will act like a dummy variable when it is included in a model because the range is 0 to 1 in increments of 1 (i.e., there are no partial fireplaces).

In addition to the continuous variables, the reformatted townhouse data set contains the following indicator variables:

- *Sale Year* (2001 through 2004)
- *Semiannual Sale Period* (First Half of 2001 through First Half of 2004)
- *Season* (winter, spring, summer, fall)
- *Occupancy* (vacant, owner, tenant)

Descriptive statistics for the indicator variables are reported in the **table on the following page**, which has been altered slightly from the standard Minitab output to include totals for each indicator variable set as a way to check for data entry errors because each should sum to $n = 129$.

Notice also that standard deviations and medians are not included in the indicator variable table. These statistics provide no useful information for indicator variables. The means are, however, quite useful because the mean of a dummy variable represents the proportion of the indicator variable set within a given category. For example, the mean of the *YR 2001* dummy variable is 0.3178, meaning that 31.78% of the sales occurred in 2001. Since 41 of the 129 sales were 2001 sales, this interpretation of the mean holds up. The table also includes minimum and maximum values for each variable. Inclusion of these statistics is another way of signaling that these are dummy variables, and they provide another way to check for data entry errors.

Minitab Descriptive Statistics Output for Indicator Variables

Variable	Total Count	Mean	Sum	Minimum	Maximum
YR 2001	129	0.3178	41	0	1
YR 2002	129	0.3411	44	0	1
YR 2003	129	0.1783	23	0	1
YR 2004	129	0.1628	21	0	1
		Total	129		
2001H1	129	0.0930	12	0	1
2001H2	129	0.2248	29	0	1
2002H1	129	0.2248	29	0	1
2002H2	129	0.1163	15	0	1
2003H1	129	0.0853	11	0	1
2003H2	129	0.0930	12	0	1
2004H1	129	0.1628	21	0	1
		Total	129		
Winter	129	0.2093	27	0	1
Spring	129	0.4341	56	0	1
Summer	129	0.2868	37	0	1
Fall	129	0.0698	9	0	1
		Total	129		
Vacant	129	0.5426	70	0	1
Owner Occupied	129	0.4186	54	0	1
Tenant Occupied	129	0.0388	5	0	1
		Total	129		

Correlations between the response variable *Sale Price* and the continuous explanatory variables are presented in the **table below**. Correlations with *Sale Price* all carry the expected signs except for *Lot Area*, which is negative. This may be an indication that townhome purchasers prefer not to have a yard to maintain, although more information would be required to confirm this conjecture. It could also be a data anomaly. If multicollinearity turns out to be an issue with a multiple regression model of these data, it is apt to reside with *Living Area*, which is positively correlated with *Bedrooms, Baths, Fireplaces*, and *Garage Stalls. Living Area* is not correlated with *Lot Area*, however.

Minitab Correlations: Sale Price vs. Numerical Variables

	Price	Time	Baths	Bedrooms	Living Area	FP	Garage	Lot SF	Age
Time (6 Mos)	0.40								
Baths	0.55	0.16							
Bedrooms	0.52	0.11	0.36						
Living Area (SF)	0.72	0.04	0.75	0.69					
Fireplaces	0.24	-0.07	0.36	0.19	0.46				
Garage	0.82	0.32	0.44	0.35	0.53	0.05			
Lot SF	-0.24	-0.16	0.01	-0.06	0.02	0.17	-.40		
Age	-0.57	0.05	-0.22	-0.04	-0.36	-0.29	-.28	-0 .01	
Days on Market	-0.26	-0.07	-0.20	-0.04	-0.24	-0.24	-.24	-0 .01	0.12

We should now look at scatter plots of sale price and each of the continuous variables that may have a nonlinear relationship to price, such as *Living Area, Lot Area*, property *Age* (depreciation proxy), and *Days on Market*. Four scatter plots are shown in Panels A, B, C, and D of the **illustration on the following page**.

Panel A may be indicative of a nonlinear inverse relationship between *Sale Price* and *Lot Area*. (We will see later, however, that the inverse relationship does not hold up after accounting for other elements of comparison.) Panel B is strongly indicative of a nonlinear inverse relationship between *Sale Price* and *Age*. Panel C suggests that the relationship between *Sale Price* and *Living Area* is probably linear for these data. Panel D could imply a nonlinear inverse relationship between *Sale Price* and *Days on Market*. It could also be interpreted as random except for higher-priced units that experience much shorter time on market than lower- and mid-priced units. (Given the rapid rise in unit prices over the study period, which we will see later in the analysis, high

Scatter Plots of Price vs. Living Area, Lot Area, Age, and Days on Market

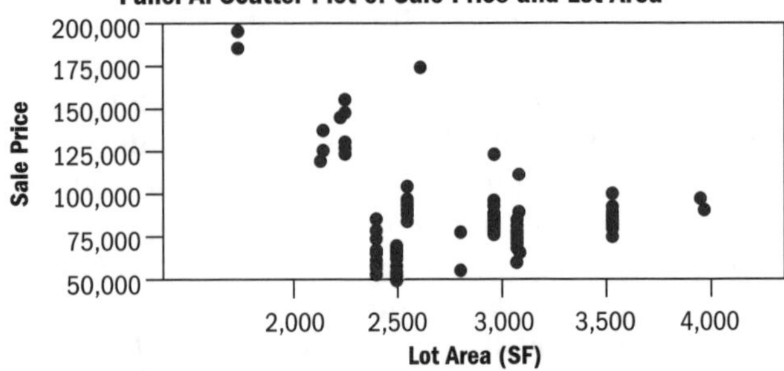

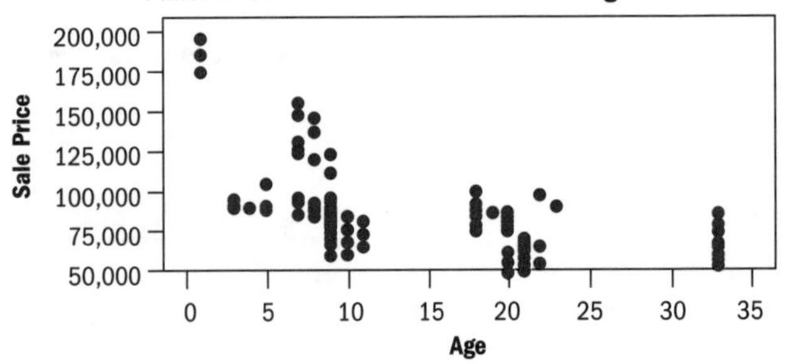

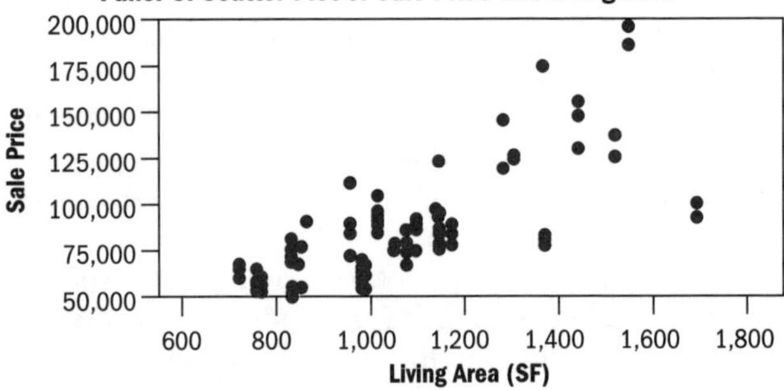

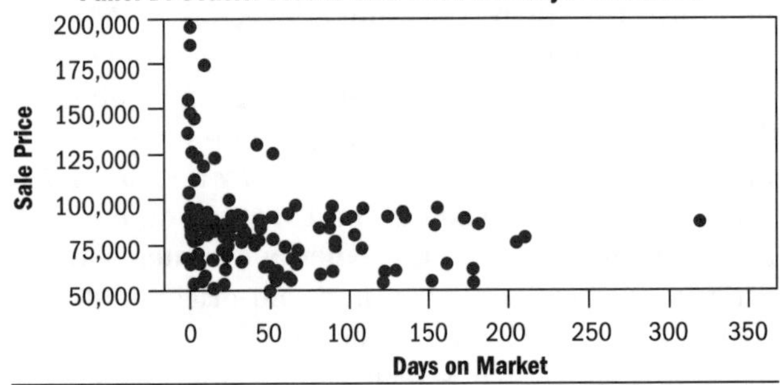

unit prices may be a proxy for improved market conditions late in the study period, which are reflected by shorter marketing periods *and* higher prices.)

If the *Days on Market* variable is included in the final model, then an estimate of future *Days on Market* must be included in an equation used for prediction using the final model (such as the valuation of the subject property in an appraisal assignment), even though the actual number of days the property might spend on the market is unknown. Generally speaking, time on market is very complex and difficult to model. In most instances time on market is correlated with how the list price is perceived by the market, measured by the difference between the list price and market value. When list price is much greater than market value, time on market is usually relatively long. When list price is less than market value, time on market is usually relatively short. Theoretically, time on market is much more likely to be related to the difference between list price and market value and less likely to be related to sale price.

The multiple regression analysis begins with selection of an initial collection of the most important numerical variables, using the best subsets procedure. We can break this procedure down into two stages: analysis of the data described in the first table of the case study and investigation of nonlinear transformations including quadratics and reciprocals.

The best subsets procedure limited to the data set's continuous variables identified the following six-variable model as "best" based on S_e, C_p, and R^2_{ADJ}:

Price = f(Living Area, Bedrooms, Garages, Time, Age, Days on Market)

$\mathbf{R^2 = 0.938}$ $\mathbf{R^2_{ADJ} = 0.935}$

Working with this base model, continuing analysis included a series of stepwise models used to test how the base model held up as interaction variables and nonlinear transformations were considered as explanatory variables. This step in the analysis resulted in the **preliminary model shown on the following page.**

Minitab Output of Variables, Expanding the Best Preliminary Model

Predictor	Coef	SE Coef	T	P
Constant	59187	4830	12.25	0.000
Spring	5011	1385	3.62	0.000
Summer	5606	1579	3.55	0.001
Fall	5902	2225	2.65	0.009
YR 2001	-20176	1808	-11.16	0.000
YR 2002	-16144	1704	-9.47	0.000
YR 2003	-10479	1806	-5.80	0.000
Living Area	29.502	5.973	4.94	0.000
Bedrooms	5104	1592	3.21	0.002
Baths	2631	1458	1.81	0.074
Garage	26652	1801	14.80	0.000
Age	-3120.6	321.4	-9.71	0.000
Age SQ	57.057	9.158	6.23	0.000
Lot Area (SF)	4.274	1.500	2.85	0.005
S = 5388.07		R-Sq = 95.9%	R-Sq(adj) = 95.5%	

Notice that prices in the spring, summer, and fall are all roughly $5,000 higher than in the winter, as measured by the coefficients of the seasonal variables. So, a *Winter/Not-Winter* dummy variable can replace the three seasonal variables in the final model, which is shown in the **table on the following page**, reducing k from 13 to 11 and increasing the ratio of $n:k$ from 9.9 to a more comfortable level of 11.7. When modeled this way, the multiple regression equation shows a significant $5,238 difference between a winter sale and sales occurring in the other three seasons.

The final model includes most of the variables first considered after running best subsets on the continuous variables. Dummy variables for year of sale replaced the linear time-trend variable, which does not restrict the functional form of price change over the study's time frame to a linear relationship. In addition, the *Days on Market* variable lost significance in the final model and was dropped. Loss of significance for this variable is not surprising given the weakness of any theoretical ties between *Days on Market* and *Sale Price* alone. In retrospect, *Days on Market* may have been inappropriately considered in the initial model.

Minitab Output of Final Model Variables

Predictor	Coef	SE Coef	T	P
Constant	64738	4807	13.47	0.000
Winter	-5238	1319	-3.97	0.000
YR 2001	-19947	1750	-11.40	0.000
YR 2002	-16016	1679	-9.54	0.000
YR 2003	-10294	1766	-5.83	0.000
Living Area	29.302	5.881	4.98	0.000
Bedrooms	5123	1580	3.24	0.002
Baths	2647	1445	1.83	0.069
Garage	26656	1786	14.93	0.000
Age	-3116.2	318.9	-9.77	0.000
Age SQ	56.822	9.085	6.25	0.000
Lot Area (SF)	4.178	1.480	2.82	0.006

S = 5350.22 R-Sq = 95.9% R-Sq(adj) = 95.5%

Analysis of Variance

Source	DF	SS	MS	F	P
Regression	11	78333602203	7121236564	248.78	0.000
Residual Error	117	3349105222	28624831		
Total	128	81682707425			

Several aspects of the final model are worthy of further consideration:

- Replacing the three seasonal variables with one seasonal variable had no effect on adjusted *R*-squared and reduced S_e slightly.
- All of the independent variables are significant at the 5% level except *Bathrooms*, which is significant at the 10% level.
- Price escalation was rapid during the time period covered by the data, showing a cumulative increase of nearly $20,000 per unit on average since 2001. This is consistent with the residential price inflation experienced in this zip code over this period of time.
- The variable *Age* is significant in quadratic form with the coefficients of *Age* and *Age-Squared* both being highly significant. This is consistent with the valuation theory that depreciation is not usually a straight-line phenomenon.
- Finally, the sign on lot size is positive, as theoretically expected. Reversing the sign found in

the correlation table once the effects of other price determining variables had been accounted for.

Indicator variables used to model changing market conditions are often misinterpreted. The coefficients are not representative of annual change in price. The -$19,947 coefficient of *YR 2001* indicates that price change by this amount on average from 2001 to 2004. The negative sign indicates that prices were lower in 2001 than in the intercept year 2004. The average annual price change for a typical unit from 2001 to 2002 is the *YR 2002* coefficient minus the *YR 2001* coefficient, which is $3,931. The average annual price change for a typical unit from 2002 to 2003 was $5,722, and it was $10,294 from 2003 to 2004.

Now that the final model has been selected, we can check a scatter plot of standardized residuals versus price estimates ($\hat{y}$) for evidence of heteroskedasticity. The **scatter plot below** shows no apparent pattern of heteroskedasticity. There are several extreme outliers, however, with standardized residuals in the vicinity of 3 or minus 3. This could suggest that a few of the sales may have included furnishings or been in substandard interior condition (i.e., variables not accounted for in the model). The outliers may also reflect randomness in the market that cannot be fully accounted for without collecting more data. In the **histogram below**, the standard errors are slightly right-skewed, which is consistent with the direction of

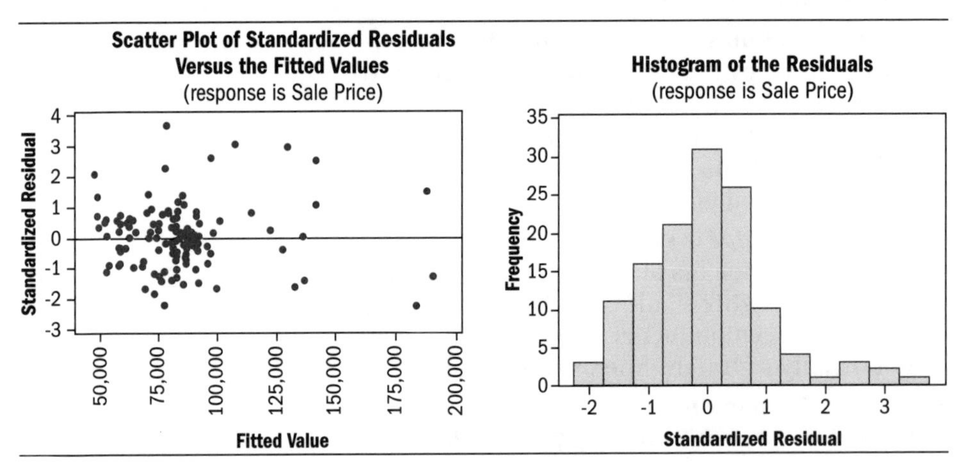

the largest outliers in the accompanying scatter plot, but otherwise the distribution appears to be approximately normal.

Muticollinearity does not appear to be a serious problem, given the significant *t* statistics on the most-correlated independent variables. (See the prior Minitab output of final model variables.) The average variance inflation factor is 2.8, excluding the high *VIF*s on the highly correlated variables *Age* and *Age-Squared.* Other than *Age* and *Age-Squared*, the highest *VIF* is associated with *Living Area*, as expected.

With this price estimation model in hand, we can use it to generate market value estimates for similar townhouses located in zip code 89015. Descriptions of three townhouses and their respective market value estimates are shown in the **table below.** Notice that all three of the prediction townhouses have independent variable values within the ranges of the independent variables in the data. If the model were to be applied for prediction outside of the range of the x variables, the tacit assumption would be that the relationship between price and the independent variables is unchanging outside of the range of the x variables. However, remember that the model cannot provide information to support this assumption.

This does not mean that prediction outside of the range is unreliable. At times, use of a regression model to predict values outside of the range of the independent variables may be justified if the analyst's market knowledge and experience is suf-

Market Value Estimates Based on the Final Model

Variable	Townhouse 1	Townhouse 2	Townhouse 3
Living Area	980 SF	1,100 SF	1,400 SF
Bedrooms	2	2	3
Baths	2	2	3
Garage Stalls	0	1	1
Age	6 yrs.	12 yrs.	8 yrs.
Age-Squared	36	144	64
Lot Size	1,800 SF	1,800 SF	2,700 SF
Expected Price	$99,863	$117,475	$145,715
95% Confidence Interval Low	$95,403	$112,514	$142,167
95% Confidence Interval High	$104,323	$122,437	$149,263

* All estimates are current as of 2004, summer season.

ficient to support the decision. For example, could the model we just developed be applied to a 20-year-old townhouse that has a floor area of 1,750 square feet, 3 bedrooms, 3 baths, and a 2-car garage, and is located on a 4,200-sq.-ft. lot in the same market area? We would have to answer, "It depends." The floor area is outside of the upper end of the data range, which is 1,696 square feet, and the lot area exceeds the data set's maximum value of 3,978 square feet. Is the townhouse market likely to differ dramatically for the larger unit? Would a 1,696-sq.-ft. unit on a 3,978-sq.-ft. lot compete effectively with a 1,750-sq.-ft. unit on a 4,200-sq.-ft. lot? If the answer to the first question is "no" and the answer to the second question is "yes," then using the model we just developed might be justified, absent a better alternative, such as data on slightly larger townhouse units.

The market value estimates based on the final model show that the width of the mean price prediction confidence interval is $8,920 for Townhouse 1, $9,923 for Townhouse 2, and $7,096 for Townhouse 3. Predictions are more precise for Townhouse 3 because the values of its elements of comparison are closer to the data centroid than the other two townhouses. The data centroid (ignoring year of sale and season, which is the same for all three predictions) is

- *Living Area* = 1,067
- *Bedrooms* = 2.11
- *Baths* = 2.01
- *Garage Stalls* = 0.12
- *Age* = 13.6
- *Age-Squared* = 185
- *Lot Area* = 2,790

It is difficult to think in multidimensional space, but this notion of a varying width confidence interval is the same as the prediction confidence interval concept developed in Chapter 9 in two-dimensional space.

Can you develop a better model? Now that you have completed the chapter, download the data from the Appraisal Institute Web site (if you have not already done so) and try to replicate what is presented here and then explore the data as a practical learning experience. As you can see, a great deal of decision-making is involved in following the path from data set to final model. It should not be surprising, therefore, that the probability is low that every analysis by a different analyst on a given data set would result in the selection of the same final model. By the way, these are actual sales from a zip code in Las Vegas, Nevada. Model fits this good are the exception rather than the rule. However, the best fits are generally found in fairly homogeneous, active residential markets such as this.

Solutions to Practice Problems

CHAPTER 2

2.1 Working with Positive and Negative Numbers

1. -2 **2.** 8 **3.** -2 **4.** -16 **5.** -16 **6.** -16 **7.** -4
8. 120 **9.** 1 **10.** $-xyzt$ **11.** 0 **12.** 0 **13.** undefined **14.** 4

2.2 Order of Operations

1. -9 **2.** 24 **3.** 17 **4.** 4 **5.** 2

2.3 Fractions, Decimals, and Percentages

1. $0.83\overline{3}$ **2.** 15 **3.** -3.25 **4.** 4.846 **5.** $0.55\overline{5}$ **6.** $10.6\overline{6}$ **7.** $\frac{x}{3}$
8. 50 **9.** 0.75 **10.** $-\frac{20}{xy}$ **11.** 9.26 **12.** $7.6\overline{6}$ **13.** 7.234 **14.** 13.4%
15. 2.56 **16.** $83.3\overline{3}$% **17.** 0.032% **18.** 1,050

2.4 Exponents and Roots

1. 49 **2.** 125 **3.** 12 **4.** x **5.** 0.64 **6.** 0.9747 **7.** 38.5
8. 75 **9.** $9x^2$ **10.** 64 **11.** 34 **12.** 64 **13.** y^5

2.5 Logarithms

1. -0.51083 **2.** 2.70805 **3.** 2.70805 **4.** 4.8283
5. 1 **6.** 2.60944 **7.** 5 **8.** -1.60944
9. 5

2.6 Summation Notation

1. 10 **2.** 8 **3.** 22 **4.** 38 **5.** 26 **6.** 100 **7.** 64
8. 108 **9.** 64 **10.** 324

2.7 Equations

1. $x = 4$ **2.** $x = 4$ **3.** $x = -75$ **4.** $x = 5$ **5.** $x = 99$

2.8 Factorials and Combinations

1. 40,320; 720; 6 **2.** 20, 1, 21 **3.** 20

2.9 Types of Data

1. discrete **2.** continuous **3.** continuous **4.** discrete **5.** nominal **6.** ordinal **7.** nominal **8.** continuous
9. continuous

CHAPTER 3

3.1 Organizing Numerical Data

1. Ordered Array: 1.1, 1.1, 1.2, 1.3, 1.3, 1.4, 1.5, 1.6, 1.7, 1.7, 1.9, 2.0, 2.0, 2.0, 2.0, 2.1, 2.2, 2.4, 2.5, 2.5, 2.6, 2.7, 2.8, 2.8, 2.8, 2.9, 3.0, 3.1, 3.3, 3.4, 3.5, 3.7, 4.3, 4.5, 4.8, 5.0, 5.6, 6.2

2. 0.158 (15.8%), 0.158 (15.8%), 0.711 (71.1%), 0.789 (78.9%)

3. 2.5 to 2.9 acres

4(a). Ordered Array: 9.1, 9.4, 9.7, 10.0, 10.2, 10.2, 10.3, 10.8, 11.1, 11.2, 11.5,11.5, 11.6, 11.6, 11.7, 11.7, 11.7, 12.2, 12.2, 12.3, 12.4, 12.8, 12.9, 13.0, 13.2

4(b). Stem and Leaf

Percent	Tenths of a Percent
9	1 4 7
10	0 2 2 3 8
11	1 2 5 5 6 6 7 7 7
12	2 2 3 4 8 9
13	0 2

4(c). 11 to 11.9 percent

3.2 Frequency and Percentage Distributions

2–4. Frequency, percentage, and cumulative percentage distributions

Class	Count	Percentage	Cumulative Percentage
80 to 84.9	2	3.6%	
85 to 89.9	5	9.1%	12.7%
90 to 94.9	8	14.5%	27.3%
95 to 99.9	15	27.3%	54.5%
100 to 104.9	9	16.4%	70.9%
105 to 109.9	8	14.5%	85.5%
110 to 114.9	5	9.1%	94.5%
115 to 119.9	3	5.5%	100.0%
Totals	55	100.0%	

5. Frequency histogram

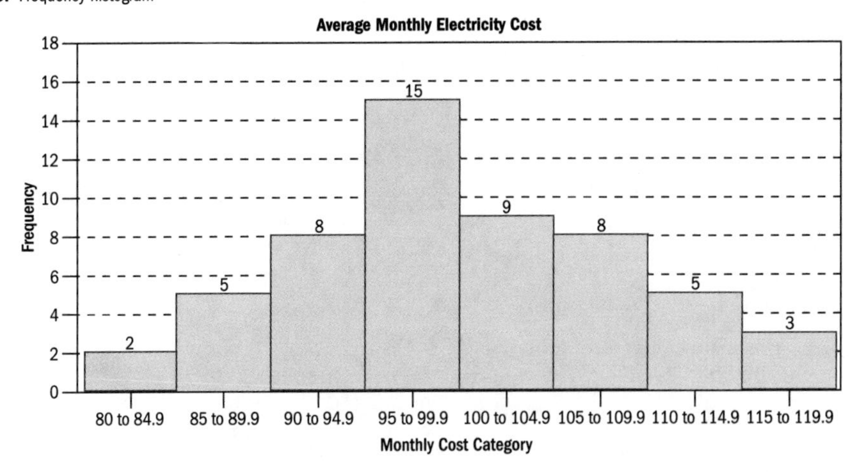

3.3 Polygons

2. Frequency and percentage distributions

Electric Bill Category	One-Bedroom Frequency	Two-Bedroom Frequency	One-Bedroom Percentage	Two-Bedroom Percentage
$70 to $74	2	0	3.6%	0.0%
$75 to $79	5	0	9.1%	0.0%
$80 to $84	8	2	14.5%	3.6%
$85 to $89	15	5	27.3%	9.1%
$90 to $94	10	8	18.2%	14.5%
$95 to $99	8	15	14.5%	27.3%
$100 to $104	5	9	9.1%	16.4%
$105 to $109	2	8	3.6%	14.5%
$110 to $114	0	5	0.0%	9.1%
$115 to $120	0	3	0.0%	5.5%
Totals	55	55	100.0%	100.0%

3. Percentage polygon overlay chart

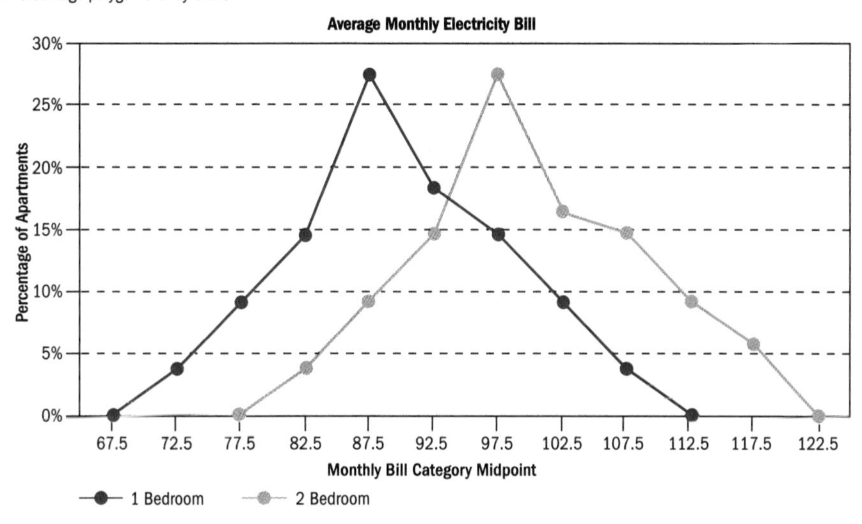

3.4 Bivariate Numerical Data

1. Water bill and floor area scatter plot shows a direct relationship (positive correlation).

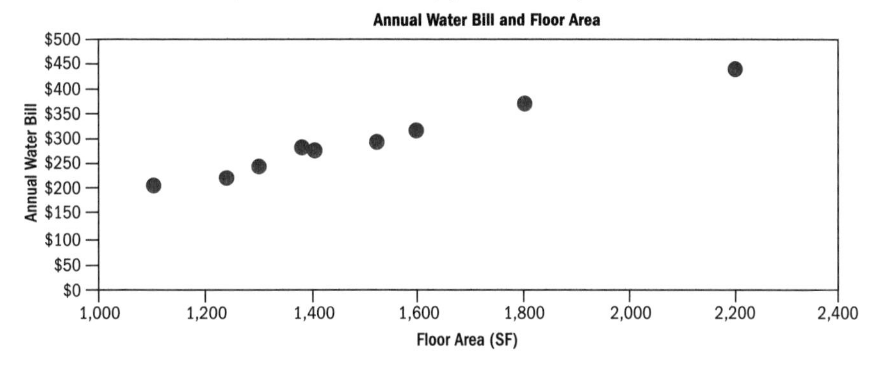

2. Population time trend shows big move in 2000 and flattening in 2004 and 2005.

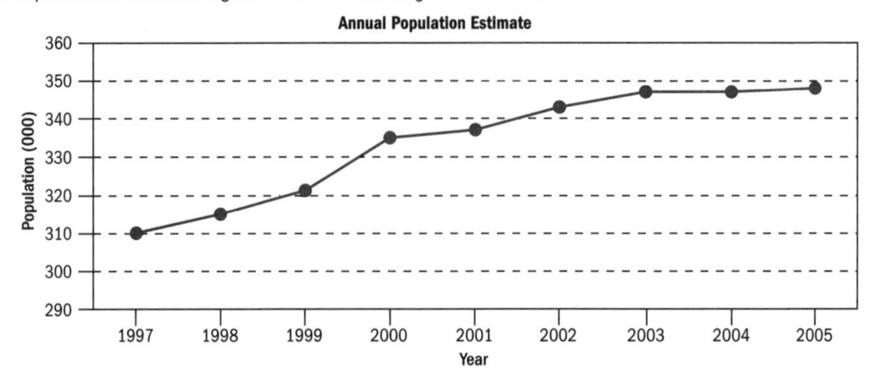

3.5 Categorical Data

1. Type of appraisal firm

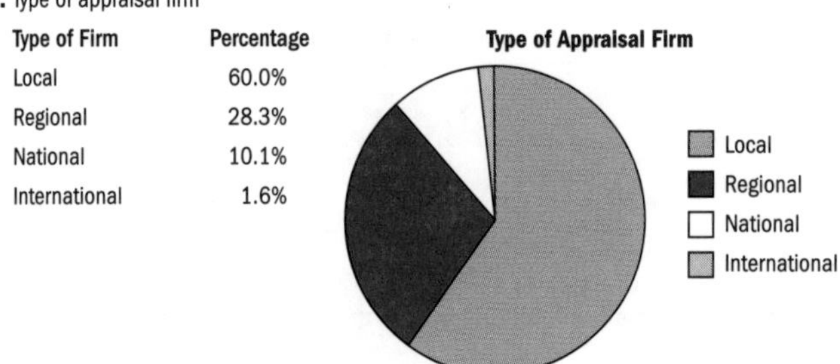

Type of Firm	Percentage
Local	60.0%
Regional	28.3%
National	10.1%
International	1.6%

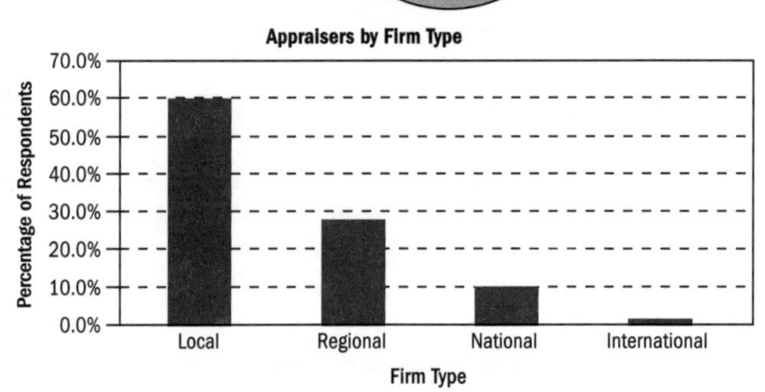

2. Apartment unit mix

	University Mkt	Mfg Market	Totals
Studio	320	38	358
1 BR	428	264	692
2 BR	560	780	1,340
3 BR	52	498	550
Totals	1,360	1,580	2,940

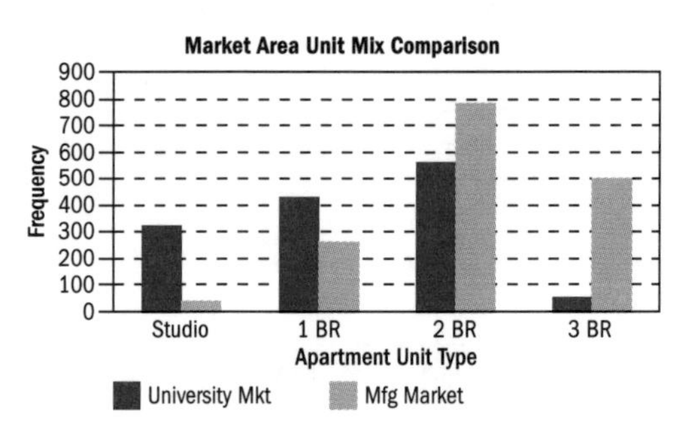

CHAPTER 4

4.1 Measures of Central Tendency

1. Mean = 1.129 Median = 1.2 Mode = 1.2 **2.** 5.34%

4.2 Measures of Dispersion

1. City A mean = 7.73%, median = 7.9%, trimodal at 7.4%, 7.9%, and 8%
City B mean = 7.32%, median = 7.2%, bimodal at 6.9% and 8%

2. City A range = 1.3%
City B range = 1.3%

Sorted Data

City A	City B
7.0	6.7
7.2	6.8
7.4	6.9
7.4	6.9
7.5	7.0
7.6	7.1
7.9	7.2
7.9	7.4
8.0	7.5
8.0	7.8
8.1	7.9
8.2	8.0
8.3	8.0

Q1 position = 3.5
Q2 position = 10.5

	City A	City B
Q1 =	7.40	6.90
Q2 =	8.05	7.85
IQR =	0.65	0.95

Note: Use of the "=QUARTILE" macro in Excel results in slightly different *Q3* values of 0.60 and 0.90 and resulting IQRs of 0.60 (City A) and 0.90 (City B).

3. $S_A = 0.41\%$ $S_B = 0.47\%$

4. $COV_A = 5.3\%$ $COV_B = 6.4\%$

5. The mode is the least reliable measure of central tendency because both are multimodal. Capitalization rates in City B are more dispersed in both a nominal and relative sense.

4.3 Box and Whisker Plots

1.

	1960	1980	2000
Mean	1,275.5	1,403.5	1,527.5
Median	1,215	1,320	1,435
Standard Deviation	163.8	180.2	227.4
COV	12.84%	12.84%	14.89%
Skewness	0.54	0.58	0.74

2.

House Living Area Comparison

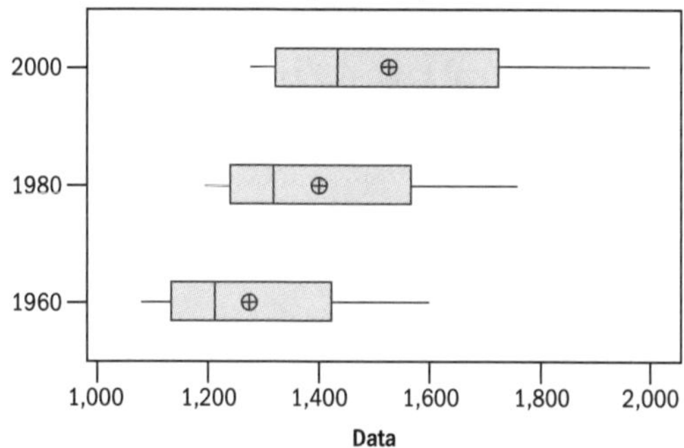

Note: The circle and cross symbol in each box plot identifies the mean.

3. Size has increased over time and become more dispersed. Newer housing is more right-skewed (extreme values tend to be larger homes).

4.4 Correlation

1.

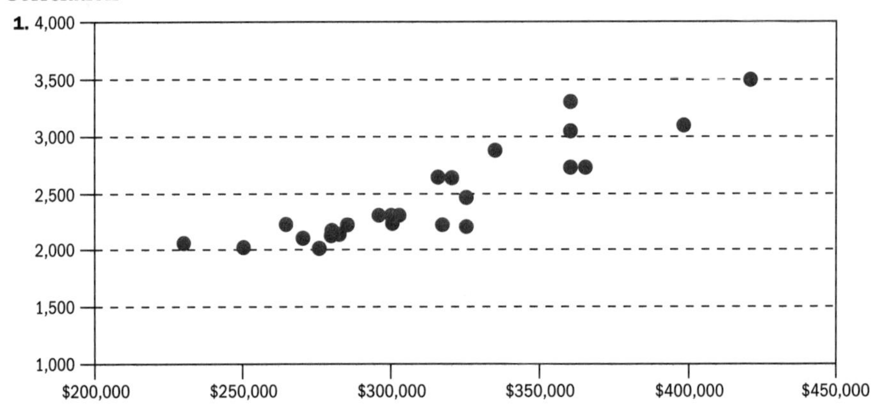

2. Pearson Correlation Coefficient = 0.905, coefficient of determination = 0.819

3. 81.9%

4.

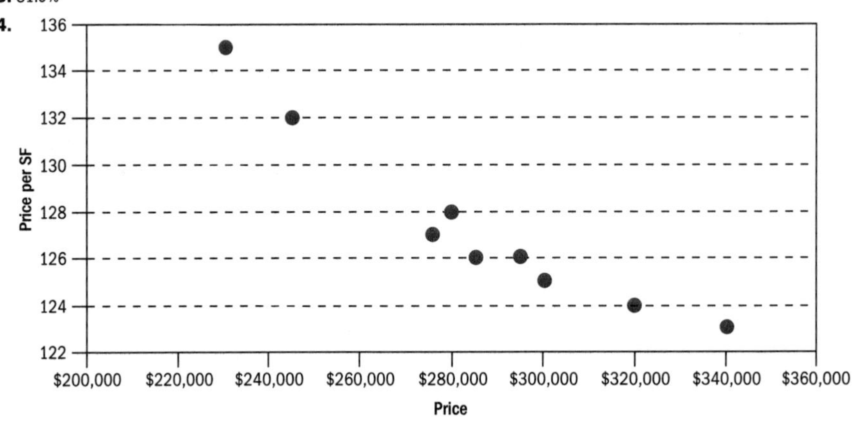

Spearman's *Rho* = -0.979

The scatter plot shows a curvilinear pattern. Therefore, Spearman's *Rho* is appropriate.

CHAPTER 5

5.1 Probability and Conditional Probability

1. $5 \div 36 = 0.139$ **2.** $1 \div 6 = 0.167$ **3.** $13 \div 52 = 0.25$

4. $4 \div 52 = 0.077$ **5.** $13 \div 51 = 0.255$ **6.** $12 \div 51 = 0.235$

5.2 Laws of Chance

1. $0.20 + 0.60 - 0.20 = 0.60$ $0.20 \div 0.60 = 0.333$

2. $0.50 \times 0.50 = 0.25$

3(a) $1{,}000 \div 2{,}200 = 0.455$ **3(b)** $250 \div 700 = 0.357$ **3(c)** No, because $0.67 \neq 0.75$

5.3 Binomial Distribution

1.

$$P(X = 4) = \frac{7!}{4! \cdot 3!} 0.40^4 \cdot 0.60^3 = 0.1935$$

2. $E[X] = np = 6(0.40) = 2.4$; $E[\text{not } X] = 6 - 2.4 = 3.6$; mean cost = 3.6($50) = $180

alternatively,

$E[\text{not } X] = n(1 - p) = 6(0.60) = 3.6$; mean cost = 3.6($50) = $180

5.4 Normal Distribution

1. 0.9772 **2.** 0.0228 **3.** 0.1075 **4.** 0.9772 – 0.8413 = 0.1359

5(a) 0.1587 **5(b)** 0.1587 **5(c)** 0.9332 **5(d)** 0.8413 – 0.3085 = 0.5328

CHAPTER 6

6.1 Research Design, Hypothesis Testing, and Sampling

1. H_0: $\mu_{(2000\text{-}2005)} - \mu_{(2006, 2007)} = 0$
H_a: $\mu_{(2000\text{-}2005)} - \mu_{(2006, 2007)} \neq 0$

2. H_0: $\mu_{(2000\text{-}2005)} - \mu_{(2006, 2007)} \leq 0$
H_a: $\mu_{(2000\text{-}2005)} - \mu_{(2006, 2007)} > 0$

3. H_0: $\mu_{(2000\text{-}2005)} - \mu_{(2006, 2007)} \geq 0$
H_a: $\mu_{(2000\text{-}2005)} - \mu_{(2006, 2007)} < 0$

4. Problem 1 is two-tailed, Problem 2 is one-tailed (right-tailed), and Problem 3 is one-tailed (left-tailed).

5. Critical value of Z is 1.64 for Problem 1 and 1.28 for Problems 2 and 3.

6. The null hypothesis is rejected at $\alpha = 0.10$ and $\alpha = 0.05$ but not at $\alpha = 0.01$. You risk Type II error as you decrease α.

7. The null hypothesis cannot be rejected at $\alpha = 0.10$ (statistical distance between $\bar{x}$ and hypothesized μ is -1.23 standard deviations and the critical value is -1.28 standard deviations).

8. A stratified random sample with 10 strata:

1) Full-time freshmen
2) Full-time sophomores
3) Full-time juniors
4) Full-time seniors
5) Full-time graduate students
6) Part-time freshmen
7) Part-time sophomores
8) Part-time juniors
9) Part-time seniors
10) Part-time graduate students

9(a). 96.04, or 97 rounded up

9(b). 242.5, or 243 rounded up

10. 747.11, or 748 rounded up

CHAPTER 7

7.1 Sampling Distribution of the Mean

1. 15, 15 15, 30 15, 7 15,45 30, 15 30, 30 30, 7 30, 45 7, 15 7, 30 7,7 7,45 45, 15 45, 30 45, 7 45, 45

Mean	Freqency
7	1
11	2
15	1
18.5	2
22.5	2
26	2
30	3
37.5	2
45	1

$\mu = 24.25$

$\mu_{\bar{x}} = 24.25$

$\sigma = 14.55$ and the standard error of the mean = 10.29

[in Excel, enter "=STDEVP(15,30,7,45)" = 14.55 and then "=14.55/SQRT(2)" = 10.29]

2.

$$Z = \frac{1{,}150 - 1{,}000}{200} = \frac{150}{200} = 0.75$$

Therefore, the probability that the rent is ≥ \$1,150 per month equals the probability that $Z \geq 0.75$, which is (1 – 0.77337) = 0.22663 (22.7%).

$$Z = \frac{1{,}150 - 1{,}000}{\frac{200}{\sqrt{25}}} = \frac{150}{40} = 3.75$$

Therefore, the probability that the sample mean is ≥ \$1,150 equals the probability that $Z \geq 3.75$, which is (1 – 0.99991) = 0.00009 (0.009%).

$$Z = \frac{1{,}150 - 1{,}000}{\frac{200}{\sqrt{9}}} = \frac{150}{66.67} = 2.25$$

Therefore, the probability that the sample mean is ≥ \$1,150 equals the probability that $Z \geq 2.25$, which is (1 – 0.98778) = 0.01222 (1.22%).

The probabilities change when sample size changes from 25 to 9 because the standard error of the mean is larger when the sample size becomes smaller.

7.2 Sampling from Non-Normal Populations

1. $t_{0.05,59} = 1.671$ $t_{0.10,59} = 1.296$

2. 0.005 0.01

3. $\bar{x} \pm t_{0.025,80}\left(\frac{S}{\sqrt{n}}\right) = \$375 \pm 1.99\left(\frac{\$28}{\sqrt{81}}\right) = \$375 \pm \$6.19$, or approximately \$369 to \$381

4.

$$t_{\frac{\alpha}{2}} = \frac{\bar{x} - 5}{\frac{S_x}{\sqrt{n}}} = \frac{5.5 - 5}{\frac{0.5}{\sqrt{98}}} = \frac{0.5}{0.0505} = 9.90$$

Critical Value: $t_{0.05,97} = 1.661$

Reject null hypothesis and conclude that the rating is biased.

5.

$$t_{\alpha} = \frac{\bar{x} - \$155}{\frac{S_x}{\sqrt{n}}} = \frac{\$157 - \$155}{\frac{\$8.50}{\sqrt{33}}} = \frac{\$2}{\$1.4797} = 1.352$$

Critical Value: $t_{0.05,32} = 1.694$

Cannot reject the null hypothesis that prices in the Better-than-Usual Middle School area are the same as everywhere else in Usual, Minn. The evidence does not support John's claim.

7.3 Inferences about the Population Proportion

1.

$$\bar{p} \pm Z\sqrt{\frac{\bar{p}(1-\bar{p})}{n}} = 0.20 \pm 1.64\sqrt{\frac{0.20(1-0.20)}{200}} = 0.20 \pm 0.0464, \text{ or approximately } 15.4\% \text{ to } 24.6\%$$

2.

$$Z \cong \frac{\bar{p} - p}{\sqrt{\frac{p(1-p)}{n}}} = \frac{64\% - 90\%}{\sqrt{\frac{90\%(1-90\%)}{50}}} = \frac{-26\%}{4.243\%} = -6.13$$

Critical Value: $-Z_{0.05} = -1.64$

Reject null hypothesis and conclude that the percentage of homes having a pool is less than 90%.

7.4 Two-Sample Tests of Means

1.

$$t_{n-1} = \frac{\bar{d} - \mu_d}{\frac{S_d}{\sqrt{n}}} = \frac{6.7333 - 0}{\frac{6.44168}{\sqrt{15}}} = \frac{6.7333}{1.663} = 4.048$$

Critical Value: $t_{0.05,14} = 1.761$

Since 4.048 ≥ 1.761, reject the null hypothesis that the mean difference was ≤ 0 and conclude that the module was effective (mean difference > 0).

Excel output for this test is reproduced on the following page. Minitab and SPSS generate a similar result. Key comparisons to the preceding arithmetic and table-dependent calculations are the Excel *t*-test statistic value of 4.048 and the Excel critical value of 1.761.

t-Test: Paired Two Sample for Means	Pre Test	Post Test
Mean	64.867	71.600
Variance	430.410	328.686
Observations	15.000	15.000
Pearson Correlation	0.954	
Hypothesized Mean Difference	0.000	
df	14.000	
t Stat	4.048	
P(T<=t) one tail	0.001	
t Critical one-tail	1.761	
P(T<=t) two tail	0.001	
t Critical two-tail	2.145	

2. Test for equal variance: $F_{35,39} = \frac{S_1^2}{S_2^2} = \frac{42.25}{39.0625} = 1.0816$

Upper *F* critical value using FINV from Excel = 1.526

Since 1.0816 < 1.526, we cannot reject variance homogeneity and use the pooled variance *t*-test.

$$S_p^2 = \frac{(n_1 - 1)S_1^2 + (n_2 - 1)S_2^2}{(n_1 - 1) + (n_2 - 1)} = \frac{(36 - 1)42.25 + (40 - 1)39.0625}{(36 - 1) + (40 - 1)} = \frac{3{,}002.1875}{74} = 40.57$$

$$t_{74} = \frac{(\bar{x}_1 - \bar{x}_2) - (\mu_1 - \mu_2)}{\sqrt{S_p^2\left(\frac{1}{n_1} + \frac{1}{n_2}\right)}} = \frac{(\$142 - \$137) - \$3.50}{\sqrt{40.57\left(\frac{1}{36} + \frac{1}{40}\right)}} = \frac{1.50}{1.463} = 1.025$$

Critical Value: $t_{0.10,74} = 1.293$

Therefore, she cannot be 90% confident that the price difference is $3.50 per square foot or more.

3. Excel solution:

F-Test Two-Sample for Variances

	Lakefront Lots	*Golf Lots*
Mean	160378.125	159471.0526
Variance	257979656.3	828670365.5
Observations	16	19
df	15	18
F	0.311317584	
P(F<=f) one-tail	0.013439914	
F Critical one-tail	0.424929425	

t-Test: Two-Sample Assuming Unequal Variances

	Lakefront Lots	*Golf Lots*
Mean	160378.125	159471.0526
Variance	257979656.3	828670365.5
Observations	16	19
Hypothesized Mean	0	
df	29	
t Stat	0.117359094	
P(T<=t) one-tail	0.453692274	
t Critical one-tail	1.699127097	
P(T<=t) two-tail	0.907384548	
t Critical two-tail	2.045230758	

Cannot reject the null hypothesis that the mean prices are equal.

7.5 Test of Means from More than Two Samples

ANOVA

Source of Variation	*SS*	*df*	*MS*	*F*	*P-value*	*F crit*
Between Groups	29.574	3	9.858	20.20772	7.52E-08	2.866265
Within Groups	17.562	36	0.487833			
Total	47.136	39				

```
Bartlett's Test (normal distribution)
Test statistic = 5.57, p-value = 0.135
Levene's Test (any continuous distribution)
Test statistic = 1.73, p-value = 0.178
```

Since 20.208 > 2.866, reject the null hypothesis that the mean cap rates of the four cities are equal.

Dependent Variable: Cap_Rate
Bonferroni

(I) 1=A, 2=B, 3=C, 4=D	(J) 1=A, 2=B, 3=C, 4=D	Mean Difference (I-J)	Std. Error	Sig.	95% Confidence Interval	
					Lower Bound	Upper Bound
1.00	2.00	-.13000	.31236	1.000	-1.0021	.7421
	3.00	-1.75000*	.31236	.000	-2.6221	-.8779
	4.00	.56000	.31236	.488	-.3121	1.4321
2.00	1.00	.13000	.31236	1.000	-.7421	1.0021
	3.00	-1.62000*	.31236	.000	-2.4921	-.7479
	4.00	.69000	.31236	.202	-.1821	1.5621
3.00	1.00	1.75000*	.31236	.000	.8779	2.6221
	2.00	1.62000*	.31236	.000	.7479	2.4921
	4.00	2.31000*	.31236	.000	1.4379	3.1821
4.00	1.00	-.56000	.31236	.488	-1.4321	.3121
	2.00	-.69000	.31236	.202	-1.5621	.1821
	3.00	-2.31000*	.31236	.000	-3.1821	-1.4379

* The mean difference is significant at the .05 level.

City C mean cap rate is different from those in Cities A, B, and D. Cannot reject the null hypothesis that cap rate pairs from Cities A, B, and D are equal.

7.6 Two-Sample Test of Proportions

$$p^* = \frac{X_1 + X_2}{n_1 + n_2} = \frac{13 + 18}{31 + 31} = \frac{31}{62} = 0.50$$

$$Z = \frac{(p_{s_1} - p_{s_2}) - (p_1 - p_2)}{\sqrt{p^*(1-p^*)\left(\frac{1}{n_1} + \frac{1}{n_2}\right)}} = \frac{(0.419) - 0.581) - 0}{\sqrt{0.50(1-0.50)\left(\frac{1}{31} + \frac{1}{31}\right)}} = \frac{-0.162}{\sqrt{0.016129}} = -1.270$$

Critical Value: $Z = -1.96$

Since -1.270 > -1.96, you cannot reject the null hypothesis that the proportions are equal.

CHAPTER 8

8.1 Chi-Square Test of Proportions

1. *p*-values are 0.107 0.094 0.015

2.

Cell	f_o	f_e	$(f_o - f_e)$	$(f_o - f_e)^2$	$\frac{(f_o - f_e)^2}{f_e}$
1	14	13.51	0.49	0.2401	0.018
2	8	8.49	-0.49	0.2401	0.028
3	21	21.49	-0.49	0.2401	0.011
4	14	13.51	0.49	0.2401	0.018
χ^2					$\Sigma = 0.075$

p-value = 0.785

Cannot reject the null hypothesis that the proportions are equal.

3.

	SAMPLE			
	School A	School B	School C	Total
Pass	86	90	92	268
Fail	14	10	8	32
Total	100	100	100	300

Cell	f_o	f_e	$(f_o - f_e)$	$(f_o - f_e)^2$	$\frac{(f_o - f_e)^2}{f_e}$
1	86	89.33	-3.33	11.09	0.124
2	14	10.67	3.33	11.09	1.039
3	90	89.33	0.67	0.45	0.005
4	10	10.67	-0.67	0.45	0.042
5	92	89.33	2.67	7.13	0.080
6	8	10.67	-2.67	7.13	0.668
χ^2					$\Sigma = 1.96$

p-value = 0.375

Cannot reject the hypothesis that pass rates are the same at the three schools (sample variation is random).

8.2 Two-Sample Test of Central Tendency: Independent Samples

1. New median expense ratio = 0.3875

Old median expense ratio = 0.41

H_0: $M_{new} \geq M_{old}$

H_a: $M_{new} < M_{old}$

Critical Value: $T_1 = 31$ ($n_1 = 6$, $n_2 = 8$), one-tailed $\alpha = 0.05$

Sorted Expense Ratio	Ratio Rank	New Sample Rank	Old Sample Rank
0.36	1	1	
0.37	2	2	
0.38	3	3	
0.39	4.5		4.5
0.39	4.5		4.5
0.395	6	6	
0.4	7		7
0.41	9.5	9.5	
0.41	9.5	9.5	
0.41	9.5		9.5
0.41	9.5		9.5
0.42	12.5		12.5
0.42	12.5		12.5
0.43	14		14
		$\Sigma T_1 = 31$	$\Sigma T_2 = 74$

Since $\Sigma T_1 \leq 31$, reject the null hypothesis and conclude that new apartment expense ratios are less than old apartment expense ratios.

2. SPSS output:

Test Statistics

	Exp_Ratio
Mann-Whitney U	10.000
Wilcoxon W	31.000
Z	-1.832
Asymp. Sig. (2-tailed)	0.67
Exact Sig. [2*(1-tailed) Sig.]	0.81[a]

a. Not corrected for ties.

Exact one-tailed *p*-value = 0.0405 (not corrected for ties). This test also rejects the null hypothesis at one-tailed $\alpha = 0.05$. Also, using the critical value table (Appendix E), the critical value of $U = 11$ for a one-tailed test with $\alpha = 0.05$, $n_1 = 6$, and $n_2 = 8$. The test statistic is $U = 31 - [6(6 + 1)]/2 = 31 - 21 = 10$. Since $10 \leq 11$, we reject the null hypothesis and conclude that the "new" expense ratio is less than the "old" expense ratio.

8.3 Two-Sample Test of Central Tendency: Paired Samples

1. $\mu = 126.5$ $\sigma = 30.8$ $Z = -2.26$ *p*-value = 0.012

Reject null hypothesis.

2. Apartment Completed	Late 2005 Vacancy Rate	Early 2003 Vacancy Rate	Differences	R_i	Difference Sign	Positive Ranks
Property 1	9%	8%	-1%	1.5	–	
Property 2	7%	7%	0%	*		
Property 3	10%	8%	-2%	3.5	–	
Property 4	5%	6%	1%	1.5	+	1.5
Property 5	9%	7%	-2%	3.5	–	
Property 6	6%	6%	0%	*		
Property 7	10%	7%	-3%	5	–	
Property 8	6%	6%	0%	*		
Property 9	8%	4%	-4%	6	–	
						$\Sigma = W = 1.5$

Critical Value: $W = 2$ (one-tailed, $\alpha = 0.05$, $n = 6$)

Since $1.5 \leq 2$, reject the null hypothesis and conclude that vacancy rates were lower in 2003.

8.4 Test of Central Tendency: More than Two Independent Samples

```
Kruskal-Wallis Test on Lot Price per Square Foot

Category   N  Median  Ave Rank
Std        8   2.500      12.0
Street     6   2.300       3.9
Green      6   2.575      15.1
Overall   20              10.5

H = 11.55  DF = 2  P = 0.003
H = 11.65  DF = 2  P = 0.003  (adjusted for ties)
```

CHAPTER 9

9.1 Deriving a Least Squares Regression Equation

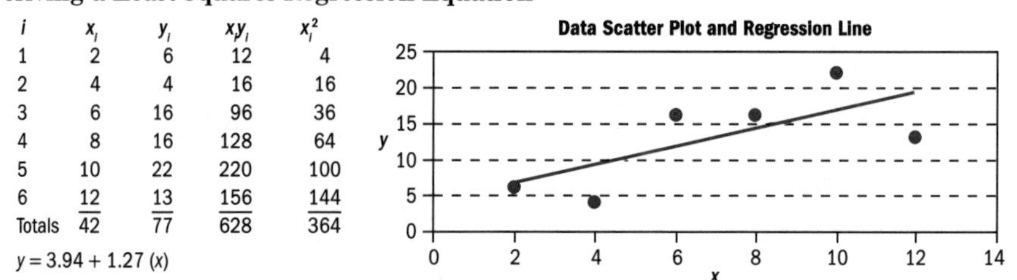

i	x_i	y_i	xy_i	x_i^2
1	2	6	12	4
2	4	4	16	16
3	6	16	96	36
4	8	16	128	64
5	10	22	220	100
6	12	13	156	144
Totals	42	77	628	364

$y = 3.94 + 1.27\,(x)$

9.2 Inferences Concerning Population Intercept and Slope

Price (\$000) = -5.376 + 105.92 (Acres)

Critical value of t: $t_{0.10,3} = 1.638$

$$t\text{-test statistic} = \frac{105.92 - 0}{13.98} = 7.58$$

Since $7.58 \geq 1.638$, reject the null hypothesis and conclude that $\beta > 0$.

The 90% confidence interval for $\beta = 105.92 \pm 2.3534(13.98)$, or 73.02 to 138.82, which is interpreted as \$73,020 to \$138,820 per 1-acre change of site area.

Note that the critical value of t for a one-tailed test differs from the value of t used in deriving the confidence interval because a confidence interval is by definition two-tailed.

9.3 Prediction Using Simple Linear Regression

Expected price when lot size equals 2.3 acres is \$238.3 thousand, or \$238,300.

Mean confidence interval when lot size = 2.3 is \$210,800 to \$265,700.

Single transaction price confidence interval when lot size = 2.3 is \$177,000 to \$299,500.

9.4 Coefficient of Determination

1.

$$0.820 = 1 - \frac{3.80}{SST}, \; SST = 21.11 \qquad SSR = SST - SSE = 17.31$$

2.

$$Price\ (\$000) = -103.76 + 0.1672(SF)$$

$$SSE = 448.4$$

$$SSR = 1{,}787.3$$

$$SST = 2{,}235.7$$

$$R^2 = \frac{1{,}787.3}{2{,}235.7} = 0.799$$

9.5 Transformations for Curvilinear Relationships

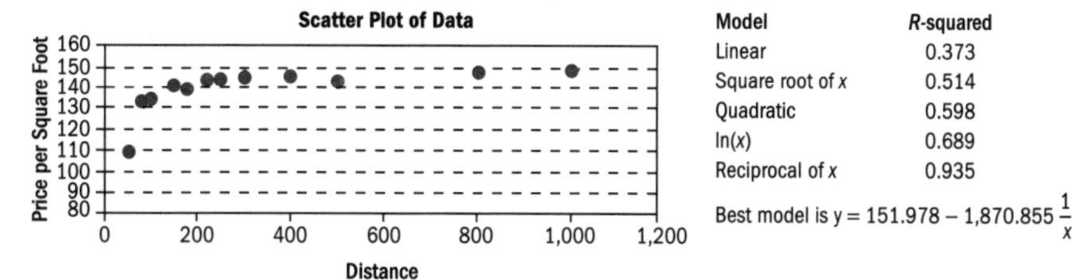

Model	R-squared
Linear	0.373
Square root of x	0.514
Quadratic	0.598
ln(x)	0.689
Reciprocal of x	0.935

Best model is $y = 151.978 - 1{,}870.855 \frac{1}{x}$

9.6 Simple Regression for Time Series Data

The compound growth model fits these data well.

The regression model is

$$\ln(Population) = 12.15486 + 0.009975(t)$$

$$a = e^{12.15486} = 190{,}015$$

$$1 + b = e^{0.009975} = 1.010025$$

Population forecast equation is $Population = 190{,}015\ (1.010025)^t$

Compound population growth rate = approximately 1% per annum

$$t = 10;\ \text{forecast population} = 209{,}946$$

$$t = 11;\ \text{forecast population} = 212{,}051$$

$$R^2 = 0.999$$

CHAPTER 10

10.1 From the Simple Linear Regression Model to the Multiple Linear Regression Model

PRICE = \$94,523 + \$280(*SF*) + \$23,108(*FLOOR*), meaning that price varies by \$280 per square foot when *Floor* is held constant, and *Price* varies by \$23,108 per floor when *Living Area* is held constant.

$F = 287.25$, p-value = 0.000, model is significantly better than mean price as a price estimator

t statistic for $SF = 7.85$, p-value = 0.000, reject H_0 that coefficient of $SF = 0$

t statistic for $FLOOR = 21.3$, p-value = 0.000, reject H_0 that coefficient of $FLOOR = 0$

$R = 0.986$, $R^2 = 0.971$, $R^2_{ADJ} = 0.968$, meaning that the correlation between actual price and predicted price for the 20 sales is 0.986, the model explains 97.1% of the variance in sale price, and most of the explanatory power is systematic rather than a function of random correlation between the dependent variable and the two independent variables.

95% Confidence interval on β_{SF} = \$205.10 to \$355.80

95% Confidence interval on β_{FLOOR} = \$20,820 to \$25,395

Predicted mean price when $SF = 1{,}240$ and $FLOOR = 11$ is \$696,461, and we are 95% confident that the mean price is between \$682,687 and \$710,735.

Beta coefficient for $SF = 0.326$ and beta coefficient for $FLOOR = 0.885$, meaning *PRICE* changes by 0.326 standard deviations with a 1 standard deviation change in *SF* and by 0.885 standard deviations with a 1 standard deviation change in *FLOOR*.

Minitab output is included below. Beta coefficients and R were computed in Excel using standardized values of the three variables.

```
The regression equation is
PRICE = 94523 + 280 SF + 23108 FLOOR


Predictor      Coef  SE Coef      T      P
Constant      94523    44518   2.12  0.049
SF           280.45    35.71   7.85  0.000
FLOOR         23108     1084  21.31  0.000


S = 15555.9   R-Sq = 97.1%   R-Sq(adj) = 96.8%


Analysis of Variance

Source          DF           SS           MS       F      P
Regression       2  1.39021E+11  69510707466  287.25  0.000
Residual Error  17   4113785067    241987357
Total           19  1.43135E+11


Predicted Values for New Observations

New
Obs     Fit  SE Fit         95% CI              95% PI
  1  696461    6529  (682687, 710235)  (660868, 732055)


Values of Predictors for New Observations

New
Obs    SF  FLOOR
  1  1240   11.0
```

10.2 Unusual Observations

Price	SF	Age	Standardized Residual	Leverage	Standardized DfFit
85,000	1,019	7	0.45144	0.04174	0.13956
80,500	1,151	9	-0.67343	0.02	-0.17596
71,500	836	9	-0.5957	0.10671	-0.26559
89,900	1,019	3	0.537	0.09049	0.22128
91,900	1,696	18	0.23518	0.55205	0.4367
74,500	1,101	20	0.70881	0.02338	0.19115
54,000	992	22	-1.99976	0.0358	-0.66093
82,500	1,151	9	-0.31286	0.02	-0.081
52,900	772	33	1.18351	0.2531	0.93444
83,950	1,151	9	-0.05144	0.02	-0.01329
83,900	1,151	9	-0.06046	0.02	-0.01562
78,000	1,056	18	1.18379	0.0078	0.28033
63,000	984	21	-0.53347	0.02906	-0.15009
56,900	984	21	-1.63321	0.02906	-0.49093
91,000	1,019	3	0.73531	0.09049	0.30516
84,000	1,019	8	0.47062	0.03263	0.13601
89,900	1,019	3	0.537	0.09049	0.22128
61,500	984	21	-0.8039	0.02906	-0.22833
84,500	1,151	9	0.04771	0.02	0.01232
54,750	984	21	-2.02082	0.02906	-0.6347
56,000	772	33	1.74239	0.2531	1.46814
84,500	1,151	9	0.04771	0.02	0.01232
85,000	1,151	9	0.13786	0.02	0.03562
82,500	1,376	20	0.66671	0.16598	0.37929

Standardized residuals more than 2 standard deviations from the mean and leverage greater than $2(k + 1)/n = 0.25$ are identified. Reexamination of three observations is warranted: Observations 9, 20, and 21. The townhouse project with 33-year-old, 772-sq.-ft units (Observations 9 and 21) has the most leverage (sample centroid = 1,070 square feet and 14 years). However, the effect of these two units on fit is not excessive. Error in estimating Observation 20 is flagged as an outlier. It sold for $54,750, whereas two other units the same size and age (Observations 13 and 14) sold for higher prices of $63,000 and $56,900 respectively. This difference could possibly be accounted for by gathering additional information (e.g., interior condition, furnishings included, seller motivation).

10.3 Model Building Issues

First, run a best subsets regression in Minitab (or stepwise in SPSS), forcing each of the alternative models to include *ACRES*, *VIEW*, and *CONSTRAINT*. Two interesting alternative models arise, both suggesting a quadratic model for view and one suggesting that *ADJ+* might become significant when *VIEW* is modeled in a nonlinear fashion.

As shown in the Minitab output on the following page, C_p is equal to the expected value of 2 with $VIEW^2$ added to the model. In addition, C_p is below the expected value of 3 when *ADJ+* is added along with $VIEW^2$. R^2_{ADJ} is 86.8% with $VIEW^2$ added and 86.9% with both $VIEW^2$ and *ADJ+* added.

Response is Price

The following variables are included in all models: Acres View Constraint

Vars	R-Sq	R-Sq(adj)	Mallows C-p	S	ADJ+	ADJ-	ViewXCons	AcresXCons	AC SQ	View SQ
1	87.7	86.8	2.0	23049						X
1	87.1	86.1	4.6	23627		X				
2	88.0	86.9	2.7	22969	X					X
2	87.8	86.7	3.4	23116					X	X
3	88.1	86.7	4.4	23102	X				X	X
3	88.0	86.7	4.5	23144	X			X		X
4	88.1	86.5	6.2	23288	X	X			X	X
4	88.1	86.5	6.2	23291	X			X	X	X
5	88.2	86.3	8.0	23477	X	X		X	X	X
5	88.1	86.2	8.2	23516	X	X	X		X	X
6	88.2	86.0	10.0	23715	X	X	X	X	X	X

PRESS and R^2_{PRESS} for the original three-variable model, the four-variable model adding $VIEW^2$, and the five-variable model including $VIEW^2$ and *ADJ+* are

	PRESS	Prediction R^2
3 variables	36,107,199,914	84.5%
4 variables	35,867,430,934	84.6%
5 variables	36,687,760,704	84.25%

Based on these statistics, transformation of the *View* variable into quadratic form offers a slight improvement in fit and prediction. The best subsets regression suggests that the two new interaction variables are unimportant.

The four-variable "best" model is

The regression equation is

Price = 101887 + 19249 Acres + 692 View + 3.16 View SQ - 37304 Constraint

Predictor	Coef	SE Coef	T	P
Constant	101887	9776	10.42	0.000
Acres	19249	6482	2.97	0.004
View	692.2	268.0	2.58	0.013
View SQ	3.161	1.728	1.83	0.073
Constraint	-37304	9477	-3.94	0.000

S = 23048.9   R-Sq = 87.7%   R-Sq(adj) = 86.8%

Analysis of Variance

Source	DF	SS	MS	F	P
Regression	4	2.04285E+11	51071313126	96.13	0.000
Residual Error	54	28687592020	531251704		
Total	58	2.32973E+11			

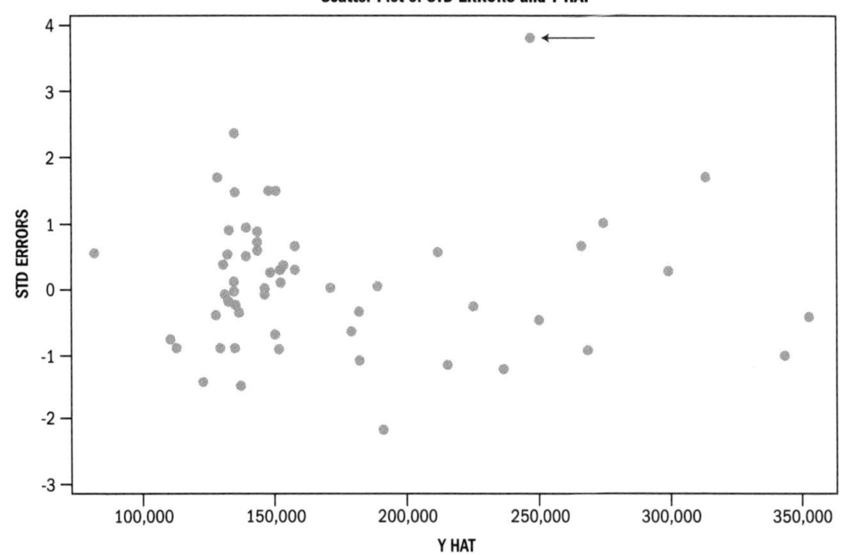

Based on this scatter plot, the model does not seem to require a test for heteroskedasticity. There is one extreme outlier, marked with an arrow, with a standardized error of 3.72 (the $325,000 sale of a 2.19-acre lot with a 100-degree view panorama and no severe developability constraint).

Useful Symbols and Formulas

Symbol	Description	Chapter
a	sample y-intercept (in linear regression)	9
AAD	average absolute deviation	4
AD	Anderson-Darling test value	8
ANOVA	analysis of variance	7
α	population intercept (in linear regression)	9
α	significance level	6
$1 - \alpha$	confidence level	6
b	sample slope (in linear regression)	9
β	population slope (in linear regression)	9
β	Type II error	6
CI	confidence interval	9
COD	coefficient of dispersion	4
COV	coefficient of variation	4
C_p	Mallow's C_p statistic	10
χ^2	chi-square statistic	8
D	sum of squared deviations between each group's average rank and the combined average rank, weighted by each group's size (in Kruskal-Wallis test)	8
df	degrees of freedom (associated with separate variance t statistic)	7
$DfFit$	test of leverage and outliers (in SPSS)	10
e	sampling error	6
e	sample error (in linear regression)	9
ε	population error (in linear regression)	9
$E[e_i]$	expected value of regression errors	9
E[X]	expected number of occurrences of Event X	5
F	F distribution	7

Symbol	Description	Chapter
F_{n_1-1, n_2-1}	F distribution with $n_1 - 1$ numerator degrees of freedom and $n_2 - 1$ denominator degrees of freedom	7
f_e	expected frequency	8
f_o	observed frequency	8
h	sum of rank weighted squared errors divided by sum of squared errors (in heteroscedasticity test)	9
H	Kruskal-Wallis statistic	8
H_a	alternative hypothesis	6
H_0	null hypothesis	6
IQR	interquartile range	4
M	median	4
MSE	mean squared error (regression)	9
MSE	estimated mean variance within each factor	7
MSF	estimated mean variance due to the factor	7
μ	population mean	4
$\mu_{\bar{x}}$	expected value of the distribution of sample means	7
$\mu_{y\|x}$	predicted mean of y given x	9
n	number of observations in the sample	4
$n+$	number of positive differences (in Wilcoxon signed ranks test)	8
N	number of observations in the population	4
p	probability of success (in binomial distribution)	5
p	p-value	7
$1 - p$	probability of failure (in binomial distribution)	5
$\bar{p}$	sample proportion	7
p^*	pooled estimate of the population proportion	7
P(A & B)	probability of Event A and Event B	5
P(A or B)	probability of Event A or Event B	5
P(X)	probability of Event X	5
$P(X\|$A)	probability of Event X given Event A	5
PDF	probability density function	7
PI	prediction interval	9
$PRESS$	prediction sum of squares	9
Q	Szroeter heteroscedasticity test statistic	9
$Q1$	first quartile	4
$Q3$	third quartile	4
r	Pearson Product-Moment Correlation Coefficient	4

Symbol	Description	Chapter
$\overline{R}_c$	sum of the combined ranks (in Kruskal-Wallis test)	8
R_i	rank of each positive difference (in Wilcoxon signed ranks test)	8
R^2	coefficient of determination	9
R^2_{ADJ}	adjusted R^2	10
R^2_{PRED}	prediction R^2	9
ρ	Spearman's *Rho*	4
S	sample standard deviation	4
S_a	standard error of a	9
S_e	regression standard error	9
$S_{\mu_{y\|x}}$	standard error of the estimate of the mean of y given x	9
$S_{y\|x}$	standard error of a single outcome of y given x	9
S^2	sample variance	4
S_p^2	pooled sample variance	7
SSE	sum of squared errors (in linear regression)	9
SSE	within-groups variation	7
SSF	between-groups variation	7
SSR	regression sum of squares	9
SST	total sample variance	7
SST	total sum of squares	9
Standardized DfFit	test of leverage and outliers more than 2 standard deviations from mean (in SPSS)	10
σ	population standard deviation	4
σ^2	population variance	4
$\sigma_{\overline{x}}$	standard error of the mean	7
T	sum of ranks (in Wilcoxon sum of ranks test)	8
t	Student's t distribution	7
t_{n-1}	t statistic at $n - 1$ degrees of freedom	7
$t_{\alpha/2, n-1}$	t statistic at $\alpha/2$ confidence level and $n - 1$ degrees of freedom	7
U	Mann-Whitney U statistic	8
VIF	variance inflation factor	10
W	Wilcoxon signed ranks test statistic	8
WLS	weighted least squares	9
$\overline{x}$	arithmetic mean of the variable x	4
$\overline{X}_G$	geometric mean of the variable X	4
Z	number of standard deviations (in normal distribution)	5

Description	Formula	Location
Adjusted R^2	$R^2_{\text{ADJ}} = 1 - \dfrac{\frac{SSE}{n-k-1}}{\frac{SST}{n-1}}$	Ch. 10, p. 296
Arithmetic Mean	$\bar{x} = \frac{\Sigma x}{n}$ (sample)	Ch. 4, p. 70
	$\mu_x = \frac{\Sigma x}{N}$ (population)	Ch. 4, p. 71
Binomial Mean	$\mu_{Binomial} = E[x] = np$	Ch. 5, p. 121
Binomial Probability	$P(X) = \frac{n!}{X!(n-X)!} p^X (1-p)^{n-X}$	Ch. 5, p. 120
Binomial Standard Deviation	$\sigma_{Binomial} = \sqrt{np(1-p)}$	Ch. 5, p. 122
χ^2 Distribution	$\chi^2_{df-1} = \sum_1^4 \frac{(f_o - f_e)^2}{f_e}$ (two proportions)	Ch. 8, p. 217
	$\chi^2_{df=k-1} = \sum_1^{2k} \frac{(f_o - f_e)^2}{f_e}$ (more than two proportions)	Ch. 8, p. 219
Coefficient of Determination (R^2)	$R^2 = \frac{SSR}{SST}$ or $R^2 = 1 - \frac{SSE}{SST}$	Ch. 9, p. 262
Coefficient of Dispersion (*COD*)	$COD = \frac{AAD}{MED(A/S)} \times 100\%$	Ch. 4, p. 86
Combinations	$\binom{n}{X} = \frac{n!}{X!(n-X)!}$	Ch. 2, p. 36
Conditional Probability	$P(A \mid B) = \frac{P(A \,\&\, B)}{P(B)}$	Ch. 5, p. 111
Covariance (*COV*)	$COV = \frac{S}{\bar{x}} \cdot 100\%$	Ch. 4, p. 84
F Distribution	$F_{n_1-1, n_2-1} = \frac{S_1^2}{S_2^2}$ (equal variance test)	Ch. 7, p. 186
	$F_{i-1, n-1} = \frac{MSF}{MSE}$ (ANOVA)	Ch. 7, p. 202
	$F = \dfrac{\frac{SSE_R - SSE_F}{k - k'}}{\frac{SSE_F}{n-k-1}}$ (partial *F* statistic)	Ch. 10, p. 333
First Quartile (position)	$Q1 = \frac{n+1}{4}$ ordered observation	Ch. 4, p. 77
General Addition Law	$P(A \text{ or } B) = P(A) + P(B) - P(A \,\&\, B)$	Ch. 5, p. 115
General Multiplication Law	$P(A \,\&\, B) = P(A \mid B) \cdot P(B)$	Ch. 5, p. 116

Description	Formula	Location
Geometric Mean	$\overline{X}_G = \sqrt[t]{(x_1 \bullet x_2 \bullet \ldots \bullet x_t)}$	Ch. 4, p. 74
Interquartile Range (*IQR*)	$IQR = Q3 - Q1$	Ch. 4, p. 78
Kruskal-Wallis H	$H = \frac{12D}{n_c(n_c + 1)}$ where $n_c = \Sigma n_i$, $D = \Sigma n_i(\overline{R}_i - \overline{R}_c)^2$, $\overline{R}_c = \frac{n_c + 1}{2}$	Ch. 8, p. 234
Mann-Whitney U	$U = T_1 - \frac{n_1(n_1 + 1)}{2}$	Ch. 8, p. 225
Mean Absolute Prediction Error	$\sum_{i=1}^{n'} \frac{\lvert y_i - \hat{y}_i \rvert}{n'}$	Ch. 9, p. 271
Mean Square Prediction Error	$\sum_{i=1}^{n'} \frac{(y_i - \hat{y}_i)^2}{n'}$	Ch. 9, p. 271
Median (position)	$\frac{n+1}{2}$ observed position	Ch. 4, p. 72
Pearson Product-Moment Correlation	$r_{XY} = \frac{n(\Sigma XY) - \Sigma X \Sigma Y}{\sqrt{(n\Sigma X^2 - (\Sigma X)^2)(n\Sigma Y^2 - (\Sigma Y)^2)}}$	Ch. 4, p. 96
Permutations	$n\_P\_X = \frac{n!}{(n - X)!}$	Ch. 2, p. 36
Prediction R^2	$R^2_{PRED} = 1 - \frac{PRESS}{SST}$	Ch. 9, p. 273
Prediction Sum of Squares (*PRESS*)	$PRESS = \sum_{i=1}^{n} (y_i - \hat{y}_i')^2$	Ch. 9, p. 273
Probability (classical)	$P(X) = \frac{X}{T}$	Ch. 5, p. 110
Sample Size (n)	$n = \frac{Z^2\sigma^2}{e^2}$ (estimating means)	Ch. 6, p. 160
	$n = \frac{Z^2 p(1 - p)}{e^2}$ (estimating proportions)	Ch. 6, p. 162

Description	Formula	Location
Simple Linear Regression	$b = \dfrac{\Sigma x_i y_i - \frac{1}{n}\Sigma x_i \Sigma y_i}{\Sigma x_i^2 - \frac{1}{n}(\Sigma x_i)^2}$	Ch. 9, p. 245
	$a = \bar{y} - b\bar{x}$	Ch. 9, p. 245
	$S_e = \sqrt{\dfrac{\Sigma(y_i - \hat{y}_i)^2}{n-2}} = \sqrt{\dfrac{SSE}{n-2}}$	Ch. 9, p. 250
	$S_a = S_e\sqrt{\dfrac{1}{n} + \dfrac{\bar{x}^2}{(n-1)S_x^2}}$	Ch. 9, p. 250
	$S_b = S_e\sqrt{\dfrac{1}{(n-1)S_x^2}}$	Ch. 9, p. 251
	$(1 - \alpha) \times 100\%$ confidence interval on $\beta = b \pm t_{\alpha/2}\, S_b$	Ch. 9, p. 254
	$S_{\mu_{y\mid x}} = S_e\sqrt{\dfrac{1}{n} + \dfrac{(x_i - \bar{x})^2}{(n-1)S_x^2}}$	Ch. 9, p. 256
	$(1 - \alpha) \times 100\%$ confidence interval on $\mu_{y\mid x} = \mu_{y\mid x} \pm t_{\alpha/2}\, S_{\mu_{y\mid x}}$	Ch. 9, p. 256
	$S_{y\mid x} = S_e\sqrt{1 + \dfrac{1}{n} + \dfrac{(x_i - \bar{x})^2}{(n-1)S_x^2}}$	Ch. 9, p. 259
	$(1 - \alpha) \times 100\%$ confidence interval on $y \mid x = y \mid x \pm t_{\alpha/2}\, S_{y\mid x}$	Ch. 9, p. 259
Skewness	$\dfrac{n}{(n-1)(n-2)}\Sigma\left(\dfrac{x_i - \bar{x}}{S}\right)^3$	Ch. 4, p. 89
Spearman's *Rho*	$\rho_{XY} = \dfrac{\frac{\Sigma d_X d_Y}{n-1}}{S_{Xr}S_{Yr}}$	Ch. 4, p. 100
Standard Deviation	$\sigma_X = \sqrt{\dfrac{\Sigma(x_i - \mu)^2}{N}}$ (population)	Ch. 4, p. 80
	$S_X = \sqrt{\dfrac{\Sigma(x_i - \bar{x})^2}{n-1}}$ (sample)	Ch. 4, p. 81
Standard Error	$\dfrac{e_i}{S_e}$	Ch. 10, p. 302
Standard Error of the Mean	$\sigma_{\bar{x}} = \dfrac{\sigma_x}{\sqrt{n}}$	Ch. 7, p. 168
Szroeter Test for Heteroskedasticity	$Q = \left(\dfrac{6n}{n^2 - 1}\right)^{1/2}\left(h - \dfrac{n+1}{2}\right)$ where $h = \dfrac{\Sigma i e_i^2}{\Sigma e_i^2}$	Ch. 9, p. 283

Description	Formula		Location
t Distribution	$t_{n-1} = \dfrac{\overline{x} - \mu_x}{S_x / \sqrt{n}}$	(means test)	Ch. 7, pp. 174, 179
	$t_{n_1+n_2-2} = \dfrac{(\overline{x}_1 - \overline{x}_2) - (\mu_1 - \mu_2)}{\sqrt{S_p^2\left[\dfrac{1}{n_1} + \dfrac{1}{n_2}\right]}}$	(two sample means with pooled variances)	Ch. 7, p. 188
	$t_{df} = \dfrac{(\overline{x}_1 - \overline{x}_2) - (\mu_1 - \mu_2)}{\sqrt{\dfrac{S_1^2}{n_1} + \dfrac{S_2^2}{n_2}}}$	(two sample means with separate variances)	Ch. 7, p. 189
	$t_{n-1} = \dfrac{\overline{d} - \mu d}{\dfrac{S\,d}{\sqrt{n}}}$	(paired samples)	Ch. 7, p. 195
	$t_{n-2} = \dfrac{b - 0}{S_b}$	(simple linear regression)	Ch. 9, pp. 253, 254
	$t_{n-k-1} = \dfrac{b - 0}{S_b}$	(multiple linear regression)	Ch. 10, p. 297
Third Quartile (position)	$Q3 = \dfrac{3(n+1)}{4}$ ordered observation		Ch. 4, p. 77
Total Sum of Squares (SST)	$SST = SSR + SSE$		Ch. 9, p. 262
Variance	$\sigma_x^2 = \dfrac{\Sigma(x_i - \mu)^2}{N}$	(population)	Ch. 4, p. 80
	$S_x^2 = \dfrac{\Sigma(x_i - \overline{x})^2}{n - 1}$	(sample)	Ch. 4, p. 81
Variance Inflation Factor (VIF)	$VIF = \dfrac{1}{(1 - R_j^2)}$		Ch. 10, p. 311
Wilcoxon Signed Ranks	$W = \sum^{n+} R_i$		Ch. 8, p. 229
Wilcoxon Sum of Ranks	$T_1 + T_2 = \dfrac{n(n+1)}{2}$		Ch. 8, p. 223
Z Distribution	$Z = \dfrac{x - \mu}{\sigma}$	(means)	Ch. 5, p. 123
	$Z \cong \dfrac{\overline{p} - p}{\sqrt{\dfrac{p(1-p)}{n}}}$	(proportions)	Ch. 7, p. 183
	$Z = \dfrac{(p_{S1} - p_{S2}) - (p_1 - p_2)}{\sqrt{p^*(1 - p^*)\left(\dfrac{1}{n_1} + \dfrac{1}{n_2}\right)}}$	(two sample proportions)	Ch. 7, p. 205

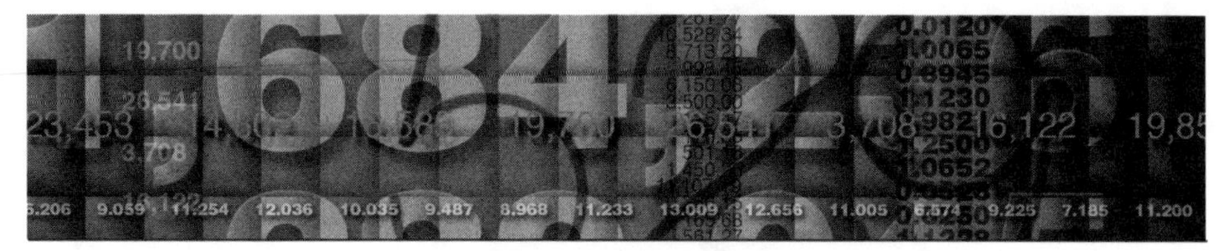

Appendix A

Standard Normal Probabilities

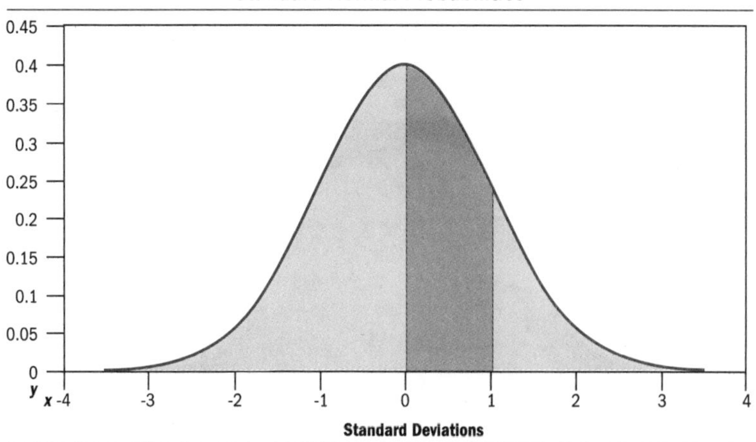

Area under the Normal Curve from 0 to *X*

X	**0.00**	**0.01**	**0.02**	**0.03**	**0.04**	**0.05**	**0.06**	**0.07**	**0.08**	**0.09**
0.0	0.00000	0.00399	0.00798	0.01197	0.01595	0.01994	0.02392	0.02790	0.03188	0.03586
0.1	0.03983	0.04380	0.04776	0.05172	0.05567	0.05962	0.06356	0.06749	0.07142	0.07535
0.2	0.07926	0.08317	0.08706	0.09095	0.09483	0.09871	0.10257	0.10642	0.11026	0.11409
0.3	0.11791	0.12172	0.12552	0.12930	0.13307	0.13683	0.14058	0.14431	0.14803	0.15173
0.4	0.15542	0.15910	0.16276	0.16640	0.17003	0.17364	0.17724	0.18082	0.18439	0.18793
0.5	0.19146	0.19497	0.19847	0.20194	0.20540	0.20884	0.21226	0.21566	0.21904	0.22240
0.6	0.22575	0.22907	0.23237	0.23565	0.23891	0.24215	0.24537	0.24857	0.25175	0.25490
0.7	0.25804	0.26115	0.26424	0.26730	0.27035	0.27337	0.27637	0.27935	0.28230	0.28524
0.8	0.28814	0.29103	0.29389	0.29673	0.29955	0.30234	0.30511	0.30785	0.31057	0.31327
0.9	0.31594	0.31859	0.32121	0.32381	0.32639	0.32894	0.33147	0.33398	0.33646	0.33891
1.0	0.34134	0.34375	0.34614	0.34849	0.35083	0.35314	0.35543	0.35769	0.35993	0.36214
1.1	0.36433	0.36650	0.36864	0.37076	0.37286	0.37493	0.37698	0.37900	0.38100	0.38298
1.2	0.38493	0.38686	0.38877	0.39065	0.39251	0.39435	0.39617	0.39796	0.39973	0.40147
1.3	0.40320	0.40490	0.40658	0.40824	0.40988	0.41149	0.41308	0.41466	0.41621	0.41774
1.4	0.41924	0.42073	0.42220	0.42364	0.42507	0.42647	0.42785	0.42922	0.43056	0.43189
1.5	0.43319	0.43448	0.43574	0.43699	0.43822	0.43943	0.44062	0.44179	0.44295	0.44408

Area under the Normal Curve from 0 to X

X	0.00	0.01	0.02	0.03	0.04	0.05	0.06	0.07	0.08	0.09
1.6	0.44520	0.44630	0.44738	0.44845	0.44950	0.45053	0.45154	0.45254	0.45352	0.45449
1.7	0.45543	0.45637	0.45728	0.45818	0.45907	0.45994	0.46080	0.46164	0.46246	0.46327
1.8	0.46407	0.46485	0.46562	0.46638	0.46712	0.46784	0.46856	0.46926	0.46995	0.47062
1.9	0.47128	0.47193	0.47257	0.47320	0.47381	0.47441	0.47500	0.47558	0.47615	0.47670
2.0	0.47725	0.47778	0.47831	0.47882	0.47932	0.47982	0.48030	0.48077	0.48124	0.48169
2.1	0.48214	0.48257	0.48300	0.48341	0.48382	0.48422	0.48461	0.48500	0.48537	0.48574
2.2	0.48610	0.48645	0.48679	0.48713	0.48745	0.48778	0.48809	0.48840	0.48870	0.48899
2.3	0.48928	0.48956	0.48983	0.49010	0.49036	0.49061	0.49086	0.49111	0.49134	0.49158
2.4	0.49180	0.49202	0.49224	0.49245	0.49266	0.49286	0.49305	0.49324	0.49343	0.49361
2.5	0.49379	0.49396	0.49413	0.49430	0.49446	0.49461	0.49477	0.49492	0.49506	0.49520
2.6	0.49534	0.49547	0.49560	0.49573	0.49585	0.49598	0.49609	0.49621	0.49632	0.49643
2.7	0.49653	0.49664	0.49674	0.49683	0.49693	0.49702	0.49711	0.49720	0.49728	0.49736
2.8	0.49744	0.49752	0.49760	0.49767	0.49774	0.49781	0.49788	0.49795	0.49801	0.49807
2.9	0.49813	0.49819	0.49825	0.49831	0.49836	0.49841	0.49846	0.49851	0.49856	0.49861
3.0	0.49865	0.49869	0.49874	0.49878	0.49882	0.49886	0.49889	0.49893	0.49896	0.49900
3.1	0.49903	0.49906	0.49910	0.49913	0.49916	0.49918	0.49921	0.49924	0.49926	0.49929
3.2	0.49931	0.49934	0.49936	0.49938	0.49940	0.49942	0.49944	0.49946	0.49948	0.49950
3.3	0.49952	0.49953	0.49955	0.49957	0.49958	0.49960	0.49961	0.49962	0.49964	0.49965
3.4	0.49966	0.49968	0.49969	0.49970	0.49971	0.49972	0.49973	0.49974	0.49975	0.49976
3.5	0.49977	0.49978	0.49978	0.49979	0.49980	0.49981	0.49981	0.49982	0.49983	0.49983
3.6	0.49984	0.49985	0.49985	0.49986	0.49986	0.49987	0.49987	0.49988	0.49988	0.49989
3.7	0.49989	0.49990	0.49990	0.49990	0.49991	0.49991	0.49992	0.49992	0.49992	0.49992
3.8	0.49993	0.49993	0.49993	0.49994	0.49994	0.49994	0.49994	0.49995	0.49995	0.49995
3.9	0.49995	0.49995	0.49996	0.49996	0.49996	0.49996	0.49996	0.49996	0.49997	0.49997
4.0	0.49997	0.49997	0.49997	0.49997	0.49997	0.49997	0.49998	0.49998	0.49998	0.49998

Source: http://www.itl.nist.gov/div898/handbook/eda/section3/eda3671.htm

Appendix B

Student's *t* Values for One-Tailed Tests

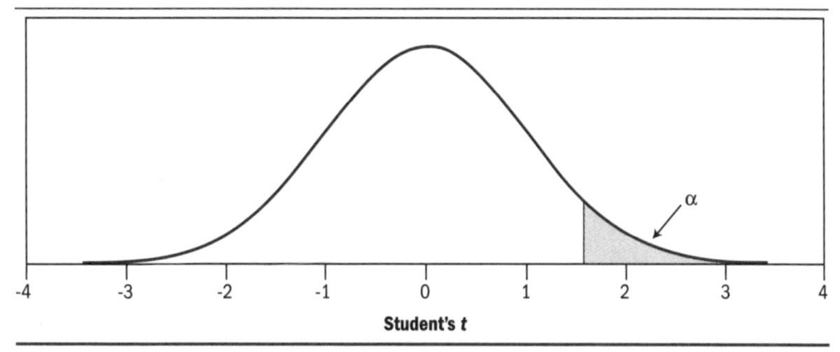

Student's *t* values associated with α probabilities

$n-1$	$\alpha = 0.10$	0.05	0.025	0.01	0.005	0.001
1.	3.078	6.314	12.706	31.821	63.657	318.313
2.	1.886	2.920	4.303	6.965	9.925	22.327
3.	1.638	2.353	3.182	4.541	5.841	10.215
4.	1.533	2.132	2.776	3.747	4.604	7.173
5.	1.476	2.015	2.571	3.365	4.032	5.893
6.	1.440	1.943	2.447	3.143	3.707	5.208
7.	1.415	1.895	2.365	2.998	3.499	4.782
8.	1.397	1.860	2.306	2.896	3.355	4.499
9.	1.383	1.833	2.262	2.821	3.250	4.296
10.	1.372	1.812	2.228	2.764	3.169	4.143
11.	1.363	1.796	2.201	2.718	3.106	4.024
12.	1.356	1.782	2.179	2.681	3.055	3.929
13.	1.350	1.771	2.160	2.650	3.012	3.852
14.	1.345	1.761	2.145	2.624	2.977	3.787
15.	1.341	1.753	2.131	2.602	2.947	3.733
16.	1.337	1.746	2.120	2.583	2.921	3.686
17.	1.333	1.740	2.110	2.567	2.898	3.646
18.	1.330	1.734	2.101	2.552	2.878	3.610
19.	1.328	1.729	2.093	2.539	2.861	3.579
20.	1.325	1.725	2.086	2.528	2.845	3.552
21.	1.323	1.721	2.080	2.518	2.831	3.527
22.	1.321	1.717	2.074	2.508	2.819	3.505

Student's *t* values associated with α probabilities

n − 1	α = **0.10**	**0.05**	**0.025**	**0.01**	**0.005**	**0.001**
23.	1.319	1.714	2.069	2.500	2.807	3.485
24.	1.318	1.711	2.064	2.492	2.797	3.467
25.	1.316	1.708	2.060	2.485	2.787	3.450
26.	1.315	1.706	2.056	2.479	2.779	3.435
27.	1.314	1.703	2.052	2.473	2.771	3.421
28.	1.313	1.701	2.048	2.467	2.763	3.408
29.	1.311	1.699	2.045	2.462	2.756	3.396
30.	1.310	1.697	2.042	2.457	2.750	3.385
31.	1.309	1.696	2.040	2.453	2.744	3.375
32.	1.309	1.694	2.037	2.449	2.738	3.365
33.	1.308	1.692	2.035	2.445	2.733	3.356
34.	1.307	1.691	2.032	2.441	2.728	3.348
35.	1.306	1.690	2.030	2.438	2.724	3.340
36.	1.306	1.688	2.028	2.434	2.719	3.333
37.	1.305	1.687	2.026	2.431	2.715	3.326
38.	1.304	1.686	2.024	2.429	2.712	3.319
39.	1.304	1.685	2.023	2.426	2.708	3.313
40.	1.303	1.684	2.021	2.423	2.704	3.307
41.	1.303	1.683	2.020	2.421	2.701	3.301
42.	1.302	1.682	2.018	2.418	2.698	3.296
43.	1.302	1.681	2.017	2.416	2.695	3.291
44.	1.301	1.680	2.015	2.414	2.692	3.286
45.	1.301	1.679	2.014	2.412	2.690	3.281
46.	1.300	1.679	2.013	2.410	2.687	3.277
47.	1.300	1.678	2.012	2.408	2.685	3.273
48.	1.299	1.677	2.011	2.407	2.682	3.269
49.	1.299	1.677	2.010	2.405	2.680	3.265
50.	1.299	1.676	2.009	2.403	2.678	3.261
51.	1.298	1.675	2.008	2.402	2.676	3.258
52.	1.298	1.675	2.007	2.400	2.674	3.255
53.	1.298	1.674	2.006	2.399	2.672	3.251
54.	1.297	1.674	2.005	2.397	2.670	3.248
55.	1.297	1.673	2.004	2.396	2.668	3.245
56.	1.297	1.673	2.003	2.395	2.667	3.242
57.	1.297	1.672	2.002	2.394	2.665	3.239
58.	1.296	1.672	2.002	2.392	2.663	3.237
59.	1.296	1.671	2.001	2.391	2.662	3.234
60.	1.296	1.671	2.000	2.390	2.660	3.232
61.	1.296	1.670	2.000	2.389	2.659	3.229
62.	1.295	1.670	1.999	2.388	2.657	3.227
63.	1.295	1.669	1.998	2.387	2.656	3.225
64.	1.295	1.669	1.998	2.386	2.655	3.223
65.	1.295	1.669	1.997	2.385	2.654	3.220
66.	1.295	1.668	1.997	2.384	2.652	3.218
67.	1.294	1.668	1.996	2.383	2.651	3.216
68.	1.294	1.668	1.995	2.382	2.650	3.214
69.	1.294	1.667	1.995	2.382	2.649	3.213
70.	1.294	1.667	1.994	2.381	2.648	3.211

Student's t values associated with α probabilities

$n - 1$	$\alpha = 0.10$	0.05	0.025	0.01	0.005	0.001
71.	1.294	1.667	1.994	2.380	2.647	3.209
72.	1.293	1.666	1.993	2.379	2.646	3.207
73.	1.293	1.666	1.993	2.379	2.645	3.206
74.	1.293	1.666	1.993	2.378	2.644	3.204
75.	1.293	1.665	1.992	2.377	2.643	3.202
76.	1.293	1.665	1.992	2.376	2.642	3.201
77.	1.293	1.665	1.991	2.376	2.641	3.199
78.	1.292	1.665	1.991	2.375	2.640	3.198
79.	1.292	1.664	1.990	2.374	2.640	3.197
80.	1.292	1.664	1.990	2.374	2.639	3.195
81.	1.292	1.664	1.990	2.373	2.638	3.194
82.	1.292	1.664	1.989	2.373	2.637	3.193
83.	1.292	1.663	1.989	2.372	2.636	3.191
84.	1.292	1.663	1.989	2.372	2.636	3.190
85.	1.292	1.663	1.988	2.371	2.635	3.189
86.	1.291	1.663	1.988	2.370	2.634	3.188
87.	1.291	1.663	1.988	2.370	2.634	3.187
88.	1.291	1.662	1.987	2.369	2.633	3.185
89.	1.291	1.662	1.987	2.369	2.632	3.184
90.	1.291	1.662	1.987	2.368	2.632	3.183
91.	1.291	1.662	1.986	2.368	2.631	3.182
92.	1.291	1.662	1.986	2.368	2.630	3.181
93.	1.291	1.661	1.986	2.367	2.630	3.180
94.	1.291	1.661	1.986	2.367	2.629	3.179
95.	1.291	1.661	1.985	2.366	2.629	3.178
96.	1.290	1.661	1.985	2.366	2.628	3.177
97.	1.290	1.661	1.985	2.365	2.627	3.176
98.	1.290	1.661	1.984	2.365	2.627	3.175
99.	1.290	1.660	1.984	2.365	2.626	3.175
100.	1.290	1.660	1.984	2.364	2.626	3.174
∞	1.282	1.645	1.960	2.326	2.576	3.090

Appendix C

Upper critical values of the F distribution for df_1 numerator degrees of freedom and df_2 denominator degrees of freedom

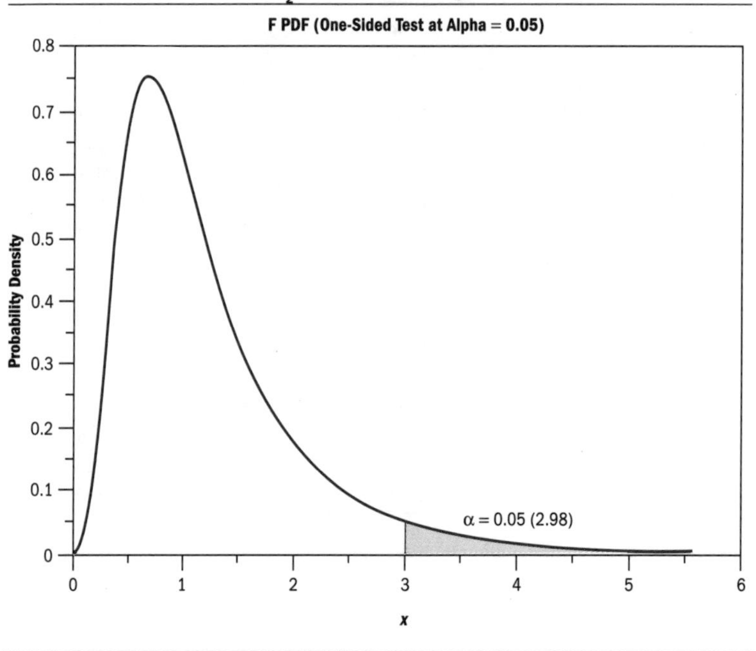

$F_{.05}(df_1, df_2)$

df_2 \ df_1	1	2	3	4	5	6	7	8	9	10
1	161.448	199.500	215.707	224.583	230.162	233.986	236.768	238.882	240.543	241.882
2	18.513	19.000	19.164	19.247	19.296	19.330	19.353	19.371	19.385	19.396
3	10.128	9.552	9.277	9.117	9.013	8.941	8.887	8.845	8.812	8.786
4	7.709	6.944	6.591	6.388	6.256	6.163	6.094	6.041	5.999	5.964
5	6.608	5.786	5.409	5.192	5.050	4.950	4.876	4.818	4.772	4.735
6	5.987	5.143	4.757	4.534	4.387	4.284	4.207	4.147	4.099	4.060
7	5.591	4.737	4.347	4.120	3.972	3.866	3.787	3.726	3.677	3.637
8	5.318	4.459	4.066	3.838	3.687	3.581	3.500	3.438	3.388	3.347

$F_{.05}(df_1, df_2)$

df_2 \ df_1	1	2	3	4	5	6	7	8	9	10
9	5.117	4.256	3.863	3.633	3.482	3.374	3.293	3.230	3.179	3.137
10	4.965	4.103	3.708	3.478	3.326	3.217	3.135	3.072	3.020	2.978
11	4.844	3.982	3.587	3.357	3.204	3.095	3.012	2.948	2.896	2.854
12	4.747	3.885	3.490	3.259	3.106	2.996	2.913	2.849	2.796	2.753
13	4.667	3.806	3.411	3.179	3.025	2.915	2.832	2.767	2.714	2.671
14	4.600	3.739	3.344	3.112	2.958	2.848	2.764	2.699	2.646	2.602
15	4.543	3.682	3.287	3.056	2.901	2.790	2.707	2.641	2.588	2.544
16	4.494	3.634	3.239	3.007	2.852	2.741	2.657	2.591	2.538	2.494
17	4.451	3.592	3.197	2.965	2.810	2.699	2.614	2.548	2.494	2.450
18	4.414	3.555	3.160	2.928	2.773	2.661	2.577	2.510	2.456	2.412
19	4.381	3.522	3.127	2.895	2.740	2.628	2.544	2.477	2.423	2.378
20	4.351	3.493	3.098	2.866	2.711	2.599	2.514	2.447	2.393	2.348
21	4.325	3.467	3.072	2.840	2.685	2.573	2.488	2.420	2.366	2.321
22	4.301	3.443	3.049	2.817	2.661	2.549	2.464	2.397	2.342	2.297
23	4.279	3.422	3.028	2.796	2.640	2.528	2.442	2.375	2.320	2.275
24	4.260	3.403	3.009	2.776	2.621	2.508	2.423	2.355	2.300	2.255
25	4.242	3.385	2.991	2.759	2.603	2.490	2.405	2.337	2.282	2.236
26	4.225	3.369	2.975	2.743	2.587	2.474	2.388	2.321	2.265	2.220
27	4.210	3.354	2.960	2.728	2.572	2.459	2.373	2.305	2.250	2.204
28	4.196	3.340	2.947	2.714	2.558	2.445	2.359	2.291	2.236	2.190
29	4.183	3.328	2.934	2.701	2.545	2.432	2.346	2.278	2.223	2.177
30	4.171	3.316	2.922	2.690	2.534	2.421	2.334	2.266	2.211	2.165
31	4.160	3.305	2.911	2.679	2.523	2.409	2.323	2.255	2.199	2.153
32	4.149	3.295	2.901	2.668	2.512	2.399	2.313	2.244	2.189	2.142
33	4.139	3.285	2.892	2.659	2.503	2.389	2.303	2.235	2.179	2.133
34	4.130	3.276	2.883	2.650	2.494	2.380	2.294	2.225	2.170	2.123
35	4.121	3.267	2.874	2.641	2.485	2.372	2.285	2.217	2.161	2.114
36	4.113	3.259	2.866	2.634	2.477	2.364	2.277	2.209	2.153	2.106
37	4.105	3.252	2.859	2.626	2.470	2.356	2.270	2.201	2.145	2.098
38	4.098	3.245	2.852	2.619	2.463	2.349	2.262	2.194	2.138	2.091
39	4.091	3.238	2.845	2.612	2.456	2.342	2.255	2.187	2.131	2.084
40	4.085	3.232	2.839	2.606	2.449	2.336	2.249	2.180	2.124	2.077
41	4.079	3.226	2.833	2.600	2.443	2.330	2.243	2.174	2.118	2.071
42	4.073	3.220	2.827	2.594	2.438	2.324	2.237	2.168	2.112	2.065
43	4.067	3.214	2.822	2.589	2.432	2.318	2.232	2.163	2.106	2.059
44	4.062	3.209	2.816	2.584	2.427	2.313	2.226	2.157	2.101	2.054
45	4.057	3.204	2.812	2.579	2.422	2.308	2.221	2.152	2.096	2.049
46	4.052	3.200	2.807	2.574	2.417	2.304	2.216	2.147	2.091	2.044
47	4.047	3.195	2.802	2.570	2.413	2.299	2.212	2.143	2.086	2.039
48	4.043	3.191	2.798	2.565	2.409	2.295	2.207	2.138	2.082	2.035
49	4.038	3.187	2.794	2.561	2.404	2.290	2.203	2.134	2.077	2.030
50	4.034	3.183	2.790	2.557	2.400	2.286	2.199	2.130	2.073	2.026
51	4.030	3.179	2.786	2.553	2.397	2.283	2.195	2.126	2.069	2.022
52	4.027	3.175	2.783	2.550	2.393	2.279	2.192	2.122	2.066	2.018
53	4.023	3.172	2.779	2.546	2.389	2.275	2.188	2.119	2.062	2.015
54	4.020	3.168	2.776	2.543	2.386	2.272	2.185	2.115	2.059	2.011
55	4.016	3.165	2.773	2.540	2.383	2.269	2.181	2.112	2.055	2.008
56	4.013	3.162	2.769	2.537	2.380	2.266	2.178	2.109	2.052	2.005

$F_{.05}(df_1, df_2)$

df_2 \ df_1	1	2	3	4	5	6	7	8	9	10
57	4.010	3.159	2.766	2.534	2.377	2.263	2.175	2.106	2.049	2.001
58	4.007	3.156	2.764	2.531	2.374	2.260	2.172	2.103	2.046	1.998
59	4.004	3.153	2.761	2.528	2.371	2.257	2.169	2.100	2.043	1.995
60	4.001	3.150	2.758	2.525	2.368	2.254	2.167	2.097	2.040	1.993
61	3.998	3.148	2.755	2.523	2.366	2.251	2.164	2.094	2.037	1.990
62	3.996	3.145	2.753	2.520	2.363	2.249	2.161	2.092	2.035	1.987
63	3.993	3.143	2.751	2.518	2.361	2.246	2.159	2.089	2.032	1.985
64	3.991	3.140	2.748	2.515	2.358	2.244	2.156	2.087	2.030	1.982
65	3.989	3.138	2.746	2.513	2.356	2.242	2.154	2.084	2.027	1.980
66	3.986	3.136	2.744	2.511	2.354	2.239	2.152	2.082	2.025	1.977
67	3.984	3.134	2.742	2.509	2.352	2.237	2.150	2.080	2.023	1.975
68	3.982	3.132	2.740	2.507	2.350	2.235	2.148	2.078	2.021	1.973
69	3.980	3.130	2.737	2.505	2.348	2.233	2.145	2.076	2.019	1.971
70	3.978	3.128	2.736	2.503	2.346	2.231	2.143	2.074	2.017	1.969
71	3.976	3.126	2.734	2.501	2.344	2.229	2.142	2.072	2.015	1.967
72	3.974	3.124	2.732	2.499	2.342	2.227	2.140	2.070	2.013	1.965
73	3.972	3.122	2.730	2.497	2.340	2.226	2.138	2.068	2.011	1.963
74	3.970	3.120	2.728	2.495	2.338	2.224	2.136	2.066	2.009	1.961
75	3.968	3.119	2.727	2.494	2.337	2.222	2.134	2.064	2.007	1.959
76	3.967	3.117	2.725	2.492	2.335	2.220	2.133	2.063	2.006	1.958
77	3.965	3.115	2.723	2.490	2.333	2.219	2.131	2.061	2.004	1.956
78	3.963	3.114	2.722	2.489	2.332	2.217	2.129	2.059	2.002	1.954
79	3.962	3.112	2.720	2.487	2.330	2.216	2.128	2.058	2.001	1.953
80	3.960	3.111	2.719	2.486	2.329	2.214	2.126	2.056	1.999	1.951
81	3.959	3.109	2.717	2.484	2.327	2.213	2.125	2.055	1.998	1.950
82	3.957	3.108	2.716	2.483	2.326	2.211	2.123	2.053	1.996	1.948
83	3.956	3.107	2.715	2.482	2.324	2.210	2.122	2.052	1.995	1.947
84	3.955	3.105	2.713	2.480	2.323	2.209	2.121	2.051	1.993	1.945
85	3.953	3.104	2.712	2.479	2.322	2.207	2.119	2.049	1.992	1.944
86	3.952	3.103	2.711	2.478	2.321	2.206	2.118	2.048	1.991	1.943
87	3.951	3.101	2.709	2.476	2.319	2.205	2.117	2.047	1.989	1.941
88	3.949	3.100	2.708	2.475	2.318	2.203	2.115	2.045	1.988	1.940
89	3.948	3.099	2.707	2.474	2.317	2.202	2.114	2.044	1.987	1.939
90	3.947	3.098	2.706	2.473	2.316	2.201	2.113	2.043	1.986	1.938
91	3.946	3.097	2.705	2.472	2.315	2.200	2.112	2.042	1.984	1.936
92	3.945	3.095	2.704	2.471	2.313	2.199	2.111	2.041	1.983	1.935
93	3.943	3.094	2.703	2.470	2.312	2.198	2.110	2.040	1.982	1.934
94	3.942	3.093	2.701	2.469	2.311	2.197	2.109	2.038	1.981	1.933
95	3.941	3.092	2.700	2.467	2.310	2.196	2.108	2.037	1.980	1.932
96	3.940	3.091	2.699	2.466	2.309	2.195	2.106	2.036	1.979	1.931
97	3.939	3.090	2.698	2.465	2.308	2.194	2.105	2.035	1.978	1.930
98	3.938	3.089	2.697	2.465	2.307	2.193	2.104	2.034	1.977	1.929
99	3.937	3.088	2.696	2.464	2.306	2.192	2.103	2.033	1.976	1.928
100	3.936	3.087	2.696	2.463	2.305	2.191	2.103	2.032	1.975	1.927

$F_{.05}(df_1, df_2)$

df_2 \ df_1	11	12	13	14	15	16	17	18	19	20
1	242.983	243.906	244.690	245.364	245.950	246.464	246.918	247.323	247.686	248.013
2	19.405	19.413	19.419	19.424	19.429	19.433	19.437	19.440	19.443	19.446
3	8.763	8.745	8.729	8.715	8.703	8.692	8.683	8.675	8.667	8.660
4	5.936	5.912	5.891	5.873	5.858	5.844	5.832	5.821	5.811	5.803
5	4.704	4.678	4.655	4.636	4.619	4.604	4.590	4.579	4.568	4.558
6	4.027	4.000	3.976	3.956	3.938	3.922	3.908	3.896	3.884	3.874
7	3.603	3.575	3.550	3.529	3.511	3.494	3.480	3.467	3.455	3.445
8	3.313	3.284	3.259	3.237	3.218	3.202	3.187	3.173	3.161	3.150
9	3.102	3.073	3.048	3.025	3.006	2.989	2.974	2.960	2.948	2.936
10	2.943	2.913	2.887	2.865	2.845	2.828	2.812	2.798	2.785	2.774
11	2.818	2.788	2.761	2.739	2.719	2.701	2.685	2.671	2.658	2.646
12	2.717	2.687	2.660	2.637	2.617	2.599	2.583	2.568	2.555	2.544
13	2.635	2.604	2.577	2.554	2.533	2.515	2.499	2.484	2.471	2.459
14	2.565	2.534	2.507	2.484	2.463	2.445	2.428	2.413	2.400	2.388
15	2.507	2.475	2.448	2.424	2.403	2.385	2.368	2.353	2.340	2.328
16	2.456	2.425	2.397	2.373	2.352	2.333	2.317	2.302	2.288	2.276
17	2.413	2.381	2.353	2.329	2.308	2.289	2.272	2.257	2.243	2.230
18	2.374	2.342	2.314	2.290	2.269	2.250	2.233	2.217	2.203	2.191
19	2.340	2.308	2.280	2.256	2.234	2.215	2.198	2.182	2.168	2.155
20	2.310	2.278	2.250	2.225	2.203	2.184	2.167	2.151	2.137	2.124
21	2.283	2.250	2.222	2.197	2.176	2.156	2.139	2.123	2.109	2.096
22	2.259	2.226	2.198	2.173	2.151	2.131	2.114	2.098	2.084	2.071
23	2.236	2.204	2.175	2.150	2.128	2.109	2.091	2.075	2.061	2.048
24	2.216	2.183	2.155	2.130	2.108	2.088	2.070	2.054	2.040	2.027
25	2.198	2.165	2.136	2.111	2.089	2.069	2.051	2.035	2.021	2.007
26	2.181	2.148	2.119	2.094	2.072	2.052	2.034	2.018	2.003	1.990
27	2.166	2.132	2.103	2.078	2.056	2.036	2.018	2.002	1.987	1.974
28	2.151	2.118	2.089	2.064	2.041	2.021	2.003	1.987	1.972	1.959
29	2.138	2.104	2.075	2.050	2.027	2.007	1.989	1.973	1.958	1.945
30	2.126	2.092	2.063	2.037	2.015	1.995	1.976	1.960	1.945	1.932
31	2.114	2.080	2.051	2.026	2.003	1.983	1.965	1.948	1.933	1.920
32	2.103	2.070	2.040	2.015	1.992	1.972	1.953	1.937	1.922	1.908
33	2.093	2.060	2.030	2.004	1.982	1.961	1.943	1.926	1.911	1.898
34	2.084	2.050	2.021	1.995	1.972	1.952	1.933	1.917	1.902	1.888
35	2.075	2.041	2.012	1.986	1.963	1.942	1.924	1.907	1.892	1.878
36	2.067	2.033	2.003	1.977	1.954	1.934	1.915	1.899	1.883	1.870
37	2.059	2.025	1.995	1.969	1.946	1.926	1.907	1.890	1.875	1.861
38	2.051	2.017	1.988	1.962	1.939	1.918	1.899	1.883	1.867	1.853
39	2.044	2.010	1.981	1.954	1.931	1.911	1.892	1.875	1.860	1.846
40	2.038	2.003	1.974	1.948	1.924	1.904	1.885	1.868	1.853	1.839
41	2.031	1.997	1.967	1.941	1.918	1.897	1.879	1.862	1.846	1.832
42	2.025	1.991	1.961	1.935	1.912	1.891	1.872	1.855	1.840	1.826
43	2.020	1.985	1.955	1.929	1.906	1.885	1.866	1.849	1.834	1.820
44	2.014	1.980	1.950	1.924	1.900	1.879	1.861	1.844	1.828	1.814
45	2.009	1.974	1.945	1.918	1.895	1.874	1.855	1.838	1.823	1.808
46	2.004	1.969	1.940	1.913	1.890	1.869	1.850	1.833	1.817	1.803
47	1.999	1.965	1.935	1.908	1.885	1.864	1.845	1.828	1.812	1.798
48	1.995	1.960	1.930	1.904	1.880	1.859	1.840	1.823	1.807	1.793
49	1.990	1.956	1.926	1.899	1.876	1.855	1.836	1.819	1.803	1.789
50	1.986	1.952	1.921	1.895	1.871	1.850	1.831	1.814	1.798	1.784

$F_{.05}(df_1, df_2)$

df_2 \ df_1	11	12	13	14	15	16	17	18	19	20
51	1.982	1.947	1.917	1.891	1.867	1.846	1.827	1.810	1.794	1.780
52	1.978	1.944	1.913	1.887	1.863	1.842	1.823	1.806	1.790	1.776
53	1.975	1.940	1.910	1.883	1.859	1.838	1.819	1.802	1.786	1.772
54	1.971	1.936	1.906	1.879	1.856	1.835	1.816	1.798	1.782	1.768
55	1.968	1.933	1.903	1.876	1.852	1.831	1.812	1.795	1.779	1.764
56	1.964	1.930	1.899	1.873	1.849	1.828	1.809	1.791	1.775	1.761
57	1.961	1.926	1.896	1.869	1.846	1.824	1.805	1.788	1.772	1.757
58	1.958	1.923	1.893	1.866	1.842	1.821	1.802	1.785	1.769	1.754
59	1.955	1.920	1.890	1.863	1.839	1.818	1.799	1.781	1.766	1.751
60	1.952	1.917	1.887	1.860	1.836	1.815	1.796	1.778	1.763	1.748
61	1.949	1.915	1.884	1.857	1.834	1.812	1.793	1.776	1.760	1.745
62	1.947	1.912	1.882	1.855	1.831	1.809	1.790	1.773	1.757	1.742
63	1.944	1.909	1.879	1.852	1.828	1.807	1.787	1.770	1.754	1.739
64	1.942	1.907	1.876	1.849	1.826	1.804	1.785	1.767	1.751	1.737
65	1.939	1.904	1.874	1.847	1.823	1.802	1.782	1.765	1.749	1.734
66	1.937	1.902	1.871	1.845	1.821	1.799	1.780	1.762	1.746	1.732
67	1.935	1.900	1.869	1.842	1.818	1.797	1.777	1.760	1.744	1.729
68	1.932	1.897	1.867	1.840	1.816	1.795	1.775	1.758	1.742	1.727
69	1.930	1.895	1.865	1.838	1.814	1.792	1.773	1.755	1.739	1.725
70	1.928	1.893	1.863	1.836	1.812	1.790	1.771	1.753	1.737	1.722
71	1.926	1.891	1.861	1.834	1.810	1.788	1.769	1.751	1.735	1.720
72	1.924	1.889	1.859	1.832	1.808	1.786	1.767	1.749	1.733	1.718
73	1.922	1.887	1.857	1.830	1.806	1.784	1.765	1.747	1.731	1.716
74	1.921	1.885	1.855	1.828	1.804	1.782	1.763	1.745	1.729	1.714
75	1.919	1.884	1.853	1.826	1.802	1.780	1.761	1.743	1.727	1.712
76	1.917	1.882	1.851	1.824	1.800	1.778	1.759	1.741	1.725	1.710
77	1.915	1.880	1.849	1.822	1.798	1.777	1.757	1.739	1.723	1.708
78	1.914	1.878	1.848	1.821	1.797	1.775	1.755	1.738	1.721	1.707
79	1.912	1.877	1.846	1.819	1.795	1.773	1.754	1.736	1.720	1.705
80	1.910	1.875	1.845	1.817	1.793	1.772	1.752	1.734	1.718	1.703
81	1.909	1.874	1.843	1.816	1.792	1.770	1.750	1.733	1.716	1.702
82	1.907	1.872	1.841	1.814	1.790	1.768	1.749	1.731	1.715	1.700
83	1.906	1.871	1.840	1.813	1.789	1.767	1.747	1.729	1.713	1.698
84	1.905	1.869	1.838	1.811	1.787	1.765	1.746	1.728	1.712	1.697
85	1.903	1.868	1.837	1.810	1.786	1.764	1.744	1.726	1.710	1.695
86	1.902	1.867	1.836	1.808	1.784	1.762	1.743	1.725	1.709	1.694
87	1.900	1.865	1.834	1.807	1.783	1.761	1.741	1.724	1.707	1.692
88	1.899	1.864	1.833	1.806	1.782	1.760	1.740	1.722	1.706	1.691
89	1.898	1.863	1.832	1.804	1.780	1.758	1.739	1.721	1.705	1.690
90	1.897	1.861	1.830	1.803	1.779	1.757	1.737	1.720	1.703	1.688
91	1.895	1.860	1.829	1.802	1.778	1.756	1.736	1.718	1.702	1.687
92	1.894	1.859	1.828	1.801	1.776	1.755	1.735	1.717	1.701	1.686
93	1.893	1.858	1.827	1.800	1.775	1.753	1.734	1.716	1.699	1.684
94	1.892	1.857	1.826	1.798	1.774	1.752	1.733	1.715	1.698	1.683
95	1.891	1.856	1.825	1.797	1.773	1.751	1.731	1.713	1.697	1.682
96	1.890	1.854	1.823	1.796	1.772	1.750	1.730	1.712	1.696	1.681
97	1.889	1.853	1.822	1.795	1.771	1.749	1.729	1.711	1.695	1.680
98	1.888	1.852	1.821	1.794	1.770	1.748	1.728	1.710	1.694	1.679
99	1.887	1.851	1.820	1.793	1.769	1.747	1.727	1.709	1.693	1.678
100	1.886	1.850	1.819	1.792	1.768	1.746	1.726	1.708	1.691	1.676

$F_{.10}(df_1, df_2)$

df_2 \ df_1	1	2	3	4	5	6	7	8	9	10
1	39.863	49.500	53.593	55.833	57.240	58.204	58.906	59.439	59.858	60.195
2	8.526	9.000	9.162	9.243	9.293	9.326	9.349	9.367	9.381	9.392
3	5.538	5.462	5.391	5.343	5.309	5.285	5.266	5.252	5.240	5.230
4	4.545	4.325	4.191	4.107	4.051	4.010	3.979	3.955	3.936	3.920
5	4.060	3.780	3.619	3.520	3.453	3.405	3.368	3.339	3.316	3.297
6	3.776	3.463	3.289	3.181	3.108	3.055	3.014	2.983	2.958	2.937
7	3.589	3.257	3.074	2.961	2.883	2.827	2.785	2.752	2.725	2.703
8	3.458	3.113	2.924	2.806	2.726	2.668	2.624	2.589	2.561	2.538
9	3.360	3.006	2.813	2.693	2.611	2.551	2.505	2.469	2.440	2.416
10	3.285	2.924	2.728	2.605	2.522	2.461	2.414	2.377	2.347	2.323
11	3.225	2.860	2.660	2.536	2.451	2.389	2.342	2.304	2.274	2.248
12	3.177	2.807	2.606	2.480	2.394	2.331	2.283	2.245	2.214	2.188
13	3.136	2.763	2.560	2.434	2.347	2.283	2.234	2.195	2.164	2.138
14	3.102	2.726	2.522	2.395	2.307	2.243	2.193	2.154	2.122	2.095
15	3.073	2.695	2.490	2.361	2.273	2.208	2.158	2.119	2.086	2.059
16	3.048	2.668	2.462	2.333	2.244	2.178	2.128	2.088	2.055	2.028
17	3.026	2.645	2.437	2.308	2.218	2.152	2.102	2.061	2.028	2.001
18	3.007	2.624	2.416	2.286	2.196	2.130	2.079	2.038	2.005	1.977
19	2.990	2.606	2.397	2.266	2.176	2.109	2.058	2.017	1.984	1.956
20	2.975	2.589	2.380	2.249	2.158	2.091	2.040	1.999	1.965	1.937
21	2.961	2.575	2.365	2.233	2.142	2.075	2.023	1.982	1.948	1.920
22	2.949	2.561	2.351	2.219	2.128	2.060	2.008	1.967	1.933	1.904
23	2.937	2.549	2.339	2.207	2.115	2.047	1.995	1.953	1.919	1.890
24	2.927	2.538	2.327	2.195	2.103	2.035	1.983	1.941	1.906	1.877
25	2.918	2.528	2.317	2.184	2.092	2.024	1.971	1.929	1.895	1.866
26	2.909	2.519	2.307	2.174	2.082	2.014	1.961	1.919	1.884	1.855
27	2.901	2.511	2.299	2.165	2.073	2.005	1.952	1.909	1.874	1.845
28	2.894	2.503	2.291	2.157	2.064	1.996	1.943	1.900	1.865	1.836
29	2.887	2.495	2.283	2.149	2.057	1.988	1.935	1.892	1.857	1.827
30	2.881	2.489	2.276	2.142	2.049	1.980	1.927	1.884	1.849	1.819
31	2.875	2.482	2.270	2.136	2.042	1.973	1.920	1.877	1.842	1.812
32	2.869	2.477	2.263	2.129	2.036	1.967	1.913	1.870	1.835	1.805
33	2.864	2.471	2.258	2.123	2.030	1.961	1.907	1.864	1.828	1.799
34	2.859	2.466	2.252	2.118	2.024	1.955	1.901	1.858	1.822	1.793
35	2.855	2.461	2.247	2.113	2.019	1.950	1.896	1.852	1.817	1.787
36	2.850	2.456	2.243	2.108	2.014	1.945	1.891	1.847	1.811	1.781
37	2.846	2.452	2.238	2.103	2.009	1.940	1.886	1.842	1.806	1.776
38	2.842	2.448	2.234	2.099	2.005	1.935	1.881	1.838	1.802	1.772
39	2.839	2.444	2.230	2.095	2.001	1.931	1.877	1.833	1.797	1.767
40	2.835	2.440	2.226	2.091	1.997	1.927	1.873	1.829	1.793	1.763
41	2.832	2.437	2.222	2.087	1.993	1.923	1.869	1.825	1.789	1.759
42	2.829	2.434	2.219	2.084	1.989	1.919	1.865	1.821	1.785	1.755
43	2.826	2.430	2.216	2.080	1.986	1.916	1.861	1.817	1.781	1.751
44	2.823	2.427	2.213	2.077	1.983	1.913	1.858	1.814	1.778	1.747
45	2.820	2.425	2.210	2.074	1.980	1.909	1.855	1.811	1.774	1.744
46	2.818	2.422	2.207	2.071	1.977	1.906	1.852	1.808	1.771	1.741
47	2.815	2.419	2.204	2.068	1.974	1.903	1.849	1.805	1.768	1.738
48	2.813	2.417	2.202	2.066	1.971	1.901	1.846	1.802	1.765	1.735
49	2.811	2.414	2.199	2.063	1.968	1.898	1.843	1.799	1.763	1.732
50	2.809	2.412	2.197	2.061	1.966	1.895	1.840	1.796	1.760	1.729

$F_{.10}(df_1, df_2)$

df_2 \ df_1	1	2	3	4	5	6	7	8	9	10
51	2.807	2.410	2.194	2.058	1.964	1.893	1.838	1.794	1.757	1.727
52	2.805	2.408	2.192	2.056	1.961	1.891	1.836	1.791	1.755	1.724
53	2.803	2.406	2.190	2.054	1.959	1.888	1.833	1.789	1.752	1.722
54	2.801	2.404	2.188	2.052	1.957	1.886	1.831	1.787	1.750	1.719
55	2.799	2.402	2.186	2.050	1.955	1.884	1.829	1.785	1.748	1.717
56	2.797	2.400	2.184	2.048	1.953	1.882	1.827	1.782	1.746	1.715
57	2.796	2.398	2.182	2.046	1.951	1.880	1.825	1.780	1.744	1.713
58	2.794	2.396	2.181	2.044	1.949	1.878	1.823	1.779	1.742	1.711
59	2.793	2.395	2.179	2.043	1.947	1.876	1.821	1.777	1.740	1.709
60	2.791	2.393	2.177	2.041	1.946	1.875	1.819	1.775	1.738	1.707
61	2.790	2.392	2.176	2.039	1.944	1.873	1.818	1.773	1.736	1.705
62	2.788	2.390	2.174	2.038	1.942	1.871	1.816	1.771	1.735	1.703
63	2.787	2.389	2.173	2.036	1.941	1.870	1.814	1.770	1.733	1.702
64	2.786	2.387	2.171	2.035	1.939	1.868	1.813	1.768	1.731	1.700
65	2.784	2.386	2.170	2.033	1.938	1.867	1.811	1.767	1.730	1.699
66	2.783	2.385	2.169	2.032	1.937	1.865	1.810	1.765	1.728	1.697
67	2.782	2.384	2.167	2.031	1.935	1.864	1.808	1.764	1.727	1.696
68	2.781	2.382	2.166	2.029	1.934	1.863	1.807	1.762	1.725	1.694
69	2.780	2.381	2.165	2.028	1.933	1.861	1.806	1.761	1.724	1.693
70	2.779	2.380	2.164	2.027	1.931	1.860	1.804	1.760	1.723	1.691
71	2.778	2.379	2.163	2.026	1.930	1.859	1.803	1.758	1.721	1.690
72	2.777	2.378	2.161	2.025	1.929	1.858	1.802	1.757	1.720	1.689
73	2.776	2.377	2.160	2.024	1.928	1.856	1.801	1.756	1.719	1.687
74	2.775	2.376	2.159	2.022	1.927	1.855	1.800	1.755	1.718	1.686
75	2.774	2.375	2.158	2.021	1.926	1.854	1.798	1.754	1.716	1.685
76	2.773	2.374	2.157	2.020	1.925	1.853	1.797	1.752	1.715	1.684
77	2.772	2.373	2.156	2.019	1.924	1.852	1.796	1.751	1.714	1.683
78	2.771	2.372	2.155	2.018	1.923	1.851	1.795	1.750	1.713	1.682
79	2.770	2.371	2.154	2.017	1.922	1.850	1.794	1.749	1.712	1.681
80	2.769	2.370	2.154	2.016	1.921	1.849	1.793	1.748	1.711	1.680
81	2.769	2.369	2.153	2.016	1.920	1.848	1.792	1.747	1.710	1.679
82	2.768	2.368	2.152	2.015	1.919	1.847	1.791	1.746	1.709	1.678
83	2.767	2.368	2.151	2.014	1.918	1.846	1.790	1.745	1.708	1.677
84	2.766	2.367	2.150	2.013	1.917	1.845	1.790	1.744	1.707	1.676
85	2.765	2.366	2.149	2.012	1.916	1.845	1.789	1.744	1.706	1.675
86	2.765	2.365	2.149	2.011	1.915	1.844	1.788	1.743	1.705	1.674
87	2.764	2.365	2.148	2.011	1.915	1.843	1.787	1.742	1.705	1.673
88	2.763	2.364	2.147	2.010	1.914	1.842	1.786	1.741	1.704	1.672
89	2.763	2.363	2.146	2.009	1.913	1.841	1.785	1.740	1.703	1.671
90	2.762	2.363	2.146	2.008	1.912	1.841	1.785	1.739	1.702	1.670
91	2.761	2.362	2.145	2.008	1.912	1.840	1.784	1.739	1.701	1.670
92	2.761	2.361	2.144	2.007	1.911	1.839	1.783	1.738	1.701	1.669
93	2.760	2.361	2.144	2.006	1.910	1.838	1.782	1.737	1.700	1.668
94	2.760	2.360	2.143	2.006	1.910	1.838	1.782	1.736	1.699	1.667
95	2.759	2.359	2.142	2.005	1.909	1.837	1.781	1.736	1.698	1.667
96	2.759	2.359	2.142	2.004	1.908	1.836	1.780	1.735	1.698	1.666
97	2.758	2.358	2.141	2.004	1.908	1.836	1.780	1.734	1.697	1.665
98	2.757	2.358	2.141	2.003	1.907	1.835	1.779	1.734	1.696	1.665
99	2.757	2.357	2.140	2.003	1.906	1.835	1.778	1.733	1.696	1.664
100	2.756	2.356	2.139	2.002	1.906	1.834	1.778	1.732	1.695	1.663

$F_{.10}(df_1, df_2)$

df_2 \ df_1	11	12	13	14	15	16	17	18	19	20
1	60.473	60.705	60.903	61.073	61.220	61.350	61.464	61.566	61.658	61.740
2	9.401	9.408	9.415	9.420	9.425	9.429	9.433	9.436	9.439	9.441
3	5.222	5.216	5.210	5.205	5.200	5.196	5.193	5.190	5.187	5.184
4	3.907	3.896	3.886	3.878	3.870	3.864	3.858	3.853	3.849	3.844
5	3.282	3.268	3.257	3.247	3.238	3.230	3.223	3.217	3.212	3.207
6	2.920	2.905	2.892	2.881	2.871	2.863	2.855	2.848	2.842	2.836
7	2.684	2.668	2.654	2.643	2.632	2.623	2.615	2.607	2.601	2.595
8	2.519	2.502	2.488	2.475	2.464	2.455	2.446	2.438	2.431	2.425
9	2.396	2.379	2.364	2.351	2.340	2.329	2.320	2.312	2.305	2.298
10	2.302	2.284	2.269	2.255	2.244	2.233	2.224	2.215	2.208	2.201
11	2.227	2.209	2.193	2.179	2.167	2.156	2.147	2.138	2.130	2.123
12	2.166	2.147	2.131	2.117	2.105	2.094	2.084	2.075	2.067	2.060
13	2.116	2.097	2.080	2.066	2.053	2.042	2.032	2.023	2.014	2.007
14	2.073	2.054	2.037	2.022	2.010	1.998	1.988	1.978	1.970	1.962
15	2.037	2.017	2.000	1.985	1.972	1.961	1.950	1.941	1.932	1.924
16	2.005	1.985	1.968	1.953	1.940	1.928	1.917	1.908	1.899	1.891
17	1.978	1.958	1.940	1.925	1.912	1.900	1.889	1.879	1.870	1.862
18	1.954	1.933	1.916	1.900	1.887	1.875	1.864	1.854	1.845	1.837
19	1.932	1.912	1.894	1.878	1.865	1.852	1.841	1.831	1.822	1.814
20	1.913	1.892	1.875	1.859	1.845	1.833	1.821	1.811	1.802	1.794
21	1.896	1.875	1.857	1.841	1.827	1.815	1.803	1.793	1.784	1.776
22	1.880	1.859	1.841	1.825	1.811	1.798	1.787	1.777	1.768	1.759
23	1.866	1.845	1.827	1.811	1.796	1.784	1.772	1.762	1.753	1.744
24	1.853	1.832	1.814	1.797	1.783	1.770	1.759	1.748	1.739	1.730
25	1.841	1.820	1.802	1.785	1.771	1.758	1.746	1.736	1.726	1.718
26	1.830	1.809	1.790	1.774	1.760	1.747	1.735	1.724	1.715	1.706
27	1.820	1.799	1.780	1.764	1.749	1.736	1.724	1.714	1.704	1.695
28	1.811	1.790	1.771	1.754	1.740	1.726	1.715	1.704	1.694	1.685
29	1.802	1.781	1.762	1.745	1.731	1.717	1.705	1.695	1.685	1.676
30	1.794	1.773	1.754	1.737	1.722	1.709	1.697	1.686	1.676	1.667
31	1.787	1.765	1.746	1.729	1.714	1.701	1.689	1.678	1.668	1.659
32	1.780	1.758	1.739	1.722	1.707	1.694	1.682	1.671	1.661	1.652
33	1.773	1.751	1.732	1.715	1.700	1.687	1.675	1.664	1.654	1.645
34	1.767	1.745	1.726	1.709	1.694	1.680	1.668	1.657	1.647	1.638
35	1.761	1.739	1.720	1.703	1.688	1.674	1.662	1.651	1.641	1.632
36	1.756	1.734	1.715	1.697	1.682	1.669	1.656	1.645	1.635	1.626
37	1.751	1.729	1.709	1.692	1.677	1.663	1.651	1.640	1.630	1.620
38	1.746	1.724	1.704	1.687	1.672	1.658	1.646	1.635	1.624	1.615
39	1.741	1.719	1.700	1.682	1.667	1.653	1.641	1.630	1.619	1.610
40	1.737	1.715	1.695	1.678	1.662	1.649	1.636	1.625	1.615	1.605
41	1.733	1.710	1.691	1.673	1.658	1.644	1.632	1.620	1.610	1.601
42	1.729	1.706	1.687	1.669	1.654	1.640	1.628	1.616	1.606	1.596
43	1.725	1.703	1.683	1.665	1.650	1.636	1.624	1.612	1.602	1.592
44	1.721	1.699	1.679	1.662	1.646	1.632	1.620	1.608	1.598	1.588
45	1.718	1.695	1.676	1.658	1.643	1.629	1.616	1.605	1.594	1.585
46	1.715	1.692	1.672	1.655	1.639	1.625	1.613	1.601	1.591	1.581
47	1.712	1.689	1.669	1.652	1.636	1.622	1.609	1.598	1.587	1.578
48	1.709	1.686	1.666	1.648	1.633	1.619	1.606	1.594	1.584	1.574
49	1.706	1.683	1.663	1.645	1.630	1.616	1.603	1.591	1.581	1.571
50	1.703	1.680	1.660	1.643	1.627	1.613	1.600	1.588	1.578	1.568

$F_{.10}(df_1, df_2)$

df_2 \ df_1	11	12	13	14	15	16	17	18	19	20
51	1.700	1.677	1.658	1.640	1.624	1.610	1.597	1.586	1.575	1.565
52	1.698	1.675	1.655	1.637	1.621	1.607	1.594	1.583	1.572	1.562
53	1.695	1.672	1.652	1.635	1.619	1.605	1.592	1.580	1.570	1.560
54	1.693	1.670	1.650	1.632	1.616	1.602	1.589	1.578	1.567	1.557
55	1.691	1.668	1.648	1.630	1.614	1.600	1.587	1.575	1.564	1.555
56	1.688	1.666	1.645	1.628	1.612	1.597	1.585	1.573	1.562	1.552
57	1.686	1.663	1.643	1.625	1.610	1.595	1.582	1.571	1.560	1.550
58	1.684	1.661	1.641	1.623	1.607	1.593	1.580	1.568	1.558	1.548
59	1.682	1.659	1.639	1.621	1.605	1.591	1.578	1.566	1.555	1.546
60	1.680	1.657	1.637	1.619	1.603	1.589	1.576	1.564	1.553	1.543
61	1.679	1.656	1.635	1.617	1.601	1.587	1.574	1.562	1.551	1.541
62	1.677	1.654	1.634	1.616	1.600	1.585	1.572	1.560	1.549	1.540
63	1.675	1.652	1.632	1.614	1.598	1.583	1.570	1.558	1.548	1.538
64	1.673	1.650	1.630	1.612	1.596	1.582	1.569	1.557	1.546	1.536
65	1.672	1.649	1.628	1.610	1.594	1.580	1.567	1.555	1.544	1.534
66	1.670	1.647	1.627	1.609	1.593	1.578	1.565	1.553	1.542	1.532
67	1.669	1.646	1.625	1.607	1.591	1.577	1.564	1.552	1.541	1.531
68	1.667	1.644	1.624	1.606	1.590	1.575	1.562	1.550	1.539	1.529
69	1.666	1.643	1.622	1.604	1.588	1.574	1.560	1.548	1.538	1.527
70	1.665	1.641	1.621	1.603	1.587	1.572	1.559	1.547	1.536	1.526
71	1.663	1.640	1.619	1.601	1.585	1.571	1.557	1.545	1.535	1.524
72	1.662	1.639	1.618	1.600	1.584	1.569	1.556	1.544	1.533	1.523
73	1.661	1.637	1.617	1.599	1.583	1.568	1.555	1.543	1.532	1.522
74	1.659	1.636	1.616	1.597	1.581	1.567	1.553	1.541	1.530	1.520
75	1.658	1.635	1.614	1.596	1.580	1.565	1.552	1.540	1.529	1.519
76	1.657	1.634	1.613	1.595	1.579	1.564	1.551	1.539	1.528	1.518
77	1.656	1.632	1.612	1.594	1.578	1.563	1.550	1.538	1.527	1.516
78	1.655	1.631	1.611	1.593	1.576	1.562	1.548	1.536	1.525	1.515
79	1.654	1.630	1.610	1.592	1.575	1.561	1.547	1.535	1.524	1.514
80	1.653	1.629	1.609	1.590	1.574	1.559	1.546	1.534	1.523	1.513
81	1.652	1.628	1.608	1.589	1.573	1.558	1.545	1.533	1.522	1.512
82	1.651	1.627	1.607	1.588	1.572	1.557	1.544	1.532	1.521	1.511
83	1.650	1.626	1.606	1.587	1.571	1.556	1.543	1.531	1.520	1.509
84	1.649	1.625	1.605	1.586	1.570	1.555	1.542	1.530	1.519	1.508
85	1.648	1.624	1.604	1.585	1.569	1.554	1.541	1.529	1.518	1.507
86	1.647	1.623	1.603	1.584	1.568	1.553	1.540	1.528	1.517	1.506
87	1.646	1.622	1.602	1.583	1.567	1.552	1.539	1.527	1.516	1.505
88	1.645	1.622	1.601	1.583	1.566	1.551	1.538	1.526	1.515	1.504
89	1.644	1.621	1.600	1.582	1.565	1.550	1.537	1.525	1.514	1.503
90	1.643	1.620	1.599	1.581	1.564	1.550	1.536	1.524	1.513	1.503
91	1.643	1.619	1.598	1.580	1.564	1.549	1.535	1.523	1.512	1.502
92	1.642	1.618	1.598	1.579	1.563	1.548	1.534	1.522	1.511	1.501
93	1.641	1.617	1.597	1.578	1.562	1.547	1.534	1.521	1.510	1.500
94	1.640	1.617	1.596	1.578	1.561	1.546	1.533	1.521	1.509	1.499
95	1.640	1.616	1.595	1.577	1.560	1.545	1.532	1.520	1.509	1.498
96	1.639	1.615	1.594	1.576	1.560	1.545	1.531	1.519	1.508	1.497
97	1.638	1.614	1.594	1.575	1.559	1.544	1.530	1.518	1.507	1.497
98	1.637	1.614	1.593	1.575	1.558	1.543	1.530	1.517	1.506	1.496
99	1.637	1.613	1.592	1.574	1.557	1.542	1.529	1.517	1.505	1.495
100	1.636	1.612	1.592	1.573	1.557	1.542	1.528	1.516	1.505	1.494

$F_{.01}(df_1, df_2)$

df_2 \ df_1	1	2	3	4	5	6	7	8	9	10
1	4052.19	4999.52	5403.34	5624.62	5763.65	5858.97	5928.33	5981.10	6022.50	6055.85
2	98.502	99.000	99.166	99.249	99.300	99.333	99.356	99.374	99.388	99.399
3	34.116	30.816	29.457	28.710	28.237	27.911	27.672	27.489	27.345	27.229
4	21.198	18.000	16.694	15.977	15.522	15.207	14.976	14.799	14.659	14.546
5	16.258	13.274	12.060	11.392	10.967	10.672	10.456	10.289	10.158	10.051
6	13.745	10.925	9.780	9.148	8.746	8.466	8.260	8.102	7.976	7.874
7	12.246	9.547	8.451	7.847	7.460	7.191	6.993	6.840	6.719	6.620
8	11.259	8.649	7.591	7.006	6.632	6.371	6.178	6.029	5.911	5.814
9	10.561	8.022	6.992	6.422	6.057	5.802	5.613	5.467	5.351	5.257
10	10.044	7.559	6.552	5.994	5.636	5.386	5.200	5.057	4.942	4.849
11	9.646	7.206	6.217	5.668	5.316	5.069	4.886	4.744	4.632	4.539
12	9.330	6.927	5.953	5.412	5.064	4.821	4.640	4.499	4.388	4.296
13	9.074	6.701	5.739	5.205	4.862	4.620	4.441	4.302	4.191	4.100
14	8.862	6.515	5.564	5.035	4.695	4.456	4.278	4.140	4.030	3.939
15	8.683	6.359	5.417	4.893	4.556	4.318	4.142	4.004	3.895	3.805
16	8.531	6.226	5.292	4.773	4.437	4.202	4.026	3.890	3.780	3.691
17	8.400	6.112	5.185	4.669	4.336	4.102	3.927	3.791	3.682	3.593
18	8.285	6.013	5.092	4.579	4.248	4.015	3.841	3.705	3.597	3.508
19	8.185	5.926	5.010	4.500	4.171	3.939	3.765	3.631	3.523	3.434
20	8.096	5.849	4.938	4.431	4.103	3.871	3.699	3.564	3.457	3.368
21	8.017	5.780	4.874	4.369	4.042	3.812	3.640	3.506	3.398	3.310
22	7.945	5.719	4.817	4.313	3.988	3.758	3.587	3.453	3.346	3.258
23	7.881	5.664	4.765	4.264	3.939	3.710	3.539	3.406	3.299	3.211
24	7.823	5.614	4.718	4.218	3.895	3.667	3.496	3.363	3.256	3.168
25	7.770	5.568	4.675	4.177	3.855	3.627	3.457	3.324	3.217	3.129
26	7.721	5.526	4.637	4.140	3.818	3.591	3.421	3.288	3.182	3.094
27	7.677	5.488	4.601	4.106	3.785	3.558	3.388	3.256	3.149	3.062
28	7.636	5.453	4.568	4.074	3.754	3.528	3.358	3.226	3.120	3.032
29	7.598	5.420	4.538	4.045	3.725	3.499	3.330	3.198	3.092	3.005
30	7.562	5.390	4.510	4.018	3.699	3.473	3.305	3.173	3.067	2.979
31	7.530	5.362	4.484	3.993	3.675	3.449	3.281	3.149	3.043	2.955
32	7.499	5.336	4.459	3.969	3.652	3.427	3.258	3.127	3.021	2.934
33	7.471	5.312	4.437	3.948	3.630	3.406	3.238	3.106	3.000	2.913
34	7.444	5.289	4.416	3.927	3.611	3.386	3.218	3.087	2.981	2.894
35	7.419	5.268	4.396	3.908	3.592	3.368	3.200	3.069	2.963	2.876
36	7.396	5.248	4.377	3.890	3.574	3.351	3.183	3.052	2.946	2.859
37	7.373	5.229	4.360	3.873	3.558	3.334	3.167	3.036	2.930	2.843
38	7.353	5.211	4.343	3.858	3.542	3.319	3.152	3.021	2.915	2.828
39	7.333	5.194	4.327	3.843	3.528	3.305	3.137	3.006	2.901	2.814
40	7.314	5.179	4.313	3.828	3.514	3.291	3.124	2.993	2.888	2.801
41	7.296	5.163	4.299	3.815	3.501	3.278	3.111	2.980	2.875	2.788
42	7.280	5.149	4.285	3.802	3.488	3.266	3.099	2.968	2.863	2.776
43	7.264	5.136	4.273	3.790	3.476	3.254	3.087	2.957	2.851	2.764
44	7.248	5.123	4.261	3.778	3.465	3.243	3.076	2.946	2.840	2.754
45	7.234	5.110	4.249	3.767	3.454	3.232	3.066	2.935	2.830	2.743
46	7.220	5.099	4.238	3.757	3.444	3.222	3.056	2.925	2.820	2.733
47	7.207	5.087	4.228	3.747	3.434	3.213	3.046	2.916	2.811	2.724
48	7.194	5.077	4.218	3.737	3.425	3.204	3.037	2.907	2.802	2.715
49	7.182	5.066	4.208	3.728	3.416	3.195	3.028	2.898	2.793	2.706
50	7.171	5.057	4.199	3.720	3.408	3.186	3.020	2.890	2.785	2.698

$F_{.01}(df_1, df_2)$

df_2 \ df_1	1	2	3	4	5	6	7	8	9	10
51	7.159	5.047	4.191	3.711	3.400	3.178	3.012	2.882	2.777	2.690
52	7.149	5.038	4.182	3.703	3.392	3.171	3.005	2.874	2.769	2.683
53	7.139	5.030	4.174	3.695	3.384	3.163	2.997	2.867	2.762	2.675
54	7.129	5.021	4.167	3.688	3.377	3.156	2.990	2.860	2.755	2.668
55	7.119	5.013	4.159	3.681	3.370	3.149	2.983	2.853	2.748	2.662
56	7.110	5.006	4.152	3.674	3.363	3.143	2.977	2.847	2.742	2.655
57	7.102	4.998	4.145	3.667	3.357	3.136	2.971	2.841	2.736	2.649
58	7.093	4.991	4.138	3.661	3.351	3.130	2.965	2.835	2.730	2.643
59	7.085	4.984	4.132	3.655	3.345	3.124	2.959	2.829	2.724	2.637
60	7.077	4.977	4.126	3.649	3.339	3.119	2.953	2.823	2.718	2.632
61	7.070	4.971	4.120	3.643	3.333	3.113	2.948	2.818	2.713	2.626
62	7.062	4.965	4.114	3.638	3.328	3.108	2.942	2.813	2.708	2.621
63	7.055	4.959	4.109	3.632	3.323	3.103	2.937	2.808	2.703	2.616
64	7.048	4.953	4.103	3.627	3.318	3.098	2.932	2.803	2.698	2.611
65	7.042	4.947	4.098	3.622	3.313	3.093	2.928	2.798	2.693	2.607
66	7.035	4.942	4.093	3.618	3.308	3.088	2.923	2.793	2.689	2.602
67	7.029	4.937	4.088	3.613	3.304	3.084	2.919	2.789	2.684	2.598
68	7.023	4.932	4.083	3.608	3.299	3.080	2.914	2.785	2.680	2.593
69	7.017	4.927	4.079	3.604	3.295	3.075	2.910	2.781	2.676	2.589
70	7.011	4.922	4.074	3.600	3.291	3.071	2.906	2.777	2.672	2.585
71	7.006	4.917	4.070	3.596	3.287	3.067	2.902	2.773	2.668	2.581
72	7.001	4.913	4.066	3.591	3.283	3.063	2.898	2.769	2.664	2.578
73	6.995	4.908	4.062	3.588	3.279	3.060	2.895	2.765	2.660	2.574
74	6.990	4.904	4.058	3.584	3.275	3.056	2.891	2.762	2.657	2.570
75	6.985	4.900	4.054	3.580	3.272	3.052	2.887	2.758	2.653	2.567
76	6.981	4.896	4.050	3.577	3.268	3.049	2.884	2.755	2.650	2.563
77	6.976	4.892	4.047	3.573	3.265	3.046	2.881	2.751	2.647	2.560
78	6.971	4.888	4.043	3.570	3.261	3.042	2.877	2.748	2.644	2.557
79	6.967	4.884	4.040	3.566	3.258	3.039	2.874	2.745	2.640	2.554
80	6.963	4.881	4.036	3.563	3.255	3.036	2.871	2.742	2.637	2.551
81	6.958	4.877	4.033	3.560	3.252	3.033	2.868	2.739	2.634	2.548
82	6.954	4.874	4.030	3.557	3.249	3.030	2.865	2.736	2.632	2.545
83	6.950	4.870	4.027	3.554	3.246	3.027	2.863	2.733	2.629	2.542
84	6.947	4.867	4.024	3.551	3.243	3.025	2.860	2.731	2.626	2.539
85	6.943	4.864	4.021	3.548	3.240	3.022	2.857	2.728	2.623	2.537
86	6.939	4.861	4.018	3.545	3.238	3.019	2.854	2.725	2.621	2.534
87	6.935	4.858	4.015	3.543	3.235	3.017	2.852	2.723	2.618	2.532
88	6.932	4.855	4.012	3.540	3.233	3.014	2.849	2.720	2.616	2.529
89	6.928	4.852	4.010	3.538	3.230	3.012	2.847	2.718	2.613	2.527
90	6.925	4.849	4.007	3.535	3.228	3.009	2.845	2.715	2.611	2.524
91	6.922	4.846	4.004	3.533	3.225	3.007	2.842	2.713	2.609	2.522
92	6.919	4.844	4.002	3.530	3.223	3.004	2.840	2.711	2.606	2.520
93	6.915	4.841	3.999	3.528	3.221	3.002	2.838	2.709	2.604	2.518
94	6.912	4.838	3.997	3.525	3.218	3.000	2.835	2.706	2.602	2.515
95	6.909	4.836	3.995	3.523	3.216	2.998	2.833	2.704	2.600	2.513
96	6.906	4.833	3.992	3.521	3.214	2.996	2.831	2.702	2.598	2.511
97	6.904	4.831	3.990	3.519	3.212	2.994	2.829	2.700	2.596	2.509
98	6.901	4.829	3.988	3.517	3.210	2.992	2.827	2.698	2.594	2.507
99	6.898	4.826	3.986	3.515	3.208	2.990	2.825	2.696	2.592	2.505
100	6.895	4.824	3.984	3.513	3.206	2.988	2.823	2.694	2.590	2.503

$F_{.01}(df_1, df_2)$

df_2 \ df_1	11	12	13	14	15	16	17	18	19	20
1	6083.35	6106.35	6125.86	6142.70	6157.28	6170.12	6181.42	6191.52	6200.58	6208.74
2	99.408	99.416	99.422	99.428	99.432	99.437	99.440	99.444	99.447	99.449
3	27.133	27.052	26.983	26.924	26.872	26.827	26.787	26.751	26.719	26.690
4	14.452	14.374	14.307	14.249	14.198	14.154	14.115	14.080	14.048	14.020
5	9.963	9.888	9.825	9.770	9.722	9.680	9.643	9.610	9.580	9.553
6	7.790	7.718	7.657	7.605	7.559	7.519	7.483	7.451	7.422	7.396
7	6.538	6.469	6.410	6.359	6.314	6.275	6.240	6.209	6.181	6.155
8	5.734	5.667	5.609	5.559	5.515	5.477	5.442	5.412	5.384	5.359
9	5.178	5.111	5.055	5.005	4.962	4.924	4.890	4.860	4.833	4.808
10	4.772	4.706	4.650	4.601	4.558	4.520	4.487	4.457	4.430	4.405
11	4.462	4.397	4.342	4.293	4.251	4.213	4.180	4.150	4.123	4.099
12	4.220	4.155	4.100	4.052	4.010	3.972	3.939	3.909	3.883	3.858
13	4.025	3.960	3.905	3.857	3.815	3.778	3.745	3.716	3.689	3.665
14	3.864	3.800	3.745	3.698	3.656	3.619	3.586	3.556	3.529	3.505
15	3.730	3.666	3.612	3.564	3.522	3.485	3.452	3.423	3.396	3.372
16	3.616	3.553	3.498	3.451	3.409	3.372	3.339	3.310	3.283	3.259
17	3.519	3.455	3.401	3.353	3.312	3.275	3.242	3.212	3.186	3.162
18	3.434	3.371	3.316	3.269	3.227	3.190	3.158	3.128	3.101	3.077
19	3.360	3.297	3.242	3.195	3.153	3.116	3.084	3.054	3.027	3.003
20	3.294	3.231	3.177	3.130	3.088	3.051	3.018	2.989	2.962	2.938
21	3.236	3.173	3.119	3.072	3.030	2.993	2.960	2.931	2.904	2.880
22	3.184	3.121	3.067	3.019	2.978	2.941	2.908	2.879	2.852	2.827
23	3.137	3.074	3.020	2.973	2.931	2.894	2.861	2.832	2.805	2.781
24	3.094	3.032	2.977	2.930	2.889	2.852	2.819	2.789	2.762	2.738
25	3.056	2.993	2.939	2.892	2.850	2.813	2.780	2.751	2.724	2.699
26	3.021	2.958	2.904	2.857	2.815	2.778	2.745	2.715	2.688	2.664
27	2.988	2.926	2.871	2.824	2.783	2.746	2.713	2.683	2.656	2.632
28	2.959	2.896	2.842	2.795	2.753	2.716	2.683	2.653	2.626	2.602
29	2.931	2.868	2.814	2.767	2.726	2.689	2.656	2.626	2.599	2.574
30	2.906	2.843	2.789	2.742	2.700	2.663	2.630	2.600	2.573	2.549
31	2.882	2.820	2.765	2.718	2.677	2.640	2.606	2.577	2.550	2.525
32	2.860	2.798	2.744	2.696	2.655	2.618	2.584	2.555	2.527	2.503
33	2.840	2.777	2.723	2.676	2.634	2.597	2.564	2.534	2.507	2.482
34	2.821	2.758	2.704	2.657	2.615	2.578	2.545	2.515	2.488	2.463
35	2.803	2.740	2.686	2.639	2.597	2.560	2.527	2.497	2.470	2.445
36	2.786	2.723	2.669	2.622	2.580	2.543	2.510	2.480	2.453	2.428
37	2.770	2.707	2.653	2.606	2.564	2.527	2.494	2.464	2.437	2.412
38	2.755	2.692	2.638	2.591	2.549	2.512	2.479	2.449	2.421	2.397
39	2.741	2.678	2.624	2.577	2.535	2.498	2.465	2.434	2.407	2.382
40	2.727	2.665	2.611	2.563	2.522	2.484	2.451	2.421	2.394	2.369
41	2.715	2.652	2.598	2.551	2.509	2.472	2.438	2.408	2.381	2.356
42	2.703	2.640	2.586	2.539	2.497	2.460	2.426	2.396	2.369	2.344
43	2.691	2.629	2.575	2.527	2.485	2.448	2.415	2.385	2.357	2.332
44	2.680	2.618	2.564	2.516	2.475	2.437	2.404	2.374	2.346	2.321
45	2.670	2.608	2.553	2.506	2.464	2.427	2.393	2.363	2.336	2.311
46	2.660	2.598	2.544	2.496	2.454	2.417	2.384	2.353	2.326	2.301
47	2.651	2.588	2.534	2.487	2.445	2.408	2.374	2.344	2.316	2.291
48	2.642	2.579	2.525	2.478	2.436	2.399	2.365	2.335	2.307	2.282
49	2.633	2.571	2.517	2.469	2.427	2.390	2.356	2.326	2.299	2.274
50	2.625	2.562	2.508	2.461	2.419	2.382	2.348	2.318	2.290	2.265

$F_{.01}(df_1, df_2)$

df_2 \ df_1	11	12	13	14	15	16	17	18	19	20
51	2.617	2.555	2.500	2.453	2.411	2.374	2.340	2.310	2.282	2.257
52	2.610	2.547	2.493	2.445	2.403	2.366	2.333	2.302	2.275	2.250
53	2.602	2.540	2.486	2.438	2.396	2.359	2.325	2.295	2.267	2.242
54	2.595	2.533	2.479	2.431	2.389	2.352	2.318	2.288	2.260	2.235
55	2.589	2.526	2.472	2.424	2.382	2.345	2.311	2.281	2.253	2.228
56	2.582	2.520	2.465	2.418	2.376	2.339	2.305	2.275	2.247	2.222
57	2.576	2.513	2.459	2.412	2.370	2.332	2.299	2.268	2.241	2.215
58	2.570	2.507	2.453	2.406	2.364	2.326	2.293	2.262	2.235	2.209
59	2.564	2.502	2.447	2.400	2.358	2.320	2.287	2.256	2.229	2.203
60	2.559	2.496	2.442	2.394	2.352	2.315	2.281	2.251	2.223	2.198
61	2.553	2.491	2.436	2.389	2.347	2.309	2.276	2.245	2.218	2.192
62	2.548	2.486	2.431	2.384	2.342	2.304	2.270	2.240	2.212	2.187
63	2.543	2.481	2.426	2.379	2.337	2.299	2.265	2.235	2.207	2.182
64	2.538	2.476	2.421	2.374	2.332	2.294	2.260	2.230	2.202	2.177
65	2.534	2.471	2.417	2.369	2.327	2.289	2.256	2.225	2.198	2.172
66	2.529	2.466	2.412	2.365	2.322	2.285	2.251	2.221	2.193	2.168
67	2.525	2.462	2.408	2.360	2.318	2.280	2.247	2.216	2.188	2.163
68	2.520	2.458	2.403	2.356	2.314	2.276	2.242	2.212	2.184	2.159
69	2.516	2.454	2.399	2.352	2.310	2.272	2.238	2.208	2.180	2.155
70	2.512	2.450	2.395	2.348	2.306	2.268	2.234	2.204	2.176	2.150
71	2.508	2.446	2.391	2.344	2.302	2.264	2.230	2.200	2.172	2.146
72	2.504	2.442	2.388	2.340	2.298	2.260	2.226	2.196	2.168	2.143
73	2.501	2.438	2.384	2.336	2.294	2.256	2.223	2.192	2.164	2.139
74	2.497	2.435	2.380	2.333	2.290	2.253	2.219	2.188	2.161	2.135
75	2.494	2.431	2.377	2.329	2.287	2.249	2.215	2.185	2.157	2.132
76	2.490	2.428	2.373	2.326	2.284	2.246	2.212	2.181	2.154	2.128
77	2.487	2.424	2.370	2.322	2.280	2.243	2.209	2.178	2.150	2.125
78	2.484	2.421	2.367	2.319	2.277	2.239	2.206	2.175	2.147	2.122
79	2.481	2.418	2.364	2.316	2.274	2.236	2.202	2.172	2.144	2.118
80	2.478	2.415	2.361	2.313	2.271	2.233	2.199	2.169	2.141	2.115
81	2.475	2.412	2.358	2.310	2.268	2.230	2.196	2.166	2.138	2.112
82	2.472	2.409	2.355	2.307	2.265	2.227	2.193	2.163	2.135	2.109
83	2.469	2.406	2.352	2.304	2.262	2.224	2.191	2.160	2.132	2.106
84	2.466	2.404	2.349	2.302	2.259	2.222	2.188	2.157	2.129	2.104
85	2.464	2.401	2.347	2.299	2.257	2.219	2.185	2.154	2.126	2.101
86	2.461	2.398	2.344	2.296	2.254	2.216	2.182	2.152	2.124	2.098
87	2.459	2.396	2.342	2.294	2.252	2.214	2.180	2.149	2.121	2.096
88	2.456	2.393	2.339	2.291	2.249	2.211	2.177	2.147	2.119	2.093
89	2.454	2.391	2.337	2.289	2.247	2.209	2.175	2.144	2.116	2.091
90	2.451	2.389	2.334	2.286	2.244	2.206	2.172	2.142	2.114	2.088
91	2.449	2.386	2.332	2.284	2.242	2.204	2.170	2.139	2.111	2.086
92	2.447	2.384	2.330	2.282	2.240	2.202	2.168	2.137	2.109	2.083
93	2.444	2.382	2.327	2.280	2.237	2.200	2.166	2.135	2.107	2.081
94	2.442	2.380	2.325	2.277	2.235	2.197	2.163	2.133	2.105	2.079
95	2.440	2.378	2.323	2.275	2.233	2.195	2.161	2.130	2.102	2.077
96	2.438	2.375	2.321	2.273	2.231	2.193	2.159	2.128	2.100	2.075
97	2.436	2.373	2.319	2.271	2.229	2.191	2.157	2.126	2.098	2.073
98	2.434	2.371	2.317	2.269	2.227	2.189	2.155	2.124	2.096	2.071
99	2.432	2.369	2.315	2.267	2.225	2.187	2.153	2.122	2.094	2.069
100	2.430	2.368	2.313	2.265	2.223	2.185	2.151	2.120	2.092	2.067

Appendix D

Wilcoxon Signed Ranks Test

Lower and Upper Critical Values Separated by Comma				
1-Tailed α	**0.050**	**0.025**	**0.010**	**0.005**
2-Tailed α	**0.100**	**0.050**	**0.020**	**0.010**
n				
5	0, 15	*	*	*
6	2, 19	0, 21	*	*
7	3, 25	2, 26	0, 28	*
8	5, 31	3, 33	1, 35	0, 36
9	8, 37	5, 40	3, 42	1, 44
10	10, 45	8, 47	5, 50	3, 52
11	13, 53	10, 56	7, 59	5, 61
12	17, 61	13, 65	10, 68	7, 71
13	21, 70	17, 74	12, 79	10, 81
14	25, 80	21, 84	16, 89	13, 92
15	30, 90	25, 95	19, 101	16, 104
16	35, 101	29, 107	233, 113	19, 117
17	41, 112	34, 119	27, 126	23, 130
18	47, 124	40, 131	32, 139	27, 144
19	53, 137	46, 144	37, 153	32, 158
20	60, 150	52, 158	43, 167	37, 173

n is the total number of positive and negative ranks, omitting zero differences.

Appendix E

Mann-Whitney U Statistic

Critical Values of Mann-Whitney U Statistic

2-Tailed α = 0.10

1-Tailed α = 0.5

n_1 \ n_2	2	3	4	5	6	7	8	9	10	11	12
2		0	0	0	0	1	1	1	2	2	2
3	0	0	1	1	2	3	3	4	5	5	6
4	0	1	2	3	4	5	6	7	8	9	10
5	0	1	3	4	5	7	8	10	11	12	14
6	0	2	4	5	7	9	11	12	14	16	18
7	1	3	5	7	9	11	13	15	18	20	22
8	1	3	6	8	11	13	16	18	21	24	26
9	1	4	7	10	12	15	18	21	24	27	30
10	2	5	8	11	14	18	21	24	28	31	34
11	2	5	9	12	16	20	24	27	31	35	39
12	2	6	10	14	18	22	26	30	34	39	43

2-Tailed α = 0.05

1-Tailed α = 0.025

n_1 \ n_2	2	3	4	5	6	7	8	9	10	11	12
2				0	0	0	0	0	1	1	1
3			0	0	1	2	2	3	3	4	4
4		0	1	2	2	3	4	5	6	7	8
5	0	0	2	3	4	5	6	7	9	10	11
6	0	1	2	4	5	7	8	10	12	13	15
7	0	2	3	5	7	9	11	13	15	17	18
8	0	2	4	6	8	11	13	15	18	20	22
9	0	3	5	7	10	13	15	18	21	23	26
10	1	3	6	9	12	15	18	21	24	27	30
11	1	4	7	10	13	17	20	23	27	30	34
12	1	4	8	11	15	18	22	26	30	34	38

Appendix F

Wilcoxon Sum of Ranks Test

Lower Critical Values of T_1

	1-tailed α	0.005	0.01	0.025	0.05	0.1
	2-tailed α	0.01	0.02	0.05	0.1	0.2
n_1	n_2					
4	4	.	.	10	11	13
4	5	.	10	11	12	14
4	6	10	11	12	13	15
4	7	10	11	13	14	16
4	8	11	12	14	15	17
4	9	11	13	14	16	19
4	10	12	13	15	17	20
4	11	12	14	16	18	21
4	12	13	15	17	19	22
5	5	15	16	17	19	20
5	6	16	17	18	20	22
5	7	16	18	20	21	23
5	8	17	19	21	23	25
5	9	18	20	22	24	27
5	10	19	21	23	26	28
5	11	20	22	24	27	30
5	12	21	23	26	28	32
6	6	23	24	26	28	30
6	7	24	25	27	29	32
6	8	25	27	29	31	34
6	9	26	28	31	33	36
6	10	27	29	32	35	38
6	11	28	30	34	37	40
6	12	30	32	35	38	42
7	7	32	34	36	39	41
7	8	34	35	38	41	44
7	9	35	37	40	43	46
7	10	37	39	42	45	49
7	11	38	40	44	47	51
7	12	40	42	46	49	54
8	8	43	45	49	51	55
8	9	45	47	51	54	58
8	10	47	49	53	56	60
8	11	49	51	55	59	63
8	12	51	53	58	62	66
9	9	56	59	62	66	70
9	10	58	61	65	69	73
9	11	61	63	68	72	76
9	12	63	66	71	75	80
10	10	71	74	78	82	87
10	11	73	77	81	86	91
10	12	76	79	84	89	94
11	11	87	91	96	100	106
11	12	90	94	99	104	110
12	12	105	109	115	120	127

Upper Critical Values of T_1

		1-tailed α	0.005	0.01	0.025	0.05	0.1
		2-tailed α	0.01	0.02	0.05	0.1	0.2

n_1	n_2	0.005 / 0.01	0.01 / 0.02	0.025 / 0.05	0.05 / 0.1	0.1 / 0.2
4	4	22	23	25	26	.
4	5	25	26	28	29	30
4	6	27	29	31	32	33
4	7	30	32	34	35	37
4	8	32	35	37	38	40
4	9	35	37	40	42	43
4	10	37	40	43	45	47
4	11	40	43	46	48	50
4	12	42	46	49	51	53
5	5	33	35	36	38	39
5	6	36	38	40	42	43
5	7	39	42	44	45	47
5	8	42	45	47	49	51
5	9	45	48	51	53	55
5	10	48	52	54	57	59
5	11	51	55	58	61	63
5	12	54	58	62	64	67
6	6	45	48	50	52	54
6	7	49	52	55	57	59
6	8	53	56	59	61	63
6	9	56	60	63	65	68
6	10	60	64	67	70	73
6	11	64	68	71	74	78
6	12	67	72	76	79	82
7	7	60	64	66	69	71
7	8	64	68	71	74	77
7	9	69	73	76	79	82
7	10	73	77	81	84	87
7	11	77	82	86	89	93
7	12	81	86	91	94	98
8	8	77	81	85	87	91
8	9	82	86	90	93	97
8	10	87	92	96	99	103
8	11	91	97	101	105	109
8	12	96	102	106	110	115
9	9	96	101	105	109	112
9	10	102	107	111	115	119
9	11	107	113	117	121	126
9	12	112	118	123	127	132
10	10	117	123	128	132	136
10	11	123	129	134	139	143
10	12	129	136	141	146	151
11	11	141	147	153	157	162
11	12	147	154	160	165	170
12	12	166	173	180	185	191

Appendix G

89015 Townhouse Data

Sale Price	Sale Date	Season	Baths	BRs	Living Area (SF)	FPs	Gar	Lot Area (SF)	Year Built	Days on Mkt	Occupancy
67500	3/18/2001	winter	1	2	850	0	0	3080	1994	4	OWN
86000	3/30/2001	winter	2	3	1148	1	0	2970	1995	46	OWN
79000	4/3/2001	spring	2	2	1151	1	0	2970	1995	35	OWN
78000	4/3/2001	spring	2	2	1376	0	0	3540	1984	8	OWN
80000	4/18/2001	spring	2	3	1376	0	0	3540	1984	5	OWN
51500	4/30/2001	spring	2	1	838	0	0	2500	1983	17	VAC
80000	5/1/2001	spring	2	2	1151	1	0	2970	1995	3	VAC
88000	5/3/2001	spring	2	2	1100	1	0	2550	1999	45	VAC
67000	5/7/2001	spring	2	3	1080	0	0	2401	1971	66	OWN
57500	5/16/2001	spring	2	2	984	0	0	2500	1983	11	OWN
77500	5/24/2001	spring	2	2	1151	1	0	2970	1995	41	OWN
85000	5/30/2001	spring	2	2	1019	0	0	2550	1997	29	VAC
80500	6/7/2001	spring	2	2	1151	1	0	2970	1995	105	VAC
71500	6/8/2001	spring	1	2	836	0	0	3080	1995	22	VAC
89900	6/10/2001	spring	2	2	1019	0	0	2550	2001	102	VAC
91900	6/11/2001	spring	3	4	1696	1	0	3540	1986	63	OWN
89400	6/13/2001	spring	2	2	1019	0	0	2550	2000	100	VAC
55000	6/21/2001	spring	2	1	858	1	0	2809	1984	9	VAC
74500	6/25/2001	spring	3	2	1101	0	0	3540	1984	25	VAC
54000	7/9/2001	summer	2	2	992	1	0	2500	1982	4	TEN
82500	7/10/2001	summer	2	2	1151	1	0	2970	1995	18	OWN
52900	7/12/2001	summer	1	2	772	1	0	2401	1971	23	VAC
83950	7/18/2001	summer	2	2	1151	1	0	2970	1995	2	OWN
83900	7/25/2001	summer	2	2	1151	1	0	2970	1995	3	VAC
78000	7/31/2001	summer	3	2	1056	1	0	3540	1986	54	OWN
80000	8/7/2001	summer	2	2	1151	1	0	2970	1995	26	VAC
119000	8/9/2001	summer	3	2	1287	1	1	2137	1996	10	OWN
76000	8/15/2001	summer	2	2	1151	1	0	2970	1995	93	VAC
63000	8/22/2001	summer	2	2	984	0	0	2500	1983	51	VAC
56900	8/22/2001	summer	2	2	984	0	0	2500	1983	55	VAC
91000	8/24/2001	summer	2	2	1019	0	0	2550	2001	28	OWN
84000	8/25/2001	summer	2	2	1019	0	0	2550	1996	83	OWN
89900	8/28/2001	summer	2	2	1019	0	0	2550	2001	89	VAC
61500	8/29/2001	summer	2	2	984	0	0	2500	1983	24	OWN

Sale Price	Sale Date	Season	Baths	BRs	Living Area (SF)	FPs	Gar	Lot Area (SF)	Year Built	Days on Mkt	Occupancy
84500	8/31/2001	summer	2	2	1151	1	0	2970	1995	34	VAC
54750	9/6/2001	summer	2	2	984	0	0	2500	1983	180	VAC
56000	9/25/2001	summer	1	2	772	0	0	2401	1971	65	VAC
84500	9/26/2001	summer	2	2	1151	1	0	2970	1995	46	OWN
85000	10/5/2001	fall	2	2	1151	1	0	2970	1995	18	VAC
82500	10/7/2001	fall	2	3	1376	0	0	3540	1984	36	OWN
72000	11/7/2001	fall	2	2	960	0	0	3080	1995	70	VAC
61000	1/2/2002	winter	2	2	992	1	0	2500	1983	57	VAC
83500	1/2/2002	winter	2	3	1148	1	0	2970	1995	11	OWN
60000	1/4/2002	winter	1	1	724	0	0	3080	1995	131	OWN
85900	1/21/2002	winter	2	3	1148	1	0	2970	1995	183	VAC
58000	1/30/2002	winter	1	2	762	0	0	2401	1971	83	VAC
80000	2/21/2002	winter	2	2	1151	1	0	2970	1995	23	VAC
79000	2/22/2002	winter	2	2	1151	1	0	2970	1995	212	VAC
89900	3/20/2002	winter	2	2	1019	1	0	2550	1996	53	OWN
75650	4/1/2002	spring	1	2	836	0	0	3080	1994	206	OWN
48000	4/4/2002	spring	2	1	838	0	0	2500	1983	165	VAC
91000	4/12/2002	spring	2	2	1019	0	0	2550	1999	34	VAC
60000	4/24/2002	spring	1	1	724	0	0	3080	1994	91	VAC
63000	4/26/2002	spring	2	2	984	0	0	2500	1983	49	OWN
64990	4/26/2002	spring	2	2	984	0	0	2500	1982	34	OWN
85000	5/9/2002	spring	2	2	1151	1	0	2970	1995	3	OWN
84000	5/15/2002	spring	2	2	1151	1	0	2970	1995	34	VAC
87900	5/23/2002	spring	2	2	1019	0	0	2550	1996	321	VAC
90000	5/24/2002	spring	2	2	1019	1	0	2550	1999	0	VAC
82900	5/24/2002	spring	2	2	1151	1	0	2970	1995	14	VAC
125000	5/27/2002	spring	3	3	1523	1	1	2151	1997	54	OWN
87500	5/30/2002	spring	2	2	1151	1	0	2970	1995	14	OWN
56000	5/31/2002	spring	1	2	762	0	0	2401	1971	56	TEN
82500	6/4/2002	spring	2	2	1151	1	0	2970	1995	3	TEN
84000	6/6/2002	spring	2	2	960	0	0	3080	1994	9	VAC
89500	6/7/2002	spring	2	2	1019	0	0	2550	1996	0	VAC
89990	6/12/2002	spring	2	2	867	0	1	3978	1981	174	VAC
83500	6/20/2002	spring	2	2	1148	1	0	2970	1995	13	VAC
99500	6/21/2002	spring	3	3	1696	1	0	3540	1986	26	OWN
77500	6/27/2002	spring	3	2	1178	1	0	3540	1986	34	OWN
90000	7/10/2002	summer	2	2	1019	0	0	2550	1999	126	OWN
69000	7/12/2002	summer	1	2	836	1	0	3080	1995	25	OWN
49900	7/23/2002	summer	2	1	838	0	0	2500	1984	52	VAC
91500	7/26/2002	summer	2	2	1100	0	0	2550	1996	32	OWN
94300	8/6/2002	summer	2	2	1019	0	0	2550	2001	7	VAC
91500	8/7/2002	summer	2	2	1019	0	0	2550	2001	10	OWN
73500	8/9/2002	summer	1	2	836	0	0	3080	1995	61	VAC
87000	8/29/2002	summer	2	2	1151	1	0	2970	1995	3	OWN
87500	9/13/2002	summer	2	2	1151	1	0	2970	1995	17	VAC
76900	9/18/2002	summer	1	2	836	0	0	3080	1995	34	VAC
64990	9/30/2002	summer	2	2	984	0	0	2500	1983	68	VAC
75000	9/30/2002	summer	3	2	1056	1	0	3540	1986	42	OWN

Sale Price	Sale Date	Season	Baths	BRs	Living Area (SF)	FPs	Gar	Lot Area (SF)	Year Built	Days on Mkt	Occupancy
88000	10/3/2002	fall	2	2	1151	1	0	2970	1995	48	VAC
66500	10/6/2002	fall	2	2	992	1	0	2500	1983	16	OWN
92900	11/5/2002	fall	2	2	1019	0	0	2550	1997	13	VAC
124000	1/15/2003	winter	3	2	1310	1	1	2254	1997	6	VAC
92500	2/18/2003	winter	2	3	1148	1	0	2970	1995	135	OWN
53700	2/28/2003	winter	1	2	762	0	0	2401	1971	123	VAC
72900	3/4/2003	winter	1	2	836	0	0	3080	1993	109	VAC
85000	3/25/2003	winter	2	3	1148	1	0	2970	1995	155	VAC
59900	4/2/2003	spring	1	2	772	0	0	2401	1971	124	VAC
95000	4/7/2003	spring	2	3	1148	1	0	2970	1995	110	OWN
61000	4/18/2003	spring	2	2	984	0	0	2500	1984	180	VAC
90000	5/1/2003	spring	2	2	1019	1	0	2550	1996	137	OWN
129900	5/2/2003	spring	3	3	1446	1	1	2254	1997	44	OWN
95000	5/30/2003	spring	2	2	1019	1	0	2550	1997	157	OWN
55000	6/5/2003	spring	2	1	838	0	0	2500	1983	154	OWN
97000	6/11/2003	spring	2	2	1144	1	0	3960	1982	68	VAC
84000	6/19/2003	spring	3	2	1178	1	0	3540	1986	89	VAC
89000	7/10/2003	summer	2	2	960	0	0	3090	1995	3	VAC
64000	7/29/2003	summer	2	2	984	0	0	2500	1983	163	VAC
95000	7/30/2003	summer	2	2	1151	1	0	2970	1995	3	VAC
95000	9/8/2003	summer	2	2	1151	1	0	2970	1995	2	OWN
95700	9/9/2003	summer	2	2	1019	0	0	2550	1997	91	VAC
81000	9/29/2003	summer	1	2	836	0	0	3080	1993	13	VAC
126000	10/2/2003	fall	3	2	1310	1	1	2254	1997	3	OWN
57000	12/1/2003	fall	1	2	762	0	0	2401	1971	63	VAC
136850	12/5/2003	fall	3	3	1523	1	1	2151	1996	0	TEN
67000	2/13/2004	winter	1	1	724	0	0	3090	1995	0	VAC
65000	2/20/2004	winter	1	1	724	0	0	3090	1993	2	VAC
85900	2/25/2004	winter	3	2	1101	0	0	3540	1984	16	VAC
104000	2/27/2004	winter	2	2	1019	1	0	2550	1999	1	OWN
155000	3/2/2004	winter	3	3	1446	1	1	2254	1997	1	OWN
147500	3/5/2004	winter	3	3	1446	1	1	2254	1997	2	OWN
174000	3/16/2004	winter	3	3	1370	0	2	2614	2003	11	OWN
74000	3/18/2004	winter	2	3	1080	0	0	2401	1971	93	OWN
88000	3/19/2004	winter	3	2	1178	1	0	3540	1986	7	VAC
68000	3/23/2004	winter	2	2	984	0	0	2500	1983	67	VAC
69500	3/26/2004	winter	3	2	984	0	0	2500	1983	7	VAC
145000	3/29/2004	winter	3	2	1287	1	1	2237	1996	5	OWN
122900	4/5/2004	spring	2	3	1148	1	0	2970	1995	18	OWN
78500	4/12/2004	spring	2	3	1080	0	0	2401	1971	45	VAC
185000	4/14/2004	spring	3	3	1553	0	2	1742	2003	2	OWN
77000	4/16/2004	spring	2	1	858	1	0	2809	1984	4	VAC
65000	4/19/2004	spring	1	2	762	0	0	2401	1971	7	TEN
86000	4/23/2004	spring	3	2	1101	0	0	3540	1985	2	VAC
195000	4/30/2004	spring	2	3	1553	0	2	1742	2003	2	OWN
85000	5/3/2004	spring	2	3	1080	0	0	2401	1971	23	VAC
111000	5/10/2004	spring	2	2	960	0	0	3090	1995	5	OWN

Bibliography

Works Cited

Chapter 2: Mathematics Review

Bittinger, Marvin L., David J. Ellenbogen, and Barbara L. Johnson. *Intermediate Algebra: Concepts and Applications.* 8th ed. Reading, Mass.: Addison-Wesley, 2009.

Chapter 3: Working with Tables and Charts

Berenson, Mark L., David M. Levine, and Timothy C. Krehbiel. *Basic Business Statistics: Concepts and Applications.* 9th ed. Upper Saddle River, N.J.: Pearson/Prentice Hall, 2004.

Tufte, Edward R. *Envisioning Information.* Cheshire, Conn.: Graphics Press, 1990.

___. *The Visual Display of Quantitative Information.* 2nd ed. Cheshire, Conn.: Graphics Press, 2001.

___. *Visual Explanations.* Cheshire, Conn.: Graphics Press, 1997.

Chapter 4: Describing Numerical Data

International Association of Assessing Officers. *Assessment Administration.* Chicago: IAAO, 2003.

Levine, David M., Timothy C. Krehbiel, and Mark L. Berenson. *Business Statistics: A First Course.* 3rd ed. Upper Saddle River, N.J.: Prentice Hall, 2003.

Chapter 5: Probability

Neter, John, William Wasserman, and Michael H. Kutner. *Applied Linear Statistical Models.* 3rd ed. Homewood, Ill.: Irwin, 1990.

Rubinfeld, Daniel L. "Reference Guide on Multiple Regression." *Reference Manual on Scientific Evidence.* 2nd ed. Washington, D.C.: Federal Judicial Center, 2000.

Chapter 6: Research Design, Hypothesis Testing, and Sampling

Babbie, Earl. *The Practice of Social Research.* 6th ed. Belmont, Calif.: Wadsworth, 1992.

Dillman, Don. *Mail and Internet Surveys: The Tailored Design Method.* Hoboken, N.J.: John Wiley & Sons, 2007.

Levine, David M., Timothy C. Krehbiel, and Mark L. Berenson. *Business Statistics: A First Course.* 3rd ed. Upper Saddle River, N.J.: Prentice Hall, 2003.

Smith, Mary L., and Gene V. Glass. *Research and Evaluation in Education and the Social Sciences.* Boston: Allyn and Bacon, 1987.

Chapter 7: Inferences About Population Means and Proportions

Levine, David M., Timothy C. Krehbiel, and Mark L. Berenson. *Business Statistics: A First Course.* 3rd ed. Upper Saddle River, N.J.: Prentice Hall, 2003.

Chapter 8: Nonparametric Tests

Conover, W. J. *Practical Nonparametric Statistics.* 3rd ed. New York: John Wiley & Sons, 1999.

Gibbons, Jean D. *Nonparametric Statistics: An Introduction.* Newbury Park, Calif.: Sage Publications, 1993.

Chapter 9: Simple Linear Regression Analysis

Box, George E. P., and D. R. Cox. "An Analysis of Transformations." *Journal of the Royal Statistical Society, Series B (Methodological)* 26, no. 2 (1964): 211-243.

Dielman, Terry. *Applied Regression: Analysis for Business and Economics.* 3rd ed. Pacific Grove, Calif.: Duxbury/Thomson Learning, 2001.

Griffiths, W., and K. Surekha. "A Monte Carlo Evaluation of the Power of Some Tests for Heteroscedasticity." *Journal of Econometrics* 31, no. 2 (1986): 219-231.

Halvorsen, Robert, and Raymond Palmquist. "The Interpretation of Dummy Variables in Semilogarithmic Equations." *American Economic Review* 70, no. 3 (1980): 474-75.

Hardin, William G., and Marvin L. Wolverton. "An Introduction to the Analysis of Covariance Model Using an Empirical Test of Foreclosure Status on Sale Price." *Assessment Journal* 6, no. 1 (1999): 50-55.

Neter, John, William Wasserman, and Michael H. Kutner. *Applied Linear Statistical Models.* 3rd ed. Homewood, Ill.: Irwin, 1990.

Chapter 10: Multiple Linear Regression Analysis

Anderson, John E. "Ridge Regression: A New Regression Technique Useful to Appraisers and Assessors." *The Real Estate Appraiser and Analyst* 45, no. 6 (November/December 1979): 48-51.

Dielman, Terry. *Applied Regression: Analysis for Business and Economics.* 3rd ed. Pacific Grove, Calif.: Duxbury/Thomson Learning, 2001.

Ferreira, Eurico J., and G. Stacy Sirmans. "Ridge Regression in Real Estate Analysis." *The Appraisal Journal* 56, no. 3 (July 1988): 311-319.

Freund, Rudolf J., and William J. Wilson. *Regression Analysis: Statistical Modeling of a Response Variable.* San Diego, Calif.: Academic Press, 1998.

Hair, Joseph F., Rolph E. Anderson, Ronald C. Tatham, and William C. Black. *Multivariate Data Analysis with Readings.* 3rd ed. New York: Macmillan, 1992.

Neter, John, William Wasserman, and Michael H. Kutner. *Applied Linear Statistical Models.* 3rd ed. Homewood, Ill.: Irwin, 1990.

Reichert, Alan K., James S. Moore, and Chien-Ching Cho. "Stabilizing Statistical Models using Ridge Regression." *The Real Estate Appraiser and Analyst* 51, no. 3 (Fall 1985): 17-22.

Additional Reading

Acock, Alan C. *A Gentle Introduction to Stata.* 2nd ed. College Station, Texas: Stata Corporation, 2006.

Anderson, David R., Dennis J. Sweeney, and Thomas A. Williams. *Essentials of Statistics for Business and Economics.* 5th ed. Mason, Ohio: Thomson Higher Education, 2009.

Asteriou, Dimitrios, and Stephen G. Hall. *Applied Econometrics: A Modern Approach Using EViews and Microfit.* revised ed. New York: Palgrave MacMillan, 2007.

Babbie, Earl R. *The Practice of Social Research.* 12th ed. Belmont, Calif.: Wadsworth Cengage, 2010.

Black, Ken. *Business Statistics for Contemporary Decision Making.* Hoboken, N.J.: John Wiley & Sons, 2008.

Conover, W. J. *Practical Nonparametric Statistics.* 3rd ed. Hoboken, N.J.: John Wiley & Sons, 1999.

Dillman, Don A., Jolene D. Smyth, and Leah Melani Christian. *Internet, Mail, and Mixed-Mode Surveys: The Tailored Design Method.* 3rd ed. Hoboken, N.J.: John Wiley & Sons, 2009.

Dretzke, Beverly J. *Statistics with Microsoft Excel.* 4th ed. Upper Saddle River, N.J.: Pearson/Prentice-Hall, 2009.

Griffith, Arthur. *SPSS for Dummies.* Hoboken, N.J.: John Wiley & Sons, 2007

Hill, R. Carter, William E. Griffiths, and Guay C. Lim. *Principles of Econometrics.* 3rd ed. Hoboken, N.J.: John Wiley & Sons, 2008.

Jacobs, Kathy, Curtis Frye, and Douglas Frye. *Excel 2007 Charts Made Easy.* New York: McGraw-Hill Professional, 2009.

Kulas, John T. *SPSS Essentials: Managing and Analyzing Social Science Data.* 2nd ed. San Francisco: Jossey-Bass, 2009.

Rosner, Bernard. *Fundamentals of Biostatistics.* 6th ed. Belmont, Calif.: Thomson-Brooks/Cole, 2005.

Weatherup, Colin. *Experimental Statistics Using MINITAB.* Bury St Edmunds, Suffolk, U.K.: Arima, 2007.